MW00709691

EMBEDDED SYSTEMS

EMBEDDED SYSTEMS

Hardware, Design, and Implementation

Edited by

Krzysztof Iniewski
CMOS Emerging Technologies Research

WILEY

A JOHN WILEY & SONS, INC., PUBLICATION

Library of Congress Cataloging-in-Publication Data:
Iniewski, Krzysztof.
    Embedded systems : hardware, design, and implementation / by Krzysztof Iniewski.
        pages cm
    Includes bibliographical references and index.
    ISBN 978-1-118-35215-1 (hardback)
    1. Embedded computer systems.   I. Title.
    TK7895.E42I526 2012
    006.2'2–dc23
                            2012034412

CONTENTS

PREFACE

Embedded computer systems surround us: smartphones, PC (personal computer) tablets, hardware embedded in cars, TVs, and even refrigerators or heating systems. In fact, embedded systems are one of the most rapidly growing segments of the computer industry today. This book offers the fundamentals of how these embedded systems can benefit anyone who has to build, evaluate, and apply these systems.

Embedded systems have become more and more prevalent over the years. Devices that we use every day have become intelligent and incorporate electronics. Data acquisition products no longer act independently. They are part of an ecosystem of interoperable communication devices. A device acquires some information, another device acquires other information, and that information is sent to a central unit for analysis. This idea of an ecosystem is powerful and flexible for the coming generation. Today, we have many examples of interoperable systems with ecosystems that address everyday problems. The ever-growing ecosystem of interoperable devices leads to increased convenience for the user and more information to solve problems. The elements of the system and the ecosystem will be explained in this book, with the understanding of the use cases and applications explained in detail.

Embedded systems are composed of hardware and software computations that are subject to physical real-time constraints. A key challenge to embedded systems design is the development of efficient methods to test and prototype realistic application configurations. This involves ultra high-speed input generation, accurate measurement and output logging, reliable environment modeling, complex timing and synchronization, hardware-in-the-loop simulation, and sophisticated analysis and visualization.

Ever since their introduction decades ago, embedded processors have undergone extremely significant transformations—ranging from relatively simple microcontrollers to tremendously complex systems-on-chip (SoCs). The best example is the iPhone, which is an embedded system platform that surpasses older personal computers or laptops. This trend will clearly continue with embedded systems virtually taking over our lives. Whoever can put together the best embedded system in the marketplace (for the given application) will clearly dominate worldwide markets. Apple is already the most valuable company in the world, surpassing Microsoft, Exxon, or Cisco.

With progress in computing power, wireless communication capabilities, and integration of various sensors and actuators, the sky is the limit for the embedded systems applications. With everyone having a smartphone in their pocket, life will be quite different from what it was 5 years ago, and the first signs are clearly visible today.

The book contains 14 carefully selected chapters. They cover areas of multicore processors, embedded graphics processing unit (GPU) and field-programmable gate array (FPGA) designs used in computing, communications, biology, industrial, and space applications. The authors are well-recognized experts in their fields and come from both academia and industry.

With such a wide variety of topics covered, I am hoping that the reader will find something stimulating to read, and discover the field of embedded systems to be both exciting and useful in science and everyday life. Books like this one would not be possible without many creative individuals meeting together in one place to exchange thoughts and ideas in a relaxed atmosphere. I would like to invite you to attend the CMOS Emerging Technologies events that are held annually in beautiful British Columbia, Canada, where many topics covered in this book are discussed. See http://www.cmoset.com for presentation slides from the previous meeting and announcements about future ones.

I would love to hear from you about this book. Please email me at kris. iniewski@gmail.com.

Let the embedded systems of the world prosper and benefit us all!

Vancouver, 2012   KRIS INIEWSKI

CONTRIBUTORS

SAMUEL ANTÃO, INESC-ID/IST, Universidade Técnica de Lisboa, Lisbon, Portugal

FUMIO ARAKAWA, Renesas Electronics Corporation, Tokyo, Japan

ROBERT J. BEYNON, Protein Function Group, Institute of Integrative Biology, University of Liverpool, Liverpool, UK

ISTVÁN BOGDÁN, Department of Automatic Control and Systems Engineering, University of Sheffield, Sheffield, UK

SAI RAHUL CHALAMALASETTI, Department of Electrical and Computer Engineering, University of Massachusetts Lowell, Lowell, MA

RICARDO CHAVES, INESC-ID/IST, Universidade Técnica de Lisboa, Lisbon, Portugal

DANIEL COCA, Department of Automatic Control and Systems Engineering, University of Sheffield, Sheffield, UK

EZZ EL-MASRY, Dalhousie University, Halifax, Nova Scotia, Canada

KAMAL EL-SANKARY, Dalhousie University, Halifax, Nova Scotia, Canada

ISSAM HAMMAD, Dalhousie University, Halifax, Nova Scotia, Canada

TAREQ HASAN KHAN, Department of Electrical and Computer Engineering, University of Saskatchewan, Saskatchewan, Canada

ANDREW LEONE, STMicroelectronics, Geneva, Switzerland

MARTIN MARGALA, Department of Electrical and Computer Engineering, University of Massachusetts Lowell, Lowell, MA

PAOLO MELONI, Department of Electrical and Electronic Engineering, University of Cagliari, Cagliari, Italy

BYEONG-GYU NAM, Chungnam National University, Daejeon, South Korea

SUDEEP PASRICHA, Department of Electrical and Computer Engineering, Colorado State University, Fort Collins, CO

SOHAN PUROHIT,   Department of Electrical and Computer Engineering, University of Massachusetts Lowell, Lowell, MA

LUIGI RAFFO,   Department of Electrical and Electronic Engineering, University of Cagliari, Cagliari, Italy

FREDERIC RISACHER,   Platform Development – Modeling, Research In Motion Limited, Waterloo, Ontario, Canada

DAVID K. RUTISHAUSER,   NASA Johnson Space Center, Houston, TX

KENNETH J. SCHULTZ,   Platform Development – Modeling, Research In Motion Limited, Waterloo, Ontario, Canada

SIMONE SECCHI,   Department of Electrical and Electronic Engineering, University of Cagliari, Cagliari, Italy

LESLEY SHANNON,   Simon Fraser University, Burnaby, British Columbia, Canada

ROBERT L. SHULER, JR.,   NASA Johnson Space Center, Houston, TX

LEONEL SOUSA,   INESC-ID/IST, Universidade Técnica de Lisboa, Lisbon, Portugal

SIDDHARTH SRINIVASAN,   Zymeworks Vancouver, British Columbia, Canada

KHAN WAHID,   Department of Electrical and Computer Engineering, University of Saskatchewan, Saskatchewan, Canada

HOI-JUN YOO,   Korea Advanced Institute of Science and Technology, Daejeon, South Korea

YONG ZOU,   Department of Electrical and Computer Engineering, Colorado State University, Fort Collins, CO

1 Low Power Multicore Processors for Embedded Systems

FUMIO ARAKAWA

1.1 MULTICORE CHIP WITH HIGHLY EFFICIENT CORES

A multicore chip is one of the most promising approaches to achieve high performance. Formerly, frequency scaling was the best approach. However, the scaling has hit the power wall, and frequency enhancement is slowing down. Further, the performance of a single processor core is proportional to the square root of its area, known as Pollack's rule [1], and the power is roughly proportional to the area. This means lower performance processors can achieve higher power efficiency. Therefore, we should make use of the multicore chip with relatively low performance processors.

The power wall is not a problem only for high-end server systems. Embedded systems also face this problem for further performance improvements [2]. MIPS is the abbreviation of million instructions per second, and a popular integer-performance measure of embedded processors. The same performance processors should take the same time for the same program, but the original MIPS varies, reflecting the number of instructions executed for a program. Therefore, the performance of a Dhrystone benchmark relative to that of a VAX 11/780 minicomputer is broadly used [3, 4]. This is because it achieved 1 MIPS, and the relative performance value is called VAX MIPS or DMIPS, or simply MIPS. Then GIPS (giga-instructions per second) is used instead of the MIPS to represent higher performance.

Figure 1.1 roughly illustrates the power budgets of chips for various application categories. The horizontal and vertical axes represent performance (DGIPS) and efficiency (DGIPS/W) in logarithmic scale, respectively. The oblique lines represent constant power (W) lines and constant product lines

Embedded Systems: Hardware, Design, and Implementation, First Edition.
Edited by Krzysztof Iniewski.

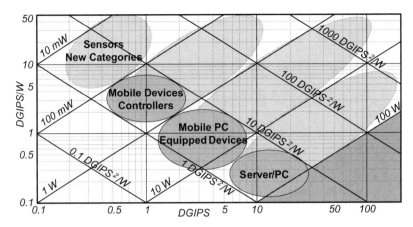

FIGURE 1.1. Power budgets of chips for various application categories.

of the power–performance ratio and the power ($DGIPS^2/W$). The product roughly indicates the attained degree of the design. There is a trade-off relationship between the power efficiency and the performance. The power of chips in the server/personal computer (PC) category is limited at around 100 W, and the chips above the 100-W oblique line must be used. Similarly, the chips roughly above the 10- or 1-W oblique line must be used for equipped-devices/mobile PCs, or controllers/mobile devices, respectively. Further, some sensors must use the chips above the 0.1-W oblique line, and new categories may grow from this region. Consequently, we must develop high $DGIPS^2/W$ chips to achieve high performance under the power limitations.

Figure 1.2 maps various processors on a graph, whose horizontal and vertical axes respectively represent operating frequency (MHz) and power–frequency ratio (MHz/W) in logarithmic scale. Figure 1.2 uses MHz or GHz instead of the DGIPS of Figure 1.1. This is because few DGIPS of the server/PC processors are disclosed. Some power values include leak current, whereas the others do not; some are under the worst conditions while the others are not. Although the MHz value does not directly represent the performance, and the power measurement conditions are not identical, they roughly represent the order of performance and power. The triangles and circles represent embedded and server/PC processors, respectively. The dark gray, light gray, and white plots represent the periods up to 1998, after 2003, and in between, respectively. The GHz^2/W improved roughly 10 times from 1998 to 2003, but only three times from 2003 to 2008. The enhancement of single cores is apparently slowing down. Instead, the processor chips now typically adopt a multi-core architecture.

Figure 1.3 summarizes the multicore chips presented at the International Solid-State Circuit Conference (ISSCC) from 2005 to 2008. All the processor chips presented at ISSCC since 2005 have been multicore ones. The axes are

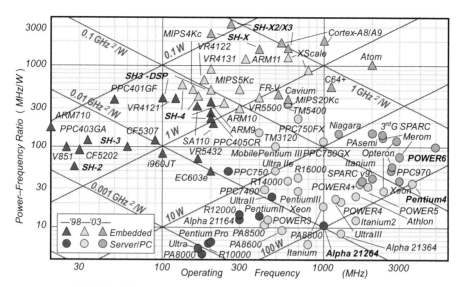

FIGURE 1.2. Performance and efficiency of various processors.

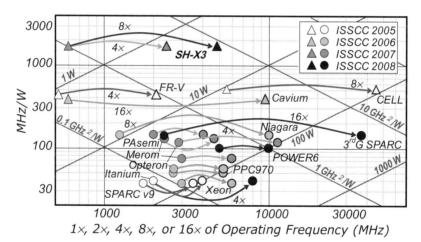

FIGURE 1.3. Some multicore chips presented at ISSCC.

similar to those of Figure 1.2, although the horizontal axis reflects the number of cores. Each plot at the start and end points of an arrow represent single core and multicore, respectively.

The performance of multicore chips has continued to improve, which has compensated for the slowdown in the performance gains of single cores in both the embedded and server/PC processor categories. There are two types of muticore chips. One type integrates multiple-chip functions into a single chip, resulting in a multicore SoC. This integration type has been popular for more than 10 years. Cell phone SoCs have integrated various types of hardware intellectual properties (HW-IPs), which were formerly integrated into multiple chips. For example, an SH-Mobile G1 integrated the function of both the application and baseband processor chips [5], followed by SH-Mobile G2 [6] and G3 [7, 8], which enhanced both the application and baseband functionalities and performance. The other type has increased number of cores to meet the requirements of performance and functionality enhancement. The RP-1, RP-2 and RP-X are the prototype SoCs, and an SH2A-DUAL [9] and an SH-Navi3 [10] are the multicore products of this enhancement type. The transition from single core chips to multicore ones seems to have been successful on the hardware side, and various multicore products are already on the market. However, various issues still need to be addressed for future multicore systems.

The first issue concerns memories and interconnects. Flat memory and interconnect structures are the best for software, but hardly possible in terms of hardware. Therefore, some hierarchical structures are necessary. The power of on-chip interconnects for communications and data transfers degrade power efficiency, and a more effective process must be established. Maintaining the external input/output (I/O) performance per core is more difficult than increasing the number of cores, because the number of pins per transistors decreases for finer processes. Therefore, a breakthrough is needed in order to maintain the I/O performance.

The second issue concerns runtime environments. The performance scalability was supported by the operating frequency in single core systems, but it should be supported by the number of cores in multicore systems. Therefore, the number of cores must be invisible or virtualized with small overhead when using a runtime environment. A multicore system will integrate different subsystems called domains. The domain separation improves system reliability by preventing interference between domains. On the other hand, the well-controlled domain interoperation results in an efficient integrated system.

The third issue relates to the software development environments. Multicore systems will not be efficient unless the software can extract application parallelism and utilize parallel hardware resources. We have already accumulated a huge amount of legacy software for single cores. Some legacy software can successfully be ported, especially for the integration type of multicore SoCs, like the SH-Mobile G series. However, it is more difficult with the enhancement type. We must make a single program that runs on multicore, or distribute functions now running on a single core to multicore. Therefore, we must improve the portability of legacy software to the multicore systems. Developing new highly parallel software is another issue. An application or parallelization specialist could do this, although it might be necessary to have specialists in both areas. Further, we need a paradigm shift in the development, for example, a higher level of abstraction, new parallel languages, and assistant tools for effective parallelization.

1.2 SUPERH™ RISC ENGINE FAMILY (SH) PROCESSOR CORES

As mentioned above, a multicore chip is one of the most promising approaches to realize high efficiency, which is the key factor to achieve high performance under some fixed power and cost budgets. Therefore, embedded systems are employing multicore architecture more and more. The multicore is good for multiplying single-core performance with maintaining the core efficiency, but does not enhance the efficiency of the core itself. Therefore, we must use highly efficient cores. SuperH™ (Renesas Electronics, Tokyo) reduced instruction set computer (RISC) engine family (SH) processor cores are highly efficient typical embedded central processing unit (CPU) cores for both single- and multicore chips.

1.2.1 History of SH Processor Cores

Since the beginning of the microprocessor history, a processor especially for PC/servers had continuously advanced its performance while maintaining a price range from hundreds to thousands of dollars [11, 12]. On the other hand, a single-chip microcontroller had continuously reduced its price, resulting in the range from dozens of cents to several dollars with maintaining its performance, and had been equipped to various products [13]. As a result, there was a situation of no demand on the processor of the middle price range from tens to hundreds of dollars.

However, with the introduction of the home game console in the late 1980s and the digitization of the home electronic appliances from the 1990s, there occurred the demands to a processor suitable for multimedia processing in this price range. Instead of seeking high performance, such a processor has attached great importance to high efficiency. For example, the performance is 1/10 of a processor for PCs, but the price is 1/100, or the performance equals to a processor for PCs for the important function of the product, but the price is 1/10. The improvement of area efficiency has become the important issue in such a processor.

In the late 1990s, a high performance processor consumed too high power for mobile devices, such as cellular phones and digital cameras, and the demand was increasing on the processor with higher performance and lower power for multimedia processing. Therefore, the improvement of the power efficiency became the important issues. Furthermore, when the 2000s begins, more functions were integrated by further finer processes, but on the other hand, the increase of the initial and development costs became a serious problem. As a result, the flexible specification and the cost reduction came to be important issues. In addition, the finer processes suffered from the more leakage current.

Under the above background, embedded processors were introduced to meet the requirements, and have improved the area, power, and development efficiencies. The SH processor cores are one of such highly efficient CPU cores.

The first SH processor was developed based on SuperH architecture as one of embedded processors in 1993. Then the SH processors have been developed as a processor with suitable performance for multimedia processing and area-and-power efficiency. In general, performance improvement causes degradation of the efficiency as Pollack's rule indicates [1]. However, we can find ways to improve both performance and efficiency. Although individually each method is a small improvement, overall it can still make a difference.

The first-generation product, SH-1, was manufactured using a 0.8-μm process, operated at 20 MHz, and achieved performance of 16 MIPS in 500 mW. It was a high performance single-chip microcontroller, and integrated a read-only memory (ROM), a random access memory (RAM), a direct memory access controller (DMAC), and an interrupt controller.

The second-generation product, SH-2, was manufactured using the same 0.8-μm process as the SH-1 in 1994 [14]. It operated at 28.5 MHz, and achieved performance of 25 MIPS in 500 mW by optimization on the redesign from the SH-1. The SH-2 integrated a cache memory and an SDRAM controller instead of the ROM and the RAM of the SH-1. It was designed for the systems using external memories. The integrated SDRAM controller did not popular at that time, but enabled to eliminate an external circuitry, and contributed to system cost reduction. In addition, the SH-2 integrated a 32-bit multiplier and a divider to accelerate multimedia processing. And it was equipped to a home game console, which was one of the most popular digital appliances. The SH-2 extend the application field of the SH processors to the digital appliances with multimedia processing.

The third-generation product SH-3 was manufactured using a 0.5-μm process in 1995 [15]. It operated at 60 MHz, and achieved performance of 60 MIPS in 500 mW. Its power efficiency was improved for a mobile device. For example, the clock power was reduced by dividing the chip into plural clock regions and operating each region with the most suitable clock frequency. In addition, the SH-3 integrated a memory management unit (MMU) for such devices as a personal organizer and a handheld PC. The MMU is necessary for a general-purpose operating system (OS) that enables various application programs to run on the system.

The fourth-generation product, SH-4, was manufactured using a 0.25-μm process in 1997 [16–18]. It operated at 200 MHz, and achieved performance of 360 MIPS in 900 mW. The SH-4 was ported to a 0.18-μm process, and its power efficiency was further improved. The power efficiency and the product of performance and the efficiency reached to 400 MIPS/W and 0.14 GIPS^2/W, respectively, which were among the best values at that time. The product roughly indicates the attained degree of the design, because there is a trade-off relationship between performance and efficiency.

The fifth-generation processor, SH-5, was developed with a newly defined instruction set architecture (ISA) in 2001 [19–21], and an SH-4A, the advanced version of the SH-4, was also developed with keeping the ISA compatibility in 2003. The compatibility was important, and the SH-4A was used for various products. The SH-5 and the SH-4A were developed as a CPU core connected to other various HW-IPs on the same chip with a SuperHyway standard internal bus. This approach was available using the fine process of 0.13 μm, and enabled to integrate more functions on a chip, such as a video codec, 3D graphics, and global positioning systems (GPS).

An SH-X, the first generation of the SH-4A processor core series, achieved a performance of 720 MIPS with 250 mW using a 0.13-μm process [22–26]. The power efficiency and the product of performance and the efficiency reached to 2,880 MIPS/W and 2.1 GIPS^2/W, respectively, which were among the best values at that time. The low power version achieved performance of 360 MIPS and power efficiency of 4,500 MIPS/W [27–29].

An SH-X2, the second-generation core, achieved 1,440 MIPS using a 90-nm process, and the low power version achieved power efficiency of 6,000 MIPS/W in 2005 [30–32]. Then it was integrated on product chips [5–8].

An SH-X3, the third-generation core, supported multicore features for both SMP and AMP [33, 34]. It was developed using a 90-nm generic process in 2006, and achieved 600 MHz and 1,080 MIPS with 360 mW, resulting in 3,000 MIPS/W and 3.2 GIPS^2/W. The first prototype chip of the SH-X3 was a RP-1 that integrated four SH-X3 cores [35–38], and the second one was a RP-2 that integrated eight SH-X3 cores [39–41]. Then, it was ported to a 65-nm low power process, and used for product chips [10].

An SH-X4, the latest fourth-generation core, was developed using a 45-nm low power process in 2009, and achieved 648 MHz and 1,717 MIPS with 106 mW, resulting in 16,240 MIPS/W and 28 GIPS^2/W [42–44].

1.2.2 Highly Efficient ISA

Since the beginning of the RISC architecture, all the RISC processor had adopted a 32-bit fixed-length ISA. However, such a RISC ISA causes larger code size than a conventional complex instruction set computer (CISC) ISA, and requires larger capacity of program memories including an instruction cache. On the other hand, a CISC ISA has been variable length to define the instructions of various complexities from simple to complicated ones. The variable length is good for realizing the compact code sizes, but requires complex decoding, and is not suitable for parallel decoding of plural instructions for the superscalar issue.

SH architecture with the 16-bit fixed-length ISA was defined in such a situation to achieve compact code sizes and simple decoding. The 16-bit fixed-length ISA was spread to other processor ISAs, such as ARM Thumb and MIPS16.

As always, there should be pros and cons of the selection, and there are some drawbacks of the 16-bit fixed-length ISA, which are the restriction of the number of operands and the short literal length in the code. For example, an instruction of a binary operation modifies one of its operand, and an extra data transfer instruction is necessary if the original value of the modified operand must be kept. A literal load instruction is necessary to utilize a longer literal than that in an instruction. Further, there is an instruction using an implicitly defined register, which contributes to increase the number of operand with no extra operand field, but requires special treatment to identify it, and spoils orthogonal characteristics of the register number decoding. Therefore, careful implementation is necessary to treat such special features.

1.2.3 Asymmetric In-Order Dual-Issue Superscalar Architecture

Since a conventional superscalar processor gave priority to performance, the superscalar architecture was considered to be inefficient, and scalar architecture was still popular for embedded processors. However, this is not always true. Since the SH-4 design, SH processors have adopted the superscalar architecture by selecting an appropriate microarchitecture with considering efficiency seriously for an embedded processor.

The asymmetric in-order dual-issue superscalar architecture is the base microarchitecture of the SH processors. This is because it is difficult for a general-purpose program to utilize the simultaneous issue of more than two instructions effectively; a performance enhancement is not enough to compensate the hardware increase for the out-of-order issue, and symmetric superscalar issue requires resource duplications. Then, the selected architecture can maintain the efficiency of the conventional scalar issue one by avoiding the above inefficient choices.

The asymmetric superscalar architecture is sensitive to instruction categorizing, because the same category instruction cannot be issued simultaneously. For example, if we categorize all floating-point instructions in the same category, we can reduce the number of floating-point register ports, but cannot issue both floating-point instructions of arithmetic and load/store/transfer operations at a time. This degrades the performance. Therefore, the categorizing requires careful trade-off consideration between performance and hardware cost.

First of all, both the integer and load/store instructions are used most frequently, and categorized to different groups of integer (INT) and load/store (LS), respectively. This categorization requires address calculation unit in addition to the conventional arithmetic logical unit (ALU). Branch instructions are about one-fifth of a program on average. However, it is difficult to use the ALU or the address calculation unit to implement the early-stage branch, which calculates the branch addresses at one-stage earlier than the other type of operations. Therefore, the branch instruction is categorized in another group of branch (BR) with a branch address calculation unit. Even a RISC processor has a special instruction that cannot fit to the superscalar issue. For example, some instruction changes a processor state, and is categorized to a group of nonsuperscalar (NS), because most of instructions cannot be issued with it.

The 16-bit fixed-length ISA frequently uses an instruction to transfer a literal or register value to a register. Therefore, the transfer instruction is categorized to the BO group to be executable on both integer and load/store

(INT and LS) pipelines, which were originally for the INT and LS groups. Then the transfer instruction can be issued with no resource conflict. A usual program cannot utilize all the instruction issue slots of conventional RISC architecture that has three operand instructions and uses transfer instructions less frequently. Extra transfer instructions of the 16-bit fixed-length ISA can be inserted easily with no resource conflict to the issue slots that would be empty for a conventional RISC.

The floating-point load/store/transfer and arithmetic instructions are categorized to the LS group and a floating-point execution (FE) group, respectively. This categorization increases the number of the ports of the floating-point register file. However, the performance enhancement deserves the increase. The floating-point transfer instructions are not categorized to the BO group. This is because neither the INT nor FE group fit to the instruction. The INT pipeline cannot use the floating-point register file, and the FE pipeline is too complicated to treat the simple transfer operation. Further, the transfer instruction is often issued with a FE group instruction, and the categorization to other than the FE group is enough condition for the performance.

The SH ISA supports floating-point sign negation and absolute value (FNEG and FABS) instructions. Although these instructions seem to fit the FE group, they are categorized to the LS group. Their operations are simple enough to execute at the LS pipeline, and the combination of another arithmetic instruction becomes a useful operation. For example, the FNEG and floating-point multiply–accumulate (FMAC) instructions became a multiply-and-subtract operation.

Table 1.1 summarizes the instruction categories for asymmetric superscalar architecture. Table 1.2 shows the ability of simultaneous issue of two instructions. As an asymmetric superscalar processor, each pipeline for the INT, LS, BR, or FE group is one, and the simultaneous issue is limited to a pair of different group instructions, except for a pair of the BO group instructions, which can be issued simultaneously using both the INT and LS pipelines. An NS group instruction cannot be issued with another instruction.

1.3   SH-X: A HIGHLY EFFICIENT CPU CORE

The SH-X has enhanced its performance by adopting superpipeline architecture to the base micro-architecture of the asymmetric in-order dual-issue superscalar architecture. The operating frequency would be limited by an applied process without fundamental change of the architecture or microarchitecture. Although conventional superpipeline architecture was thought inefficient as was the conventional superscalar architecture before applying to the SH-4, the SH-X core enhanced the operating frequency with maintaining the high efficiency.

TABLE 1.1. Instruction Categories for Asymmetric Superscalar Architecture

INT	FE
ADD; ADDC; ADDV;	FADD; FSUB; FMUL;
SUB; SUBC; SUBV;	FDIV; FSQRT; FCMP;
MUL; MULU; MULS;	FLOAT; FTRC;
DMULU; DMULS;	FCNVSD; FCNVDS;
DIV0U; DIV0S; DIV1;	FMAC; FIPR; FTRV;
CMP; NEG; NEGC; NOT;	FSRRA; FSCA;
DT; MOVT; CLRT; SETT;	FRCHG; FSCHG; FPCHG
CLRMAC; CLRS; SETS;	
TST Rm, Rn; TST imm, R0;	**BO**
AND Rm, Rn; AND imm, R0;	MOV imm, Rn;
OR Rm, Rn; OR imm, R0;	MOV Rm, Rn; NOP
XOR Rm, Rn; XOR imm, R0;	
ROTL; ROTR; ROTCL; ROTCR;	**BR**
SHAL; SHAR; SHAD; SHLD;	BRA; BSR; BRAF; BSRF;
SHLL; SHLL2; SHLL8; SHLL16;	BT; BF; BT/S; BF/S;
SHLR; SHLR2; SHLR8; SHLR16;	JMP; JSR; RTS
EXTU; EXTS; SWAP; XTRCT	
	NS
LS	AND imm, @(R0,GBR);
	OR imm, @(R0,GBR);
MOV (load/store);	XOR imm, @(R0,GBR);
MOVA; MOVCA;	TST imm, @(R0,GBR);
FMOV; FLDI0; FLDI1;	MAC; SYNCO;
FABS; FNEG;	MOVLI; MOVCO;
FLDS; FSTS; LDS; STS;	LDC (SR/SGR/DBR);
LDC (except SR/SGR/DBR);	STC (SR); RTE;
STC (except SR);	LDTLB; ICBI; PREFI;
OCBI; OCBP; OCBWB; PREF	TAS; TRAPA; SLEEP

TABLE 1.2. Simultaneous Issue of Instructions

		Second Instruction Category					
		BO	INT	LS	BR	FE	NS
First Instruction Category	BO	✓	✓	✓	✓	✓	
	INT	✓		✓	✓	✓	
	LS	✓	✓		✓	✓	
	BR	✓	✓	✓		✓	
	FE	✓	✓	✓	✓		
	NS						

TABLE 1.3. Microarchitecture Selections of SH-X

	Selections	Other Candidates	Merits
Pipeline stages	7	5, 6, 8, 10, 15, 20	1.4 times frequency enhancement
Branch acceleration	Out-of-order issue	BTB, branch with plural instructions	Compatibility, small area, for low frequency branch
Branch prediction	Dynamic (BHT, global history)	Static (fixed direction, hint bit in instruction)	
Latency concealing	Delayed execution, store buffers	Out-of-order issue	Simple, small

1.3.1 Microarchitecture Selections

The SH-X has seven-stage superpipeline to maintain the efficiency among various numbers of stages applied to various processors up to highly super-pipelined 20 stages [45]. The conventional seven-stage pipeline degraded the cycle performance compared with the five-stage one that is popular for efficient embedded processors. Therefore, appropriate methods were chosen to enhance and recover the cycle performance with the careful trade-off judgment of performance and efficiency. Table 1.3 summarizes the selection result of the microarchitecture.

An out-of-order issue is the popular method used by a high-end processor to enhance the cycle performance. However, it requires much hardware and is too inefficient especially for general-purpose register handling. The SH-X adopts an in-order issue except branch instructions using no general-purpose register.

The branch penalty is the serious problem for the superpipeline architecture. The SH-X adopts a branch prediction and an out-of-order branch issue, but does not adopt a more expensive way with a branch target buffer (BTB) and an incompatible way with plural instructions. The branch prediction is categorized to static and dynamic ones, and the static ones require the architecture change to insert the static prediction result to the instruction. Therefore, the SH-X adopts a dynamic one with a branch history table (BHT) and a global history.

The load/store latencies are also a serious problem, and the out-of-order issue is effective to hide the latencies, but too inefficient to adopt as mentioned above. The SH-X adopts a delayed execution and a store buffer as more efficient methods.

The selected methods are effective to reduce the pipeline hazard caused by the superpipeline architecture, but not effective to avoid a long-cycle stall caused by a cache miss for an external memory access. Such a stall could be avoided by an out-of-order architecture with large-scale buffers, but is not a serious problem for embedded systems.

1.3.2 Improved Superpipeline Structure

Figure 1.4 illustrates a conventional seven-stage superpipeline structure. The seven stages consist of 1st and 2nd instruction fetch (I1 and I2) stages and an instruction decoding (ID) stage for all the pipelines, 1st to 4th execution (E1, E2, E3, and E4) stages for the INT, LS, and FE pipelines. The FE pipeline has nine stages with two extra execution stages of E5 and E6.

A conventional seven-stage pipeline has less performance than a five-stage one by 20%. This means the performance gain of the superpipeline architecture is only $1.4 \times 0.8 = 1.12$ times, which would not compensate the hardware increase. The branch and load-use-conflict penalties increase by the increase of the instruction-fetch and data-load cycles, respectively. They are the main reason of the 20% performance degradation.

Figure 1.5 illustrates the seven-stage superpipeline structure of the SH-X with delayed execution, store buffer, out-of-order branch, and flexible forwarding. Compared with the conventional pipeline shown in Figure 1.4, the INT pipeline starts its execution one-cycle later at the E2 stage, a store data is buffered to the store buffer at the E4 stage and stored to the data cache at the E5 stage, the data transfer of the floating-point unit (FPU) supports

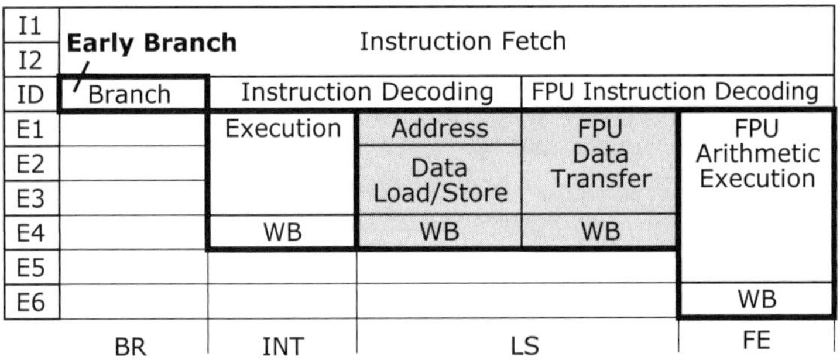

FIGURE 1.4. Conventional seven-stage superpipeline structure.

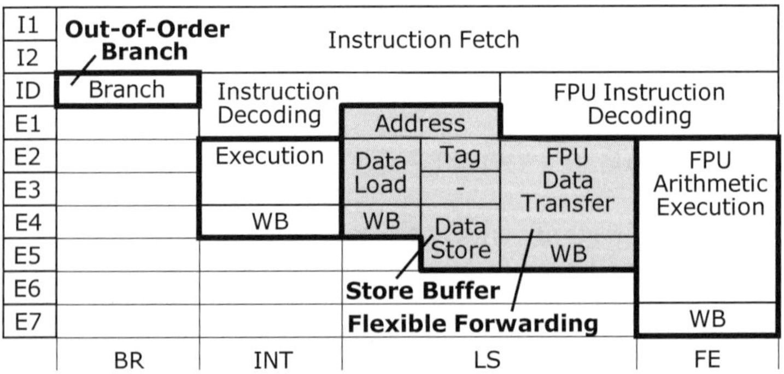

FIGURE 1.5. Seven-stage superpipeline structure of SH-X.

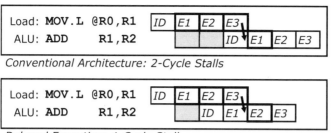

FIGURE 1.6. Load-use conflict reduction by delayed execution.

flexible forwarding. The BR pipeline starts at the ID stage, but is not synchronized to the other pipelines for an out-of-order branch issue.

The delayed execution is effective to reduce the load-use conflict as Figure 1.6 illustrates. It also lengthens the decoding stages into two except for the address calculation, and relaxes the decoding time. With the conventional architecture shown in Figure 1.4, a load instruction, MOV.L, set ups an R0 value at the ID stage, calculates a load address at the E1 stage, loads a data from the data cache at the E2 and E3 stages, and the load data is available at the end of the E3 stage. An ALU instruction, ADD, setups R1 and R2 values at the ID stage, adds the values at the E1 stage. Then the load data is forwarded from the E3 stage to the ID stage, and the pipeline stalls two cycles. With the delayed execution, the load instruction execution is the same, and the add instruction setups R1 and R2 values at E1 stage, adds the values at the E2 stage. Then the load data is forwarded from the E3 stage to the E1 stage, and the pipeline stalls only one cycle. This is the same cycle as those of conventional five-stage pipeline structures.

As illustrated in Figure 1.5, a store instruction performs an address calculation, TLB (translation lookaside buffer) and cache-tag accesses, a store-data latch, and a data store to the cache at the E1, E2, E4, and E5 stages, respectively, whereas a load instruction performs a cache access at the E2 stage. This means the three-stage gap of the cache access timing between the E2 and the E5 stages of a load and a store. However, a load and a store use the same port of the cache. Therefore, a load instruction gets the priority to a store instruction if the access is conflicted, and the store instruction must wait the timing with no conflict. In the N-stage gap case, N entries are necessary for the store buffer to treat the worst case, which is a sequence of N consecutive store issues followed by N consecutive load issues, and the SH-X implemented three entries.

1.3.3   Branch Prediction and Out-of-Order Branch Issue

Figure 1.7 illustrates a branch execution sequence of the SH-X before branch acceleration with a program sequence consisting of compare, conditional-branch, delay-slot, and branch-target instructions.

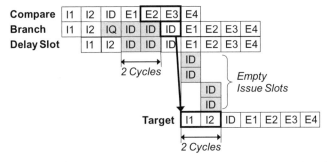

FIGURE 1.7. Branch execution sequence before branch acceleration.

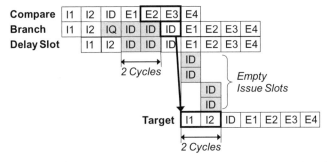

FIGURE 1.8. Branch execution sequence of SH-X.

The conditional-branch and delay-slot instructions are issued three cycles after the compare instruction issue, and the branch-target instruction is issued three cycles after the branch issue. The compare operation starts at the E2 stage by the delayed execution, and the result is available at the middle of the E3 stage. Then the conditional-branch instruction checks the result at the latter half of the ID stage, and generates the target address at the same ID stage, followed by the I1 and I2 stages of the target instruction. As a result, eight empty issue slots or four stall cycles are caused as illustrated. This means only one-third of the issue slots are used for the sequence.

Figure 1.8 illustrates the execution sequence of the SH-X after branch acceleration. The branch operation can start with no pipeline stall by a branch prediction, which predicts the branch direction that the branch is taken or not taken. However, this is not early enough to make the empty issue slots zero. Therefore, the SH-X adopted an out-of-order issue to the branches using no general-purpose register.

The SH-X fetches four instructions per cycle, and issues two instructions at most. Therefore, Instructions are buffered in an instruction queue (IQ) as illustrated. A branch instruction is searched from the IQ or an instruction-cache output at the I2 stage and provided to the ID stage of the branch pipeline for the out-of-order issue earlier than the other instructions provided to the ID stage in order. Then the conditional branch instruction is issued right after it is fetched while the preceding instructions are in the IQ, and the issue becomes early enough to make the empty issue slots zero. As a result, the target instruction is fetched and decoded at the ID stage right after the delay-slot instruction. This means no branch penalty occurs in the sequence when the preceding or delay-slot instructions stay two or more cycles in the IQ.

The compare result is available at the E3 stage, and the prediction is checked if it is hit or miss. In the miss case, the instruction of the correct flow is decoded at the ID stage right after the E3 stage, and two-cycle stall occurs. If the correct flow is not held in the IQ, the miss-prediction recovery starts from the I1 stage, and takes two more cycles.

Historically, the dynamic branch prediction method started from a BHT with 1-bit history per entry, which recorded a branch direction of taken or not for the last time, and predicted the same branch direction. Then, a BHT with 2-bit history per entry became popular, and the four direction states of strongly taken, weekly taken, weekly not taken, and strongly not taken were used for the prediction to reflect the history of several times. There were several types of the state transitions, including a simple up-down transition. Since each entry held only one or two bits, it is too expensive to attach a tag consisting of a part of the branch-instruction address, which was usually about 20 bits for a 32-bit addressing. Therefore, we could increase the number of entries about ten or twenty times without the tag. Although the different branch instructions could not be distinguished without the tag and there occurred a false hit, the merit of the entry increase exceeded the demerit of the false hit. A global history method was also popular for the prediction, and usually used with the 2-bit/entry BHT.

The SH-X stalled only two cycles for the prediction miss, and the performance was not so sensitive to the hit ratio. Further, the one-bit method required a state change only for a prediction miss, and it could be done during the stall. Therefore, the SH-X adopted a dynamic branch prediction method with a 4K-entry 1-bit/entry BHT and a global history. The size was much smaller than the instruction and data caches of 32 kB each.

1.3.4 Low Power Technologies

The SH-X achieved excellent power efficiency by using various low-power technologies. Among them, hierarchical clock gating and pointer-controlled

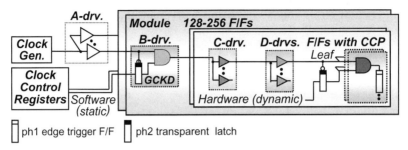

FIGURE 1.9. Conventional clock-gating method. CCP, control clock pin; GCKD, gated clock driver cell.

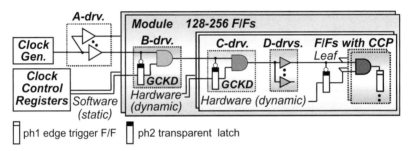

FIGURE 1.10. Clock-gating method of SH-X. CCP, control clock pin; GCKD, gated clock driver cell.

pipeline are explained in this section. Figure 1.9 illustrates a conventional clock-gating method. In this example, the clock tree has four levels with A-, B-, C-, and D-drivers. The A-driver receives the clock from the clock generator, and distributes the clock to each module in the processor. Then, the B-driver of each module receives the clock and distributes it to various submodules, including 128–256 flip-flops (F/Fs). The B-driver gates the clock with the signal from the clock control register, whose value is statically written by software to stop and start the modules. Next, the C- and D-drivers distribute the clock hierarchically to the leaf F/Fs with a Control Clock Pin (CCP). The leaf F/Fs are gated by hardware with the CCP to avoid activating them unnecessarily. However, the clock tree in the module is always active while the module is activated by software.

Figure 1.10 illustrates the clock-gating method of the SH-X. In addition to the clock gating at the B-driver, the C-drivers gate the clock with the signals dynamically generated by hardware to reduce the clock tree activity. As a result, the clock power is 30% less than that of the conventional method.

The superpipeline architecture improved operating frequency, but increased number of F/Fs and power. Therefore, one of the key design considerations

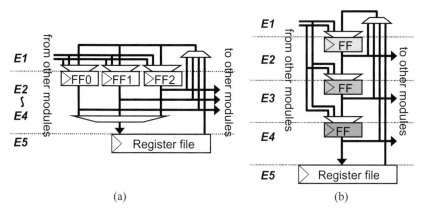

FIGURE 1.11. F/Fs of (a) pointer-controlled and (b) conventional pipelines.

TABLE 1.4. Relationship of F/Fs and Pipeline Stages

Pointer	FF0	FF1	FF2
0	E2	E4	E3
1	E3	E2	E4
2	E4	E3	E2

was to reduce the activity ratio of the F/Fs. To address this issue, a pointer-controlled pipeline was developed. It realizes a pseudo pipeline operation with a pointer control. As shown in Figure 1.11a, three pipeline F/Fs are connected in parallel, and the pointer is used to show which F/Fs correspond to which stages. Then, only one set of F/Fs is updated in the pointer-controlled pipeline, while all pipeline F/Fs are updated every cycle in the conventional pipeline, as shown in Figure 1.11b.

Table 1.4 shows the relationship between F/Fs FF0-FF2 and pipeline stages E2-E4 for each pointer value. For example, when the pointer indexes zero, the FF0 holds an input value at E2 and keeps it for three cycles as E2, E3, and E4 latches until the pointer indexes zero again and the FF0 holds a new input value. This method is good for a short latency operation in a long pipeline. The power of pipeline F/Fs decreases to 1/3 for transfer instructions, and decreases by an average of 25% as measured using Dhrystone 2.1.

1.3.5   Performance and Efficiency Evaluations

The SH-X performance was measured using the Dhrystone 2.1 benchmark, as well as those of the SH-3 and the SH-4. The Dhrystone is a popular benchmark for evaluating integer performance of embedded processors. It is small enough to fit all the program and data into the caches, and to use at the beginning of the processor development. Therefore, only the processor core architecture can be evaluated without the influence from the system level architecture, and the evaluation result can be fed back to the architecture design. On the contrary, the system level performance cannot be measured considering cache miss rates, external memory access throughput and latencies, and so on. The evaluation result includes compiler performance because the Dhrystone benchmark is described in C language.

Figure 1.12 shows the evaluated result of the cycle performance, architectural performance, and actual performance. Starting from the SH-3, five major enhancements were adopted to construct the SH-4 microarchitecture. The SH-3 achieved 1.0 MIPS/MHz when it was released, and the SH-4 compiler enhanced its performance to 1.1. The cycle performance of the SH-4 was enhanced to 1.81 MIPS/MHz by Harvard architecture, superscalar architecture, adding BO group, early-stage branch, and zero-cycle MOV operation. The SH-4 enhanced the cycle performance by 1.65 times form the SH-3, excluding the compiler contribution. The SH-3 was a 60-MHz processor in a 0.5-μm process, and estimated to be a 133-MHz processor in a 0.25-μm process. The SH-4 achieved 200 MHz in the same 0.25-μm process. Therefore, SH-4 enhanced the frequency by 1.5 times form the SH-3. As a result, the

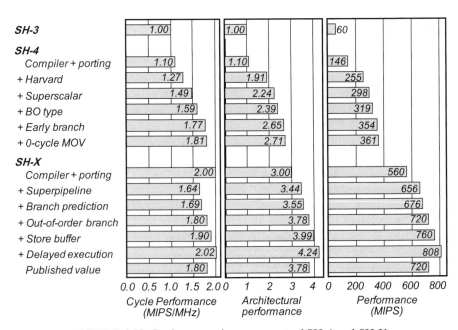

FIGURE 1.12. Performance improvement of SH-4 and SH-X.

architectural performance of the SH-4 is $1.65 \times 1.5 = 2.47$ times as high as that of the SH-3.

With adopting a conventional seven-stage superpipeline, the performance was decreased by 18% to 1.47 MIPS/MHz. Branch prediction, out-of-order branch issue, store buffer and delayed execution of the SH-X improve the cycle performance by 23%, and recover the 1.8 MIPS/MHz. Since 1.4 times high operating frequency was achieved by the superpipeline architecture, the architectural performance of the SH-X was also 1.4 times as high as that of the SH-4. The actual performance of the SH-X was 720 MIPS at 400 MHz in a 0.13-μm process, and improved by two times from the SH-4 in a 0.25-μm process.

Figures 1.13 and 1.14 show the area and power efficiency improvements, respectively. The upper three graphs of both the figures show architectural performance, relative area/power, and architectural area–/power–performance ratio. The lower three graphs show actual performance, area/power, and area–/power–performance ratio.

The area of the SH-X core was 1.8 mm² in a 0.13-μm process, and the area of the SH-4 was estimated as 1.3 mm² if it was ported to a 0.13-μm process. Therefore, the relative area of the SH-X was 1.4 times as much as that of the SH-4, and 2.26 times as much as the SH-3. Then, the architectural area efficiency of the SH-X was nearly equal to that of the SH-4, and 1.53 times as high as the SH-3. The actual area efficiency of the SH-X reached 400 MIPS/mm², which was 8.5 times as high as the 74 MIPS/ mm² of the SH-4.

SH-4 was estimated to achieve 200 MHz, 360 MIPS with 140 mW at 1.15 V, and 280 MHz, 504 MIPS with 240 mW at 1.25 V. The power efficiencies were 2,500 and 2,100 MIPS/W, respectively. On the other hand, SH-X achieved 200 MHz, 360 MIPS with 80 mW at 1.0 V, and 400 MHz, 720 MIPS with

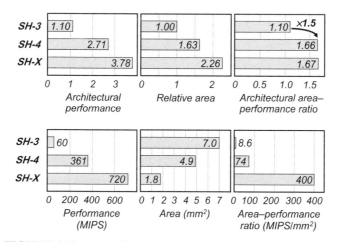

FIGURE 1.13. Area efficiency improvement of SH-4 and SH-X.

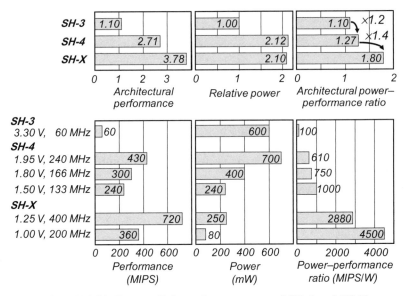

FIGURE 1.14. Power efficiency improvement of SH-4 and SH-X.

250 mW at 1.25 V. The power efficiencies were 4,500 and 2,880 MIPS/W, respectively. As a result, the power efficiency of the SH-X improved by 1.8 times from that of the SH-4 at the same frequency of 200 MHz, and by 1.4 times at the same supply voltage with enhancing the performance by 1.4 times. These were architectural improvements, and actual improvements were multiplied by the process porting.

1.4   SH-X FPU: A HIGHLY EFFICIENT FPU

The floating-point architecture and microarchitecture of the SH processors achieve high multimedia performance and efficiency. An FPU of the SH processor is highly parallel with keeping the efficiency for embedded systems in order to compensate the insufficient parallelism of the dual-issue superscalar architecture for highly parallel applications like 3D graphics.

In late 1990s, it became difficult to support higher resolution and advanced features of the 3D graphics. It was especially difficult to avoid overflow and underflow of fixed-point data with small dynamic range, and there was a demand to use floating-point data. Since it was easy to implement a four-way parallel operation with fixed-point data, equivalent performance had to be realized to change the data type to the floating-point format at reasonable costs.

Since an FPU was about three times as large as a fixed-point unit, and a four-way SMID required four times as large a datapath, it was too expensive

to integrate a four-way SMID FPU. The latency of the floating-point operations was long, and required more number of registers than the fixed-point
operations. Therefore, efficient parallelization and latency-reduction methods
had to be developed.

1.4.1 FPU Architecture of SH Processors

Sixteen is the limit of the number of registers directly specified by the 16-
bit fixed-length ISA, but the SH FPU architecture defines 32 registers as
two banks of 16 registers. The two banks are front and back banks, named
FR0-FR15 and XF0-XF15, respectively, and they are switched by changing a
control bit FPSCR.FR in a floating-point status and control register (FPSCR).
Most of instructions use only the front bank, but some instructions use both
the front and back banks. The front bank registers are used as eight pairs or
four length-4 vectors as well as 16 registers, and the back bank registers are
used as eight pairs or a four-by-four matrix. They are defined as follows:

$$DRn = (FRn, FR[n+1])(n: 0, 2, 4, 6, 8, 10, 12, 14),$$

$$FV0 = \begin{pmatrix} FR0 \\ FR1 \\ FR2 \\ FR3 \end{pmatrix}, FV4 = \begin{pmatrix} FR4 \\ FR5 \\ FR6 \\ FR7 \end{pmatrix}, FV8 = \begin{pmatrix} FR8 \\ FR9 \\ FR10 \\ FR11 \end{pmatrix}, FV12 = \begin{pmatrix} FR12 \\ FR13 \\ FR14 \\ FR15 \end{pmatrix},$$

$$XDn = (XFn, XF[n+1])(n: 0, 2, 4, 6, 8, 10, 12, 14),$$

$$XMTRX = \begin{pmatrix} XF0 & XF4 & XF8 & XF12 \\ XF1 & XF5 & XF9 & XF13 \\ XF2 & XF6 & XF10 & XF14 \\ XF3 & XF7 & XF11 & XF15 \end{pmatrix}.$$

Since an ordinary SIMD architecture of an FPU is too expensive for an
embedded processor as described above, another parallelism is applied to
the SH processors. The large hardware of an FPU is for a mantissa alignment
before the calculation and normalization and rounding after the calculation.
Further, a popular FPU instruction, FMAC, requires three read and one write
ports. The consecutive FMAC operations are a popular sequence to accumulate plural products. For example, an inner product of two length-4 vectors
is one of such sequences, and popular in a 3D graphics program. Therefore, a
floating-point inner-product instruction (FIPR) is defined to accelerate the
sequence with smaller hardware than that for the SIMD. It uses the two of
four length-4 vectors as input operand, and modifies the last register of one
of the input vectors to store the result. The defining formula is as follows:

$$FR[n+3] = FVm \cdot FVn(m, n: 0, 4, 8, 12).$$

This modifying-type definition is similar to the other instructions. However, for a length-3 vector operation, which is also popular, you can get the result without destroying the inputs, by setting one element of the input vectors to zero.

The FIPR produces only one result, which is one-fourth of a four-way SIMD, and can save the normalization and rounding hardware. It requires eight input and one output registers, which are less than the 12 input and four output registers for a four-way SIMD FMAC. Further, the FIPR takes much shorter time than the equivalent sequence of one FMUL and three FMACs, and requires small number of registers to sustain the peak performance. As a result, the hardware is about half of the four-way SIMD.

The rounding rule of the conventional floating-point operations is strictly defined by an American National Standards Institute/Institute of Electrical and Electronics Engineers (ANSI/IEEE) 754 floating-point standard. The rule is to keep accurate values before rounding. However, each instruction performs the rounding, and the accumulated rounding error sometimes becomes very serious. Therefore, a program must avoid such a serious rounding error without relying to hardware if necessary. The sequence of one FMUL and three FMACs can also cause a serious rounding error. For example, the following formula results in zero if we add the terms in the order of the formula by FADD instructions:

$$1.0 \times 2^{127} + 1.FFFFFE \times 2^{102} + 1.FFFFFE \times 2^{102} - 1.0 \times 2^{127}.$$

However, the exact value is $1.FFFFFE \times 2^{103}$, and the error is $1.FFFFFE \times 2^{103}$ for the formula, which causes the worst error of 2^{-23} times of the maximum term. We can get the exact value if we change the operation order properly. The floating-point standard defines the rule of each operation, but does not define the result of the formula, and either of the result is fine for the conformance. Since the FIPR operation is not defined by the standard, we defined its maximum error as "2^{E-25} + rounding error of result" to make it better than or equal to the average and worst-case errors of the equivalent sequence that conforms the standard, where E is the maximum exponent of the four products.

A length-4 vector transformation is also popular operation of a 3D graphics, and a floating-point transform vector instruction (FTRV) is defined. It requires 20 registers to specify the operands in a modification type definition. Therefore, the defining formula is as follows, using a four-by-four matrix of all the back bank registers, XMTRX, and one of the four front-bank vector registers, FV0-FV3:

$$FVn = XMTRX \cdot FVn(n: 0, 4, 8, 12).$$

Since a 3D object consists of a lot of polygons expressed by the length-4 vectors, and one XMTRX is applied to a lot of the vectors of a 3D object, the

XMTRX is not so often changed, and is suitable for using the back bank. The FTRV operation is implemented as four inner-product operations by dividing the XMTRX into four vectors properly, and its maximum error is the same as the FIPR.

The newly defined FIPR and FTRV can enhance the performance, but data transfer ability becomes a bottleneck to realize the enhancement. Therefore, a pair load/store/transfer mode is defined to double the data move ability. In the pair mode, floating-point move instructions (FMOVs) treat 32 front- and back-bank floating-point registers as 16 pairs, and directly access all the pairs without the bank switch controlled by the FPSCR.FR bit. The mode switch between the pair and normal modes is controlled by a move-size bit FPSCR. SZ in the FPSCR.

The 3D graphics requires high performance but uses only a single precision. On the other hand, a double precision format is popular for server/PC market, and would eases a PC application porting to a handheld PC. Although the performance requirement is not so high as the 3D graphics, software emulation is too slow compared with hardware implementation. Therefore, the SH architecture has single- and double-precision modes, which are controlled by a precision bit FPSCR.PR of the FPSCR. Further, a floating-point register-bank, move-size, and precision change instructions (FPCRG, FSCHG, and FRCHG) were defined for fast changes of the modes defined above. This definition can save the small code space of the 16-bit fixed length ISA. Some conversion operations between the precisions are necessary, but not fit to the mode separation. Therefore, SH architecture defines two conversion instructions in the double-precision mode. An FCNVSD converts a single-precision data to a double-precision one, and an FCNVDS converts vice versa. In the double-precision mode, eight pairs of the front-bank registers are used for double-precision data, and one 32-bit register, FPUL, is used for a single-precision or integer data, mainly for the conversion.

The FDIV and floating-point square-root instruction (FSQRT) are long latency instructions, and could cause serious performance degradations. The long latencies are mainly from the strict operation definitions by the ANSI/IEEE 754 floating-point standard. We have to keep accurate value before rounding. However, there is another way if we allow proper inaccuracies.

A floating-point square-root reciprocal approximate (FSRRA) is defined as an elementary function instruction to replace the FDIV, FSQRT, or their combination. Then we do not need to use the long latency instructions. 3D graphics applications especially require a lot of reciprocal and square-root reciprocal values, and the FSRRA is highly effective. Further, 3D graphics require less accuracy, and the single-precision without strict rounding is enough accuracy. The maximum error of the FSRRA is $\pm 2^{E-21}$, where E is the exponent value of an FSRRA result. The FSRRA definition is as follows:

$$FRn = \frac{1}{\sqrt{FRn}}.$$

A floating-point sine and cosine approximate (FSCA) is defined as another popular elementary function instruction. Once the FSRRA is introduced, extra hardware is not so large for the FSCA. The most poplar definition of the trigonometric function is to use radian for the angular unit. However, the period of the radian is 2π, and cannot be expressed by a simple binary number. Therefore, the FSCA uses fixed-point number of rotations as the angular expression. The number consists of 16-bit integer and 16-bit fraction parts. Then the integer part is not necessary to calculate the sine and cosine values by their periodicity, and the 16-bit fraction part can express enough resolution of $360/65{,}536 = 0.0055°$. The angular source operand is set to a CPU-FPU communication register FPUL because the angular value is a fixed-point number. The maximum error of the FSCA is $\pm 2^{-22}$, which is an absolute value and not related to the result value. Then the FSCA definition is as follows:

$$FRn = \sin(2\pi \cdot FPUL),\ FR[n+1] = \cos(2\pi \cdot FPUL).$$

1.4.2 Implementation of SH-X FPU

Table 1.5 shows the pitches and latencies of the FE-category instructions of the SH-3E, SH-4, and SH-X. As for the SH-X, the simple single-precision instructions of FADD, FSUB, FLOAT, and FTRC, have three-cycle latencies. Both single- and double-precision FCMPs have two-cycle latencies. Other single-precision instructions of FMUL, FMAC, and FIPR, and the double-precision instructions, except FMUL, FCMP, FDIV, and FSQRT, have five-cycle latencies. All the above instructions have one-cycle pitches.

The FTRV consists of four FIPR like operations resulting in four-cycle pitch and eight-cycle latency. The FDIV and FSQRT are out-of-order completion instructions having two-cycle pitches for the first and last cycles to initiate a special resource operation and to perform postprocesses of the result. Their pitches of the special resource expressed in the parentheses are about halves of the mantissa widths, and the latencies are four cycles more than the special-resource pitches. The FSRRA has one-cycle pitch, three-cycle pitch of the special resource, and five-cycle latency. The FSCA has three-cycle pitch, five-cycle pitch of the special resource, and seven-cycle latency. The double-precision FMUL has three-cycle pitch and seven-cycle latency.

Multiply–accumulate (MAC) is one of the most frequent operations in intensive computing applications. The use of four-way SIMD can achieve the same throughput as the FIPR, but the latency is longer and the register file has to be larger. Figure 1.15 illustrates an example of the differences according to the pitches and latencies of the FE-category SH-X instructions shown in Table 1.5. In this example, each box shows an operation issue slot. Since FMUL and FMAC have five-cycle latencies, we must issue 20 independent

TABLE 1.5. Pitch/Latency of FE-Category Instructions

Single Precision	SH-3E	SH-4	SH-X
FADD FRm, FRn	1/2	1/3	1/3
FSUB FRm, FRn	1/2	1/3	1/3
FMUL FRm, FRn	1/2	1/3	1/5
FDIV FRm, FRn	13/14	2 (10) /12	2 (13) /17
FSQRT FRn	13/14	2 (9) /11	2 (13) /17
FCMP/EQ FRm, FRn	1/1	1/2	1/2
FCMP/GT FRm, FRn	1/1	1/2	1/2
FLOAT FPUL, FRn	1/2	1/3	1/3
FTRC FRm, FPUL	1/2	1/3	1/3
FMAC FR0, FRm, FRn	1/2	1/3	1/5
FIPR FVm, FVn, FRn + 3	–	1/4	1/5
FTRV XMTRX, FVn	–	4/7	4/8
FSRRA FRn	–	–	1 (3) /5
FSCA FPUL, DRn	–	–	3 (5) /7

Double Precision	–	SH-4	SH-X
FADD DRm, DRn	–	6/8	1/5
FSUB DRm, DRn	–	6/8	1/5
FMUL DRm, DRn	–	6/8	3/7
FDIV DRm, DRn	–	5 (23) /25	2 (28) /32
FSQRT DRm, DRn	–	5 (22) /24	2 (28) /32
FCMP/EQ DRm,DRn	–	2/2	1/2
FCMP/EQ DRm,DRn	–	2/2	1/2
FLOAT DRn	–	2/4	1/5
FTRC DRm, FPUL	–	2/4	1/5
FCNVSD FPUL, FRn	–	2/4	1/5
FCNVDS DRm, FPUL	–	2/4	1/5

operations for peak throughput in the case of four-way SIMD. The result is available 20 cycles after the FMUL issue. On the other hand, five independent operations are enough to get the peak throughput of a program using FIPRs. Therefore, FIPR requires one-quarter of the program's parallelism and registers.

Figure 1.16 compares the pitch and latency of an FSRRA and the equivalent sequence of an FSQRT and an FDIV according to Table 1.5. Each of the FSQRT and FDIV occupies 2 and 13 cycles of the MAIN FPU and special resources, respectively, and takes 17 cycles to get the result, and the result is available 34 cycles after the issue of the FSQRT. In contrast, the pitch and latency of the FSRRA are one and five cycles that are only one-quarter and approximately one-fifth of those of the equivalent sequences,

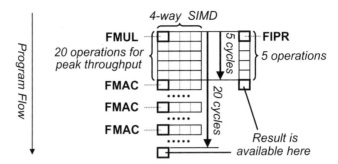

FIGURE 1.15. Four-way SIMD versus FIPR.

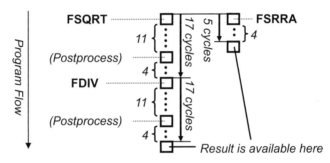

FIGURE 1.16. FSRRA versus equivalent sequence of FSQRT and FDIV.

respectively. The FSRRA is much faster using a similar amount of the hardware resource.

The FSRRA can compute a reciprocal as shown in Figure 1.17. The FDIV occupies 2 and 13 cycles of the MAIN FPU and special resources, respectively, and takes 17 cycles to get the result. On the other hand, the FSRRA and FMUL sequence occupies 2 and 3 cycles of the MAIN FPU and special resources, respectively, and takes 10 cycles to get the result. Therefore, the FSRRA and FMUL sequence is better than using the FDIV if an application does not require a result conforming to the IEEE standard, and 3D graphics is one of such applications.

Figure 1.18 illustrates the FPU arithmetic execution pipeline. With the delayed execution architecture, the register-operand read and forwarding are done at the E1 stage, and the arithmetic operation starts at E2. The short arithmetic pipeline treats three-cycle latency instructions. All the arithmetic pipelines share one register write port to reduce the number of ports. There are four forwarding source points to provide the specified latencies for any

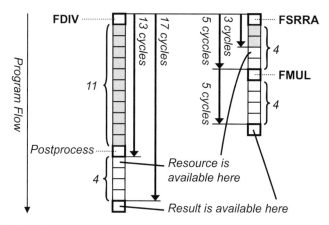

FIGURE 1.17. FDIV versus equivalent sequence of FSRRA and FMUL.

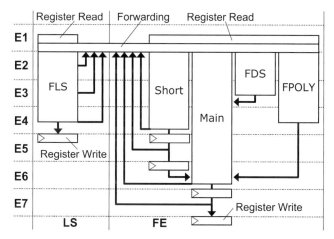

FIGURE 1.18. Arithmetic execution pipeline of SH-X FPU.

cycle distance of the define-and-use instructions. The FDS pipeline is occupied by 13/28 cycles to execute a single/double FDIV or FSQRT, and these instructions cannot be issued frequently. The FPOLY pipeline is three-cycles long and is occupied three or five times to execute an FSRRA or FSCA instruction. Therefore, the third E4 stage and E6 stage of the main pipeline are synchronized for the FSRRA, and the FPOLY pipeline output merges with the main pipeline at this point. The FSCA produce two outputs, and the first output is produced at the same timing of the FSRRA, and the second one is produced two-cycle later, and the main pipeline is occupied for three cycles, although

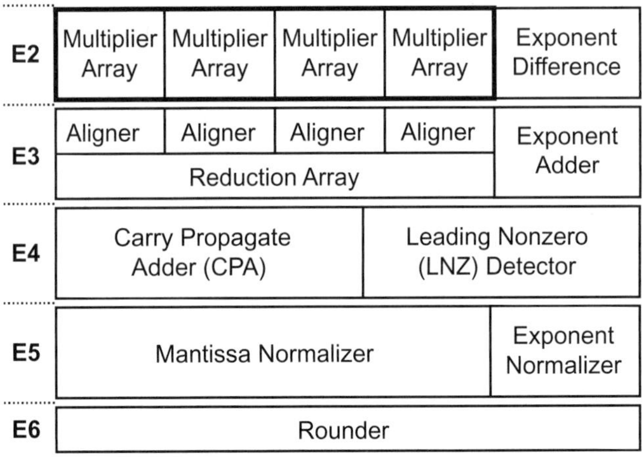

E2	Multiplier Array	Multiplier Array	Multiplier Array	Multiplier Array	Exponent Difference
E3	Aligner	Aligner	Aligner	Aligner	Exponent Adder
	Reduction Array				
E4	Carry Propagate Adder (CPA)		Leading Nonzero (LNZ) Detector		
E5	Mantissa Normalizer				Exponent Normalizer
E6	Rounder				

FIGURE 1.19. Main pipeline of SH-X FPU.

the second cycle is not used. The FSRRA and FSCA are implemented by calculating the cubic polynomials of the properly divided periods. The width of the third-order term is 8 bits, which adds only a small area overhead, while enhancing accuracy and reducing latency.

Figure 1.19 illustrates the structure of the main FPU pipeline. There are four single-precision multiplier arrays at E2 to execute FIPR and FTRV and to emulate double-precision multiplication. Their total area is less than that of a double-precision multiplier array. The calculation of exponent differences is also done at E2 for alignment operations by four aligners at E3. The four aligners align eight terms consisting of four sets of sum and carry pairs of four products generated by the four multiplier arrays, and a reduction array reduces the aligned eight terms to two at E3. The exponent value before normalization is also calculated by an exponent adder at E3. A carry propagate adder (CPA) adds two terms from the reduction array, and a leading nonzero (LNZ) detector searches the LNZ position of the absolute value of the CPA result from the two CPA inputs precisely and with the same speed as the CPA at E4. Therefore, the result of the CPA can be normalized immediately after the CPA operation with no correction of position errors, which is often necessary when using a conventional 1-bit error LNZ detector. Mantissa and exponent normalizers normalize the CPA and exponent-adder outputs at E5 controlled by the LNZ detector output. Finally, the rounder rounds the normalized results into the ANSI/IEEE 754 format. The extra hardware required for the special FPU instructions of the FIPR, FTRV, FSRRA and FSCA is about 30% of the original FPU hardware, and the FPU area is about 10–20% of the processor core depending on the size of the first and second on-chip memories. Therefore, the extra hardware is about 3–6% of the processor core.

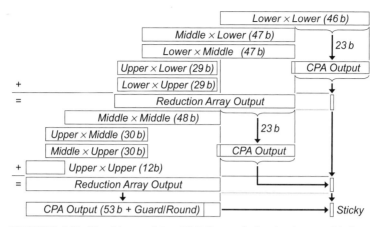

FIGURE 1.20. Double-precision FMUL emulation by four multipliers.

The SH-X FPU can use four 24-by-24 multipliers for the double-precision FMUL emulation. Since the double-precision mantissa width is more than twice of the single-precision one, we have to divide a multiplication into nine parts. Then we need three cycles to emulate the nine partial multiplications by four multipliers. Figure 1.20 illustrates the flow of the emulation. At the first step, a lower-by-lower product is produced, and its lower 23 bits are added by the CPA. Then the CPA output is ORed to generate a sticky bit. At the second step, four products of middle-by-lower, lower-by-middle, upper-by-lower, and lower-by-upper are produced and accumulated to the lower-by-lower product by the reduction array, and its lower 23 bits are also used to generate a sticky bit. At the third step, the remaining four products of middle-by-middle, upper-by-middle, middle-by-upper, and upper-by-upper are produced and accumulated to the already accumulated intermediate values. Then, the CPA adds the sum and carry of the final product, and 53-bit result and guard/round/sticky bits are produced. The accumulated terms of the second and third steps are 10 because each product consists of sum and carry, but the bitwise position of some terms are not overlapped. Therefore, the eight-term reduction array is enough to accumulate them.

1.4.3 Performance Evaluations with 3D Graphics Benchmark

The floating-point architecture was evaluated by a simple 3D graphics benchmark shown in Figure 1.21. It consists of coordinate transformations, perspective transformations, and intensity calculations of a parallel beam of light in Cartesian coordinates. A 3D-object surface is divided into triangular polygons to be treated by the 3D graphics. The perspective transformation assumes a flat screen expressed as $z = 1$. A strip model is used, which is a 3D object expression method to reduce the number of vertex vectors. In the model, each

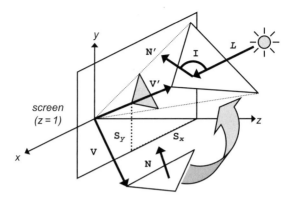

FIGURE 1.21. Simple 3D graphics benchmark.

triangle has three vertexes, but each vertex is shared by three triangles, and the number of vertex per triangle is one. The benchmark is expressed as follows, where T represents a transformation matrix, V and N represent vertex and normal vectors of a triangle before the coordinate transformations, respectively, N′ and V′ represent the ones after the transformations, respectively, S_x and S_y represent x and y coordinates of the projection of V′, respectively, L represents a vector of the parallel beam of light, I represents a intensity of a triangle surface, and V″ is an intermediate value of the coordinate transformations.

$$V'' = TV, V' = \frac{V''}{V''_w}, S_x = \frac{V'_x}{V'_z}, S_y = \frac{V'_y}{V'_z}, N' = TN, I = \frac{(L, N')}{\sqrt{(N', N')}},$$

$$T = \begin{pmatrix} T_{xx} & T_{xy} & T_{xz} & T_{xw} \\ T_{yx} & T_{yy} & T_{yz} & T_{yw} \\ T_{zx} & T_{zy} & T_{zz} & T_{zw} \\ T_{wx} & T_{wy} & T_{wz} & T_{ww} \end{pmatrix}, V = \begin{pmatrix} V_x \\ V_y \\ V_z \\ 1 \end{pmatrix}, V' = \begin{pmatrix} V'_x \\ V'_y \\ V'_z \\ 1 \end{pmatrix}, V'' = \begin{pmatrix} V''_x \\ V''_y \\ V''_z \\ V''_w \end{pmatrix},$$

$$N = \begin{pmatrix} N_x \\ N_y \\ N_z \\ 0 \end{pmatrix}, N' = \begin{pmatrix} N'_x \\ N'_y \\ N'_z \\ 0 \end{pmatrix}, L = \begin{pmatrix} L_x \\ L_y \\ L_z \\ 0 \end{pmatrix}.$$

The coordinate and perspective transformations require 7 FMULs, 12 FMACs, and 2 FDIVs without FTRV, FIPR, and FSRRA, and 1 FTRV, 5 FMULs, and 2 FSRRAs with them. The intensity calculation requires 7 FMULs, 12 FMACs, 1 FSQRT, and 1 FDIV without them, and 1 FTRV, 2 FIPRs, 1 FSRRA, and 1 FMUL with them.

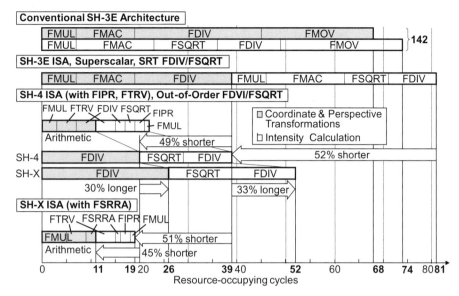

FIGURE 1.22. Resource occupying cycles for a 3D benchmark.

Figure 1.22 shows the resource-occupying cycles of the SH-3E, SH-4, and SH-X. After program optimization, no register conflict occurs, and performance is restricted only by the floating-point resource-occupying cycles. The gray areas of the graph represent the cycles of the coordinate and perspective transformations.

The Conventional SH-3E architecture takes 68 cycles for coordinate and perspective transformations, and 142 cycles when intensity is also calculated. Applying superscalar architecture and SRT method for FDIV/FSQRT with keeping the SH-3E ISA, they become 39 and 81 cycles, respectively. The SH-4 architecture having the FIPR/FTRV and the out-of-order FDIV/FSQRT makes them 20 and 39 cycles, respectively. The performance is good, but only the FDIV/FSQRT resource is busy in this case. Further, applying the super-pipeline architecture with keeping the SH-4 ISA, they become 26 and 52 cycles, respectively. Although the operating frequency grows higher by the super-pipeline architecture, the cycle performance degradation is serious, and almost no performance gain is achieved. In the SH-X ISA case with the FSRRA, they become 11 and 19 cycles, respectively. Clearly, the FSRRA solves the long pitch problem of the FDIV/FSQRT.

Since we emphasized the importance of the efficiency, we evaluated the area and power efficiencies. Figure 1.23 shows the area efficiencies. The upper half shows architectural performance, relative area, and architectural area–performance ratio to compare the area efficiencies with no process porting effect. According to the above cycles, the relative cycle performance of the coordinate and perspective transformations of the SH-4 and SH-X to the

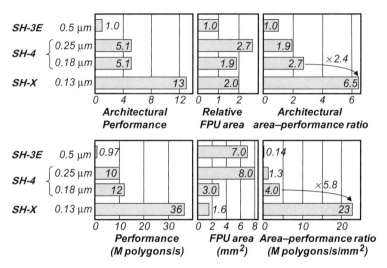

FIGURE 1.23. Area efficiencies of SH-3E, SH-4, and SH-X.

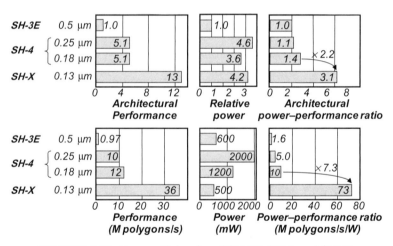

FIGURE 1.24. Power efficiencies of SH-3E, SH-4, and SH-X.

SH-3E are 68/20 = 3.4 and 68/11 = 6.2, respectively. As explained in Section 1.3.5, the relative frequency of the SH-4 and SH-X are 1.5 and 2.1, respectively. Then the architectural performance of the SH-4 and SH-X are 3.4 × 1.5 = 5.1 and 6.2 × 2.1 = 13, respectively. Although the relative areas increased, the performance improvements are much higher, and the efficiency is greatly enhanced. The lower half shows real performance, area, and area–performance ratio. The frequencies of the SH-3E, SH-4 in 0.25- and 0.18-μm and SH-X are 66, 200, 240, and 400 MHz, respectively. The efficiency is further enhanced using the finer process. Similarly, the power efficiency is also enhanced greatly, as shown in Figure 1.24.

1.5 SH-X2: FREQUENCY AND EFFICIENCY ENHANCED CORE

An SH-X2 was developed as the second-generation core, and achieved performance of 1,440 MIPS at 800 MHz using a 90-nm process. The low power version achieved the power efficiency of 6,000 MIPS/W. The performance and efficiency are greatly enhanced from the SH-X by both the architecture and micro-architecture tuning and the process porting.

1.5.1 Frequency Enhancement

According to the SH-X analyzing, the ID stage was the most critical timing part, and the branch acceleration successfully reduced the branch penalty. Therefore, we added the third instruction fetch stage (I3) to the SH-X2 pipeline to relax the ID stage timing. The cycle performance degradation was negligible small by the successful branch architecture, and the SH-X2 achieved the same cycle performance of 1.8 MIPS/MHz as the SH-X.

Figure 1.25 illustrates the pipeline structure of the SH-X2. The I3 stage was added, and performs branch search and instruction predecoding. Then the ID stage timing was relaxed, and the achievable frequency increased.

Another critical timing path was in first-level (L1) memory access logic. SH-X had L1 memories of a local memory and I- and D-caches, and the local memory was unified for both instruction and data accesses. Since all the memories could not be placed closely, a memory separation for instruction and data was good to relax the critical timing path. Therefore, the SH-X2 separated the unified L1 local memory of the SH-X into instruction and data local memories (ILRAM and OLRAM). With the other various timing tuning, the SH-X2 achieved 800 MHz using a 90-nm generic process from the SH-X's 400 MHz using a 130-nm process. The improvement was far higher than the process porting effect.

I1	Out-of-Order	Instruction Fetch				
I2	Branch					
I3	Branch Search / Instruction Predecoding					
ID	Branch	Instruction Decoding		FPU Instruction Decoding		
E1			Address			
E2		Execution	Data Load	Tag	FPU Data Transfer	FPU Arithmetic Execution
E3				-		
E4		WB	WB	Data Store		
E5					WB	
E6			Store Buffer			
E7			Flexible Forwarding			WB
	BR	INT	LS		FE	

FIGURE 1.25. Eight-stage superpipeline structure of SH-X2.

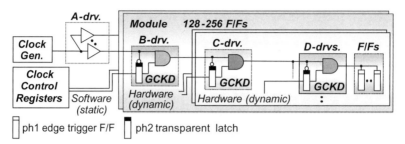

FIGURE 1.26. Clock-gating method of SH-X2. GCKD, gated clock driver cell.

1.5.2 Low Power Technologies

The SH-X2 enhanced the low power technologies from that of the SH-X. Figure 1.26 shows the clock gating method of the SH-X2. The D-drivers also gate the clock with the signals dynamically generated by hardware, and the leaf F/Fs requires no CCP. As a result, the clock tree and total powers are 14 and 10% lower, respectively, than in the SH-X method.

The SH-X2 adopted a way prediction method to the instruction cache. The SH-X2 aggressively fetched the instructions using branch prediction and early branch techniques to compensate branch penalty caused by long pipeline. The power consumption of the instruction cache reached 17% of the SH-X2, and the 64% of the instruction cache power was consumed by data arrays. The way prediction misses were less than 1% in most cases, and was 0% for the Dhrystone 2.1. Then, the 56% of the array access was eliminated by the prediction for the Dhrystone. As a result, the instruction cache power was reduced by 33%, and the SH-X2 power was reduced by 5.5%.

1.6 SH-X3: MULTICORE ARCHITECTURE EXTENSION

Continuously, the SH cores has achieved high efficiency as described above. The SH-X3 core is the third generation of the SH-4A processor core series to achieve higher performance with keeping the high-efficiency maintained in all the SH core series. The multicore architecture is the next approach for the series.

1.6.1 SH-X3 Core Specifications

Table 1.6 shows the specifications of an SH-X3 core designed based on the SH-X2 core. The most of the specifications are the same as that of the SH-X2 core as the successor of it. In addition to such succeeded specifications,

TABLE 1.6. SH-X3 Processor Core Specifications

ISA	SuperH 16-Bit Encoded ISA
Pipeline structure	Dual-issue superscalar 8-stage pipeline
Operating frequency	600 MHz (90-nm generic CMOS process)
Performance	
Dhrystone 2.1	1080 MIPS
FPU (Peak)	4.2/0.6 GFLOPS (single/double)
Caches	8–64 KB I/D each
Local memories	
1st/2nd level	4–128 KB I/D each/128 KB to 1 MB
Power/power efficiency	360 mW/3,000 MIPS/W
Multiprocessor support	
SMP support	Coherency for data caches (up to 4 cores)
AMP support	DTU for local memories
Interrupt	Interrupt distribution and Inter-processor interrupt
Low power modes	Light sleep, sleep, and resume standby
Power management	Operating frequency and low power mode can be different for each core.

the core supports both symmetric and asymmetric multiprocessor (SMP and AMP) features with interrupt distribution and interprocessor interrupt, in corporate with an interrupt controller of such SoCs as RP-1 and RP-2. Each core of the cluster can be set to one of the SMP and AMP modes individually.

It also supports three low power modes of light sleep, sleep, and resume standby. The new light-sleep mode is to respond to a snoop request from the SNC while the core is inactive. In this mode, the data cache is active for the snoop operation, but the other modules are inactive.

In a chip multiprocessor, the core loads are not equal, and each SH-X3 core can operate at a different operating frequency and in a different low power mode to minimize the power consumption for the load. The core can support the SMP features even such heterogeneous operation modes of the cores.

1.6.2 Symmetric and Asymmetric Multiprocessor Support

The four SH-X3 cores constitute a cluster sharing an SNC and a DBG to support symmetric multiprocessor (SMP) and multicore debug features. The SNC has a duplicated address array (DAA) of data caches of all the four cores, and is connected to the cores by a dedicated snoop bus separated from the SuperHyway to avoid both deadlock and interference by some cache coherency protocol operations. The DAA minimizes the number of data cache accesses of the cores for the snoop operations, resulting in the minimum coherency maintenance overhead.

The supported SMP data cache coherency protocols are standard MESI (modified, exclusive, shared, invalid) and ESI modes for copyback and write-through modes, respectively. The copyback and MESI modes are good for performance, and the write-through and ESI modes are suitable to control some accelerators that cannot control the data cache of the SH-X3 cores properly.

The SH-X3 outputs one of the following snoop requests of the cache line to the SNC with the line address and write-back data, if any:

1. Invalidate request for write and shared case.
2. Fill-data request for read and cache-miss case.
3. Fill-data and invalidate request for write and cache-miss case.
4. Write-back request to replace a dirty line.

The SNC transfers a request other than a write-back one to proper cores by checking its DAA, and the requested core processes the requests.

The on-chip RAMs and the data transfer among the various memories are the key features for the AMP support. The use of on-chip RAM makes it possible to control the data access latency, which cannot be controlled well in systems with on-chip caches. Therefore, each core integrates L1 instruction and data RAMs, and a second-level (L2) unified RAM. The RAMs are globally addressed to transfer data to/from the other globally addressed memories. Then, application software can place data in proper timing and location. The SH-X3 integrates a data transfer unit (DTU) to accelerate the data transfer to/from the other modules.

1.6.3   Core Snoop Sequence Optimization

Each core should operate at the proper frequency for its load, but in some cases of the SMP operation, a low frequency core can cause a long stall of a high frequency core. We optimized the cache snoop sequences for the SMP mode to minimize such stalls. Table 1.7 summarizes the coherency overhead cycles. These cycles vary according to various conditions; the table indicates a typical case.

Figure 1.27a,b show examples of core snoop sequences before and after the optimization. The case shown is a "write access to a shared line," which is the third case in the table. The operating frequencies of core #0, #1, and #2 are 600, 150, and 600 MHz, respectively. Initially, all the data caches of the cores hold a common cache line, and all the cache-line states are "shared."

Sequence (a) is as follows:

1. *Core Snoop Request*: Core #0 stores data in the cache, changes the stored-line state from "Shared" to "Modified," and sends a "Core Snoop Request" of the store address to the SNC.

TABLE 1.7. Coherency Overhead Cycles

Cache Line State			Overhead (SCLK Cycles)			
			Snooped Core: 600 MHz		Snooped Core: 150 MHz	
Access Type	Accessed Core	Snooped Core	Not Optimized	Optimized	Not Optimized	Optimized
Read	S, E, M	–	0	0	0	0
Write	E, M	–	0	0	0	0
	S	S	10	**4**	19	**4**
Read or Write	Miss	Miss	5	5	5	5
		S	10	**5**	19	**5**
		E	10	10	19	19
		M	13	13	22	22

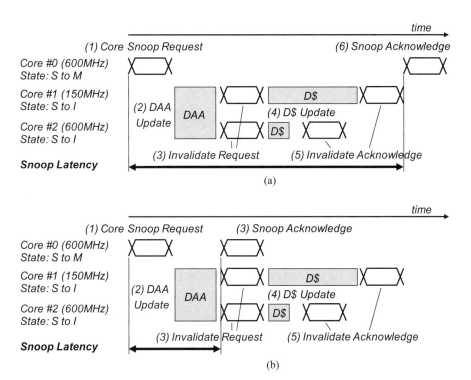

FIGURE 1.27. Core snoop sequences (a) before and (b) after optimization.

2. *DAA Update*: The SNC searches the DAA of all the cores, and changes the states of the hit lines from "Shared" to "Modified" for core #0 and "Invalid" for cores #1 and #2. The SNC runs at SCLK frequency (300 MHz).

3. *Invalidate Request*: The SNC sends "Invalidate Request" to cores #1 and #2.

4. *Data Cache Update*: Cores #1 and #2 change the states of the corresponding cache lines from "Shared" to "Invalid." The processing time depends on each core's ICLK.

5. *Invalidate Acknowledge*: Cores #1 and #2 return "Invalidate Acknowledge" to the SNC.

6. *Snoop Acknowledge*: The SNC returns "Snoop Acknowledge" to core #0.

As shown in Figure 1.27a, the return from core #1 is late due to its low frequency, resulting in long snoop latency.

Sequence (b) is as follows by the optimization:

1. Core Snoop Request
2. DAA Update
3. Snoop Acknowledge and Invalidate Request
4. Data Cache Update
5. Invalidate Acknowledge

The "Snoop Acknowledge" is moved from the 6th to the 3rd step by eliminating the wait of the "Invalidate Acknowledge," and the late response of the slow core does not affect the operation of the fast core. In the optimized sequence, the SNC is busy for some cycles after the "Snoop Acknowledge," and the next "Core Snoop Request" must wait if the SNC is still busy. However, this is rare for ordinary programs.

The sequence of another case, a "read miss and hit to another core's modified line," which is the last case in the table, is as follows:

1. *Core Snoop Request*: A data read of core #0 misses its cache and sends a "Core Snoop Request" of the access address to the SNC.

2. *DAA Update*: The SNC searches the DAA of all the cores, and changes the states of the hit lines from "Modified" to "Shared."

3. *Data Transfer Request*: The SNC sends a "Data Transfer Request" to the core of the hit line for the cache fill data of core #0.

4. *Data Cache Update*: The requested core reads the requested data and changes the states of the corresponding line of the DAA to "Shared." The processing time depends on each core's ICLK.

5. *Data Transfer Response and Write-Back Request*: The requested core returns the requested data and requests a write back to the SNC.
6. *Snoop Acknowledge and Write-Back Request*: The SNC returns "Snoop Acknowledge" to core #0 with the fill data, and requests a write-back of the returned data to the main memory.
7. *Data Cache Update 2*: Core #0 completes the "Read" operation by replacing a cache line with the fill data.

In this case, core #0 must wait for the fill data, and the early "Snoop Acknowledge" is impossible.

1.6.4   Dynamic Power Management

Each core can operate at different CPU clock (ICLK) frequencies and can stop individually while other processors are running with a short switching time in order to achieve both the maximum processing performance and the minimum operating power for various applications. A data cache coherency is maintained during operations at different frequencies, including frequencies lower than the on-chip system bus clock (SCLK). The following four schemes make it possible to change each ICLK frequency individually while maintaining data cache coherency.

1. Each core has its own clock divider for an individual clock frequency change.
2. A handshake protocol is executed before the frequency change to avoid conflicts in bus access, while keeping the other cores running.
3. Each core supports various ICLK frequency ratios to SCLK, including a lower frequency than that of SCLK.
4. Each core has a light-sleep mode to stop its ICLK while maintaining data cache coherency.

The global ICLK and the SCLK that run up to 600 and 300 MHz, respectively, are generated by a global clock pulse generator (GCPG) and distributed to each core. Both the global ICLK and SCLK are programmable by setting the frequency control register in the GCPG. Each local ICLK is generated from the global ICLK by the clock divider of each core. The local CPG (LCPG) of a core executes a handshake sequence dynamically when the frequency control register of the LCPG is changed, so that it can keep the other cores running and can maintain coherency in data transfers of the core. The previous approach assumed a low frequency in a clock frequency change, and it stopped all the cores when a frequency was changed. The core supports "light-sleep mode" to stop its ICLK except for its data cache in order to maintain the data cache coherency. This mode is effective for reducing the power of an SMP system.

1.6.5 RP-1 Prototype Chip

The RP-1 is the first multicore chip with four SH-X3 CPU cores. It supports both symmetric and asymmetric multiprocessor (SMP and AMP) features for embedded applications. The SMP and AMP modes can be mixed to construct a hybrid system of the SMP and AMP. Each core can operate at different frequencies and can stop individually with maintaining its data-cache coherency while the other processors are running in order to achieve both the maximum processing performance and the minimum operating power for various applications.

1.6.5.1 RP-1 Specifications Table 1.8 summarizes the RP-1 specifications. It was fabricated as a prototype chip using a 90-nm CMOS process to accelerate the research and development of various embedded multicore systems. The RP-1 achieved a total of 4,320 MIPS at 600 MHz by the four SH-X3 cores measured using the Dhrystone 2.1 benchmark. The RP-1 integrates four SH-X3 cores with a snoop controller (SNC) to maintain the data-cache coherency among the cores, DDR2-SDRAM and SRAM memory interfaces, a PCI-Express interface, some HW-IPs for various types of processing, and some peripheral modules. The HW-IPs include a DMA controller, a display unit, and accelerators. Each SH-X3 core includes a 32-kB four-way set-associative instruction and data caches, an 8-kB instruction local RAM (ILRAM), a 16-kB operand local RAM (OLRAM), and a 128-kB unified RAM (URAM).

Figure 1.28 illustrates a block diagram of the RP-1. The four SH-X3 cores, a snoop controller (SNC), and a debug module (DBG) constitute a cluster.

TABLE 1.8. RP-1 Specifications

Process technology	90-nm, 8-layer Cu, triple-Vth, CMOS
Chip size/power	97.6 mm² (9.88 mm × 9.88 mm)/3 W (typical, 1.0 V)
Supply voltage/clock frequency	1.0 V (internal), 1.8/3.3 V(I/O)/600 MHz
SH-X3 core	
Size	2.60 mm × 2.80 mm
I/D-cache	32-kB four-way set-associative (each)
ILRAM/OLRAM/URAM	8/16/128 KB (unified)
Snoop controller (SNC)	Duplicated Address Array (DAA) of four D-caches
Centralized shared memory (CSM)	128 kB
External interfaces	DDR2-SDRAM, SRAM, PCI-Express
Performance	
CPU	4,320 MIPS (Dhrystone 2.1, 4 core total)
FPU	16.8 GFLOPS (peak, 4 core total)
Package	FCBGA 554 pin, 29 × 29 mm

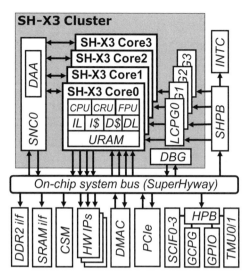

FIGURE 1.28. Block diagram of RP-1. SNC, snoop controller (cntl.); DAA, duplicated address array; CRU, cache RAM control unit; I$/D$, instruction (inst.)/data cache; IL/DL, Inst./data local memory; URAM, unified RAM; DBG, debug module; GCPG/LCPG, global/local CPG; INTC, interrupt cntl.; SHPB,HPB, peripheral bus bridge; CSM, centralized shared memory; DMAC, direct memory access cntl.; PCIe, PCIexpress interface (i/f); SCIF, serial communication i/f; GPIO, general purpose IO; TMU, timer unit.

The HW-IPs are connected to an on-chip system bus (SuperHyway). The arrows to/from the SuperHyway indicate connections from/to initiator/target ports, respectively.

1.6.5.2 Chip Integration and Evaluations   Figure 1.29 shows the chip micrograph of the RP-1. The chip was integrated in two steps to minimize the design period of the physical integration, and successfully fabricated: (1) First, a single core was laid out as a hard macro and completed timing closure of the core, and (2) the whole chip was laid out with instancing the core four times.

We evaluated the processing performance and power reduction in parallel processing on the RP-1. Figure 1.30 plots the time required to execute the SPLASH-2 suite [46] depending on the number of threads on an SMP Linux system. The RP-1 reduced the processing time to 50.5–52.6% and 27.1–36.9% with two and four threads, respectively. The time should be 50 and 25% for ideal performance scalability. The major overhead was synchronization and snoop time. The SNC improved cache coherency performance, and the performance overhead by snoop transactions was reduced to up to 0.1% when SPLASH-2 was executed.

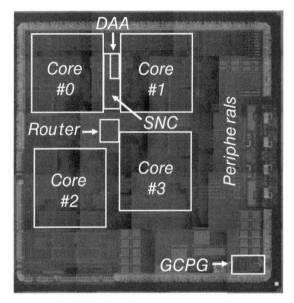

FIGURE 1.29. Chip micrograph of RP-1.

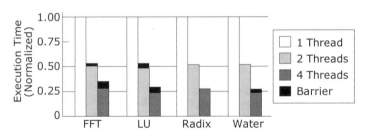

FIGURE 1.30. Execution time of SPLASH-2 suite.

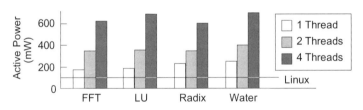

FIGURE 1.31. Active power of SPLASH-2 suite.

Figure 1.31 shows the power consumption of the SPLASH-2 suite. The suite ran at 600 MHz and at 1.0 V. The average power consumption of one, two, and four threads was 251, 396, and 675 mW, respectively. This included 104 mW of active power for the idle tasks of SMP Linux. The results of the performance and power evaluation showed that the power efficiency was maintained or enhanced when the number of threads increased.

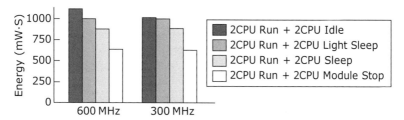

FIGURE 1.32. Energy consumption with low power modes.

Figure 1.32 shows the energy consumption with low power modes. These modes were implemented to save power when fewer threads were running than available on CPU cores. As a benchmark, two threads of fast Fourier transform (FFT) were running on two CPU cores, and two CPU cores were idle. The energy consumed in the light sleep, sleep, and module stop modes at 600 MHz was 4.5, 22.3, and 44.0% lower than in the normal mode, respectively, although these modes took some time to stop and start the CPU core and to save and return the cache. The execution time increased by 79.5% at 300 MHz, but the power consumption decreased, and the required energy decreased by 5.2%.

1.6.6 RP-2 Prototype Chip

The RP-2 is a prototype multicore chip with eight SH-X3 CPU cores. It was fabricated in a 90-nm CMOS process that was the same process used for the RP-1. The RP-2 achieved a total of 8,640 MIPS at 600 MHz by the eight SH-X3 cores measured with the Dhrystone 2.1 benchmark. Because it is difficult to lay out the eight cores close to each other, we did not select a tightly coupled cluster of eight cores. Instead, the RP-2 consists of two clusters of four cores, and the cache coherency is maintained in each cluster. Therefore, the intercluster cache coherency must be maintained by software if necessary.

1.6.6.1 RP-2 Specifications Table 1.9 summarizes the RP-2 specifications. The RP-2 integrates eight SH-X3 cores as two clusters of four cores, DDR2-SDRAM and SRAM memory interfaces, DMA controllers, and some peripheral modules. Figure 1.33 illustrates a block diagram of the RP-2. The arrows to/from the SuperHyway indicate connections from/to initiator/target ports, respectively.

1.6.6.2 Power Domain and Partial Power Off Power-efficient chip design for embedded applications requires several independent power domains where the power of unused domains can be turned off. The power domains were initially introduced to an SoC for mobile phones [5], which defined 20 hierarchical power domains. In contrast, high performance multicore chips use leaky low-Vt transistors for CPU cores, and reducing the leakage power of such cores is the primary goal.

TABLE 1.9. RP-2 Specifications

Process technology	90-nm, 8-layer Cu, triple-Vth, CMOS
Chip size/power	104.8 mm²/2.8 W (typical, 1.0 V, Dhrystone 2.1)
Supply voltage/clock frequency	1.0 V (internal), 1.8/3.3 V(I/O)/600 MHz
SH-X3 core	
Size	6.6 mm² (3.36 × 1.96 mm)
I/D-cache	16-kB four-way set-associative (each)
ILRAM/OLRAM/URAM	8/32/64 kB (unified)
CSM/external interfaces	128 kB/DDR2-SDRAM, SRAM
Performance	
CPU	8,640 MIPS (Dhrystone 2.1, 8 core total)
FPU	33.6 GFLOPS (peak, 8 core total)

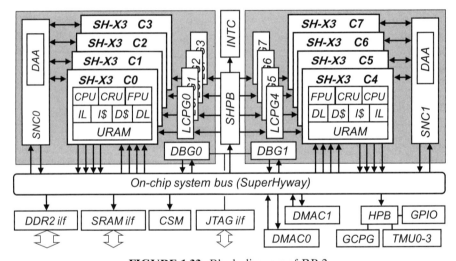

FIGURE 1.33. Block diagram of RP-2.

The RP-2 was developed for target use in power-efficient high performance embedded applications. Sixteen power domains were defined so that they can be independently powered off. A resume-standby mode was also defined for fast resume operation, and the power levels of the CPU and the URAM of a core are off and on, respectively. Each processor core can operate at a different frequency or even dynamically stop the clock to maintain processing performance while reducing the average operating power consumption.

Figure 1.34 illustrates the power domain structure of eight CPU cores with eight URAMs. Each core is allocated to a separate power domain so that

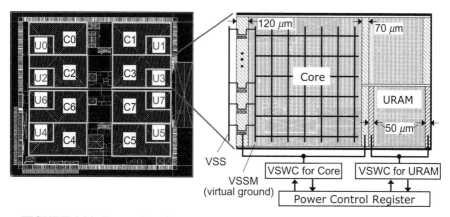

FIGURE 1.34. Power domain structure of eight CPU cores with eight URAMs.

TABLE 1.10. Power Modes of CPU Cores

CPU Power Modes	Normal	Light Sleep	Sleep	Resume	Power Off
Clock for CPU and URAM	On	Off	Off	Off	Off
Clock for I/D Cache	On	On	Off	Off	Off
Power supply for CPU	On	On	On	Off	Off
Power supply for URAM	On	On	On	On	Off
Leakage current (mA)[a]	162	162	162	22	0

[a] Measured at room temperature at 1.0 V, eight-core total.

the power supply can be cut off while unused. Two power domains (Cn and Un, for n ranging from 0 to 7) are assigned to each core, where Un is allocated only for URAM. By keeping the power of Un on, the CPU status is saved to URAM before the Cn power is turned off, and restored from URAM after Cn power is turned on. This shortens the restart time compared with a power-off mode in which both Cn and Un are powered off together. Each power domain is surrounded by power switches and controlled by a power switch controller (VSWC).

Table 1.10 summarizes the power modes of each CPU. Light sleep mode is suitable for dynamic power saving while cache coherency is maintained. In sleep mode, almost all clocks for the CPU core are stopped. In resume standby mode, the leakage current for eight cores is reduced to 22 from 162 mA in sleep mode, and leakage power is reduced by 86%.

1.6.6.3 Synchronization Support Hardware   The RP-2 has barrier registers to support CPU core synchronization for multiprocessor systems. Software can

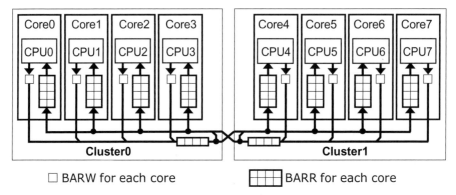

FIGURE 1.35. Barrier registers for synchronization.

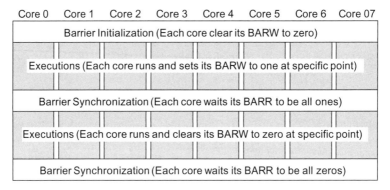

FIGURE 1.36. Synchronization example using barrier registers.

use these registers for fast synchronization between the cores. In the synchronization, one core waits for other cores to reach a specific point in a program. Figure 1.35 illustrates the barrier registers for the synchronization. In a conventional software solution, the cores have to test and set a specific memory location, but this requires long cycles. We provide three sets of barrier registers to accelerate the synchronization. Each CPU core has a 1-bit barrier write (BARW) register that it notifies when it reaches a specific point. The BARW values of all the cores are gathered by hardware to form an 8-bit barrier read (BARR) register of each core so that each core can obtain all the BARW values from its BARR register with a single instruction. As a result, the synchronization is fast and does not disturb other transactions on the Super-Hyway bus.

Figure 1.36 shows an example of the barrier register usage. In the beginning, all the BARW values are initialized to zero. Then each core inverts its BARW

TABLE 1.11. Eight-Core Synchronization Cycles

	Conventional Method (via External Memory)	RP-2 Method (via BARW/BARR registers)
Average clock cycles	52,396	8,510
Average difference	20,120	10

value when it reaches a specific point, and it checks and waits until all its BARR values are ones reflecting the BARW values. The synchronization is complete when all the BARW values are inverted to ones. The next synchronization can start immediately with the BARWs being ones, and is complete when all the BARW values are inverted to zeros.

Table 1.11 compares the results of eight-core synchronizations with and without the barrier registers. The average number of clock cycles required for a certain task to be completed with and without barrier registers is 8,510 and 52,396 cycles, respectively. The average differences in the synchronizing cycles between the first and last cores are 10 and 20,120 cycles, with and without the barrier registers, respectively. These results show that the barrier registers effectively improve the synchronization.

1.6.6.4 Chip Integration and Evaluations   The RP-2 was fabricated using the same 90-nm CMOS process as that for the RP-1. Figure 1.37 is the chip micrograph of the RP-2. It achieved a total of 8,640 MIPS at 600 MHz by the eight SH-X3 cores measured with the Dhrystone 2.1 benchmark, and consumed 2.8 W at 1.0 V, including leakage power.

The fabricated RP-2 chip was evaluated using the SPLASH-2 benchmarks on an SMP Linux OS. Figure 1.38 plots the RP-2 execution time on one cluster based on the number of POSIX threads. The processing time was reduced to 51–63% with two threads and to 41–27% with four or eight threads running on one cluster. Since there were fewer cores than threads, the eight-thread case showed similar performance to the four-thread one. Furthermore, in some cases, the increase in the number of threads resulted in an increase in the processing time due to the synchronization overhead.

1.7 SH-X4: ISA AND ADDRESS SPACE EXTENSION

Continuously, embedded systems expand their application fields, and enhance their performance and functions in each field. As a key component of the system, embedded processors must enhance their performance and functions with maintaining or enhancing their efficiencies. As the latest SH processor core, the SH-X4 extended its ISA and address space efficiently for this purpose.

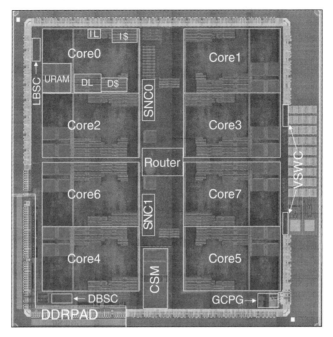

FIGURE 1.37. Chip micrograph of RP-2.

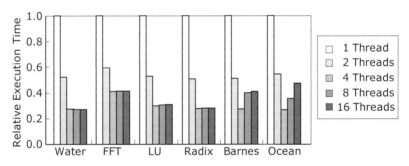

FIGURE 1.38. RP-2 execution time according to number of POSIX threads.

The SH-X4 was integrated on the RP-X heterogeneous multicore chip as two 4-core clusters with four Flexible Engine/Generic ALU Arrays (FE-GAs) [47, 48], two MX-2 matrix processors [49], a Video Processing Unit 5 (VPU5) [50, 51], and various peripheral modules.

1.7.1 SH-X4 Core Specifications

Table 1.12 shows the specifications of an SH-X4 core designed based on the SH-X3 core. The most of the specifications are the same as those of the SH-X3

TABLE 1.12. SH-X4 Processor Core Specifications

ISA	SuperH 16-Bit ISA with Prefix Extension
Operating frequency	648 MHz (45-nm low power CMOS process)
Performance	
Dhrystone 2.1	1,717 MIPS (2.65 MIPS/MHz)
FPU (peak)	4.5/0.6 GFLOPS (single/double)
Power/power efficiency	106 mW/16 GIPS/W
Address space	
Logical	32 bits, 4 GB
Physical	40 bits, 1 TB

core, and the same ones are not shown. The SH-X4 extended the ISA with some prefixes, and the cycle performance is enhanced from 2.23 to 2.65 MIPS/ MHz. As a result, the SH-X4 achieved 1,717 MIPS at 648 MHz. The 648 MHz is not so high compared with the 600 MHz of the SH-X3, but the SH-X4 achieved the 648 MHz in a low power process. Then, the typical power consumption is 106 mW, and the power efficiency reached as high as 16 GIPS/W.

1.7.2 Efficient ISA Extension

The 16-bit fixed-length ISA of the SH cores is an excellent feature enabling a higher code density than that of 32-bit fixed-length ISAs of conventional RISCs. However, we made some trade-off to establish the 16-bit ISA. Operand fields are carefully shortened to fit the instructions into the 16 bits according to the code analysis of typical embedded programs in the early 1990s. The 16-bit ISA was the best choice at that time and the following two decades. However, required performance grew higher and higher, program size and treating data grew larger and larger. Therefore, we decided to extend the ISA by some prefix codes.

The week points of the 16-bit ISA are (1) short-immediate operand, (2) lack of three-operand operation instructions, and (3) implicit fixed-register operand. The short-immediate ISA uses a two-instruction sequence of a long-immediate load and a use of the loaded data, instead of a long immediate instruction. A three-operand operation becomes a two-instruction sequence of a move instruction and a two-operand instruction. The implicit fixed-register operand makes register allocation difficult, and causes inefficient register allocations.

The popular ISA extension from the 16-bit ISA is a variable-length ISA. For example, an IA-32 is a famous variable-length ISA, and ARM Thumb-2 is a variable-length ISA of 16 and 32 bits. However, a variable-length instruction consists of plural unit-length codes, and each unit-length code has plural meaning depending on the preceding codes. Therefore, the variable-length ISA causes complicated, large, and slow parallel-issue logic with serial code analysis.

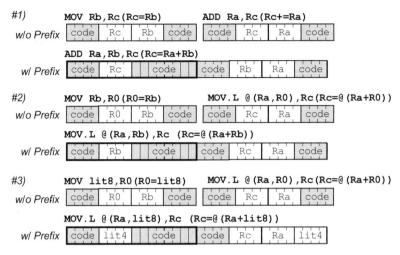

FIGURE 1.39. Examples of ISA extension.

Another way is using prefix codes. The IA-32 uses some prefixes, as well as the variable-length instructions, and using prefix codes is one of the conventional ways. However, if we use the prefix codes but not use the variable-length instructions, we can implement a parallel instruction decoding easily. The SH-X4 introduced some 16-bit prefix codes to extend the 16-bit fixed-length ISA.

Figure 1.39 shows some examples of the ISA extension. The first example (#1) is an operation "Rc = Ra + Rb (Ra, Rb, Rc: registers)", which requires a two-instruction sequence of "MOV Ra, Rc (Rc = Ra)" and "ADD Rb, Rc (Rc + = Rb)" before extension, but only one instruction "ADD Ra, Rb, Rc" after the extension. The new instruction is made of the "ADD Ra, Rb" by a prefix to change a destination register operand Rb to a new register operand Rc. The code sizes are the same, but the number of issue slots reduces from two to one. Then the next instruction can be issued simultaneously if there is no other pipeline stall factor.

The second example (#2) is an operation "Rc = @(Ra + Rb)," which requires a two-instruction sequence of "MOV Rb, R0 (R0 = Rb)" and "MOV.L @(Ra, R0), Rc (Rc = @(Ra + R0))" before extension, but only an instruction "MOV.L @(Ra, Rb), Rc" after the extension. The new instruction is made of the "MOV @(Ra, R0), Rc" by a prefix to change the R0 to a new register operand. Then we do not need to use the R0, which is the third implicit fixed operand with no operand field to specify. It makes the R0 busy and register allocation inefficient to use the R0-fixed operand, but the above extension solve the problem.

The third example (#3) is an operation "Rc = @(Ra + lit8) (lit8: 8-bit literal)," which requires a two-instruction sequence of "MOV lit8, R0

(R0 = lit8)" and "MOV.L @(Ra, R0), Rc (Rc = @(Ra + R0))" before extension, but only an instruction "MOV.L @(Ra, lit8), Rc" after the extension. The new instruction is made of the "MOV.L @(Ra, lit4), Rc (lit4: 4-bit literal)" by a prefix to extend the lit4 to lit8. The prefix can specify the loaded data size in memory and the extension type of signed or unsigned if the size is 8 or 16 bits, as well as the extra 4-bit literal.

Figure 1.40 illustrates the instruction decoder of the SH-X4 enabling a dual issue, including extended instructions by prefix codes. The gray parts are the extra logic for the extended ISA. Instruction registers at the I3 stage hold first four 16-bit codes, which was two codes for the conventional 16-bit fixed-length ISA. The simultaneous dual-issue of the instructions with prefixes consumes the four codes per cycle at peak throughput. Then, a predecoder checks each code in parallel if it is a prefix or not, and outputs control signals of multiplexers MUX to select the inputs of prefix and normal decoders properly.

The Table 1.13 summarizes all cases of the input patterns and corresponding selections. A code after the prefix code is always a normal code, and hardware

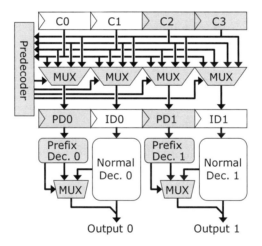

FIGURE 1.40. Instruction decoder of SH-X4.

TABLE 1.13. Input Patterns and Selections

Input				Output			
C0	C1	C2	C3	PD0	ID0	PD1	ID1
N	N	–	–	–	C0	–	C1
N	P	–	–	–	C0	C1	C2
P	–	N	–	C0	C1	–	C2
P	–	P	–	C0	C1	C2	C3

P, prefix; n, normal; –, arbitrary code.

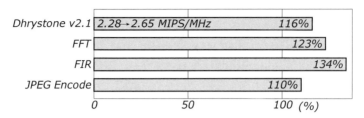

FIGURE 1.41. Performance improvement ratio by prefix codes.

need not check it. Each prefix decoder decodes a provided prefix code, and overrides the output of the normal decoder appropriately. As a result, the instruction decoder performs the dual issue of instructions with prefixes.

Figure 1.41 shows evaluation results of the extended ISA with four benchmark programs. The performance of Dhrystone 2.1 was accelerated from 2.24 to 2.65 MIPS/MHz by 16%. The performance of FFT, finite impulse response (FIR), and JPEG encoding were improved by 23, 34, and 10%, respectively. On the other hand, area overhead of the prefix code implementation was less than 2% of the SH-X4. This means the ISA extension by the prefix codes enhanced both performance and efficiency.

1.7.3 Address Space Extension

The 32-bit address can define an address space of 4 GB. The space consists of main memory, on-chip memories, various IO spaces, and so on. Then the maximum linearly addressed space is 2 GB for the main memory. However, the total memory size is continuously increasing, and will soon exceed 2 GB even in an embedded system. Therefore, we extended the number of physical address bits to 40 bits, which can define 1-TB address space. The logical address space remains 32-bit, and the programming model is unchanged. Then the binary compatibility is maintained. The logical address space extension would require the costly 32- to 64-bit extensions of register files, integer executions, branch operations, and so on.

Figure 1.42 illustrates an example of the extension. The 32-bit logical address space is compatible to the predecessors of the SH-X4. The MMU translates the logical address to a 32/40-bit physical address by TLB or privileged mapping buffer (PMB) in 32/40-bit physical address mode, respectively. The TLB translation is a well-known dynamic method, but the original PMB translation is a static method to avoid exceptions possible for the TLB translation. Therefore, the PMB page sizes are larger than that of the TLB to cover the PMB area efficiently.

The logical space is divided into five regions, and the attribute of each region can be specified as user-mode accessible or inaccessible, translated by TLB or PMB, and so on. In the example, the P0/U0 region is user-mode

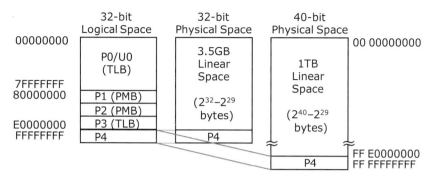

FIGURE 1.42. An example of logical and physical address spaces of SH-X4.

accessible and translated by TLB, the P1 and P2 region are user-mode inaccessible and translated by PMB, and the P3 region is user-mode inaccessible and translated by TLB. The P4 region includes a control register area that is mapped on the bottom of physical space so that the linear physical space is not divided by the control register area.

1.7.4 Data Transfer Unit

High-speed and efficient data transfer is one of the key features for multicore performance. The SH-X4 core integrates a DTU for this purpose. A DMAC is conventional hardware for the data transfer. However, the DTU has some advantage to the DMAC, because the DTU is a part of an SH-X4 core. For example, when a DMAC transfer the data between a memory in an SH-X4 core and a main memory, the DMAC must initiate two SuperHyway bus transactions between the SH-X4 core and the DMAC and between the DMAC and the main memory. On the other hand, the DTU can perform the transfer with one SuperHyway bus transaction between the SH-X4 core and the main memory. In addition, the DTU can use the initiator port of the SH-X4 core, whereas the DMAC must have its own initiator port, and even if all the SH-X4 cores have a DTU, no extra initiator port is necessary. Another merit is that the DTU can share the unified TLB (UTLB) of the SH-X4 core, and the DTU can handle a logical address.

Figure 1.43 shows an example of a data transfer between an SH-X4 core and an FE-GA. The DTU has a transfer TLB (TTLB) as a micro TLB that caches UTLB entries of the CPU for independent executions. The DTU can get a UTLB entry when the translation misses the TTLB. The DTU action is defined by a command chain in a local memory. The DTU can execute the command chain of plural commands without CPU control. In the example, the DTU transfers data in a local memory of the SH-X4 to a memory in the FE-GA. The source data specified by the source address from the command

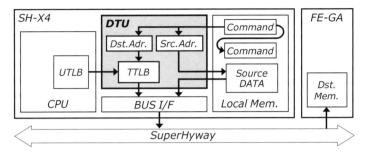

FIGURE 1.43. An example of DTU transfer between SH-X4 and FE-GA.

is read from the local memory, and the destination address specified by the command is translated by the TTLB. Then the address and data are output to the SuperHyway via the bus interface, and the data are transferred to the destination memory of the FE-GA.

1.7.5 RP-X Prototype Chip

A heterogeneous multicore is one of the most promising approaches to attain high performance with low frequency and power, for consumer electronics or scientific applications. The RP-X is the latest prototype multicore chip with eight SH-X4 cores, four FE-GAs, two MX-2s, a VPU5, and various peripheral modules. It was fabricated using 45-nm CMOS process. The RP-X achieved 13.7 GIPS at 648 MHz by the eight SH-X4 cores measured using the Dhrystone 2.1 benchmark, and a total of 114.7 GOPS with 3.07 W. It attained a power efficiency of 37.3 GOPS/W.

1.7.5.1 RP-X Specifications The RP-X specifications are summarized in Table 1.14. It was fabricated using a 45-nm CMOS process, integrating eight SH-X4 cores, four FE-GAs, two MX-2s, one VPU5, one SPU, and various peripheral modules as a heterogeneous multicore SoC, which is one of the most promising approaches to attain high performance with low frequency and power, for consumer electronics or scientific applications.

The eight SH-X4 cores achieved 13.7 GIPS at 648 MHz measured using the Dhrystone 2.1 benchmark. Four FE-GAs, dynamically reconfigurable processors, were integrated and attained a total performance of 41.5GOPS and a power consumption of 0.76 W. Two 1024-way MX-2s were integrated and attained a total performance of 36.9GOPS and a power consumption of 1.10 W. Overall, the efficiency of the RP-X was 37.3 GOPS/W at 1.15 V, excluding special-purpose cores of a VPU5 and an SPU. This was the highest among comparable processors. The operation granularity of the SH-X4, FE-GA and MX-2 processors are 32, 16, and 4 bits, respectively, and thus, we can assign the appropriate processor cores for each task in an effective manner.

TABLE 1.14. RP-X Specifications

Process technology	45-nm, 8-layer Cu, triple-Vth, CMOS
Chip size	153.76 mm^2 (12.4 mm × 12.4 mm)
Supply voltage	1.0–1.2 V (internal), 1.2/1.5/1.8/2.5/3.3 V (I/O)
Clock frequency	648 MHz (SH-X4), 324 MHz (FE-GA, MX-2)
Total power consumption	3.07 W (648 MHz, 1.15 V)
Processor cores and performances	
8× SH-X4 CPU	13.7 GIPS (Dhrystone 2.1, 8-core total)
FPU	36.3 GFLOPS (8-core total)
4× FE-GA	41.5 GOPS (4-core total)
2× MX-2	36.9 GOPS (2-core total)
Programmable special purpose cores	VPU5 (video processing unit) for MPEG2, H.264, VC-1
	SPU (sound processing unit) for AAC, MP3
Total performances and power	114.7 GOPS, 3.07 W, 37.3 GOPS/W (648 MHz, 1.15 V)
External interfaces	2× DDR3-SDRAM (32-bit, 800 MHz), SRAM, PCIexpress (rev 2.0, 2.5 GHz, 4 lanes), serial ATA

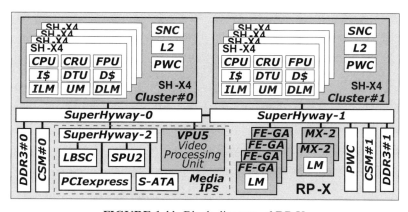

FIGURE 1.44. Block diagram of RP-X.

Figure 1.44 illustrates the structure of the RP-X. The processor cores of the SH-X4, FE-GA, and MX-2, the programmable special purpose cores of the VPU5 and SPU, and the various modules are connected by three SuperHyway buses to handle high-volume and high-speed data transfers. SuperHyway-0 connects the modules for an OS, general tasks, and video processing, SuperHyway-1 connects the modules for media acceleration, and SuperHyway-2 connects media IPs except for the VPU5. Some peripheral buses and modules are not shown in the figure.

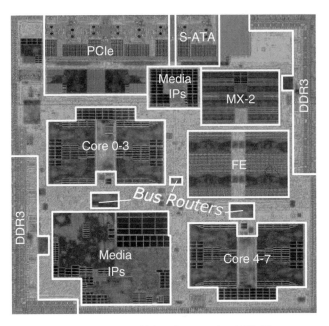

FIGURE 1.45. Chip micrograph of RP-X.

A DTU was implemented in each SH-X4 core to transfer data to and from the special-purpose cores or various memories without using CPU instructions. In this kind of system, multiple OSes are used to control various functions, and thus high-volume and high-speed memories are required.

1.7.5.2 Chip Integration and Evaluations The RP-X was fabricated using a 45-nm low power CMOS process. A chip micrograph of the RP-X is in Figure 1.45. It achieved a total of 13,738 MIPS at 648 MHz by the eight SH-X4 cores measured using the Dhrystone 2.1 benchmark, and consumed 3.07 W at 1.15 V including leakage power.

The RP-X is a prototype chip for consumer electronics or scientific applications. As an example, we produced a digital TV prototype system with IP networks (IP-TV), including image recognition and database search. Its system configuration and memory usage are shown in Figure 1.46. The system is capable of decoding 1080i audio/video data using a VPU and an SPU on the OS#1. For image recognition, the MX-2s are used for image detection and feature quantity calculation, and the FE-GAs are used for optical flow calculation of a VGA (640 × 480) video at 15 fps on the OS#2. These operations required 30.6 and 0.62 GOPS of the MX-2 and FE-GA, respectively. The SH-X4 cores are used for database search using the results of the above operations on the OS#3, as well as supporting of all the processing, including OS#1, OS#2, OS#3,

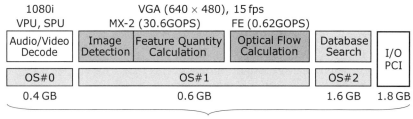

FIGURE 1.46. System configuration and memory usage of a digital TV.

TABLE 1.15. Performance and Power Consumption of RP-X

	Operating Frequency	Performance	Power	Power Efficiency
SH-X4	648 MHz	36.3 GFLOPS	0.74 W	49.1 GFLOPS/W
MX-2	324 MHz	36.9 GOPS	0.81 W	45.6 GOPS/W
FE-GA	324 MHz	41.5 GOPS	1.12 W	37.1 GOPS/W
Others	324/162/81 MHz	–	0.40 W	–
Total	–	114.7 GOPS	3.07 W	37.3 GOPS/W

and data transfers between the cores. The main memories of 0.4, 0.6, 1.6, and 1.8 GB are assigned to OS#1, OS#2, OS#3, and PCI, respectively, for a total of 4.4 GB.

Table 1.15 lists the total performance and power consumption at 1.15 V when eight CPU cores, four FE-GAs, and two MX-2s are used at the same time. The power efficiency of the CPU cores, FE-GAs, and MX-2s reached 42.9 GFLOPS/W, 41.5 GOPS/W, and 36.9 GOPS/W, respectively. The power consumption of the other components was reduced to 0.40 W by clock gating of 31 out of 44 modules. In total, if we count 1 GFLOPS as 1 GOPS, the RP-X achieved 37.3 GOPS/W at 1.15 V, excluding I/O area power consumption.

REFERENCES

[1] P.P. Gelsinger, "Microprocessors for the new millennium challenges, opportunities, and new frontiers," in ISSCC Digest of Technical Papers, Session 1.3, Feb. 2001.

[2] F. Arakawa, "Multicore SoC for embedded systems," in International SoC Design Conference (ISOCC) 2008, Nov. 2008, pp. I-180–I-183.

[3] R.P. Weicker, "Dhrystone: a synthetic programming benchmark," Communications of ACM, 27(10), Oct. 1984, pp. 1013–1030.

[4] R.P. Weicker, "Dhrystone benchmark: rationale for version 2 and measurement rules," ACM SIGPLAN Notices, 23(8), Aug. 1988, pp. 49–62.

[5] T. Hattori et al., "A power management scheme controlling 20 power domains for a sin-gle-chip mobile processor," in ISSCC Dig. Tech. Papers, Session 29.5, Feb., 2006.

[6] M. Ito et al., "A 390 MHz single-chip application and dual-mode baseband processor in 90 nm triple-Vt CMOS," in ISSCC Dig. Tech. Papers, Session 15.3, Feb. 2007.

[7] M. Naruse et al., "A 65 nm single-chip application and dual-mode baseband processor with partial clock activation and IP-MMU," in ISSCC Dig. Tech. Papers, Session 13.3, Feb. 2008.

[8] M. Ito et al., "A 65 nm single-chip application and dual-mode baseband processor with par-tial clock activation and IP-MMU," IEEE Journal of Solid-State Circuits, 44(1), Jan. 2009, pp. 83–89.

[9] K. Hagiwara et al., "High performance and low power SH2A-DUAL core for embedded microcontrollers," in COOL Chips XI Proceedings, Session XI, no. 2, April 2008. 1.

[10] H. Kido et al., "SoC for car navigation systems with a 53.3 GOPS image recognition engine," in HOT CHIPS 21, Session 6, no 3, Aug. 2009.

[11] R.G. Daniels, "A participant's perspective," IEEE Micro, 16(2), Apr. 1996, pp. 8–15.

[12] L. Gwennap, "CPU technology has deep roots," Microprocessor Report, 10(10), Aug. 1996, pp. 9–13.

[13] H. Nakamura et al., "A circuit methodology for CMOS microcomputer LSIs," in ISSCC Dig. Tech. Papers, Feb. 1983, pp. 134–135.

[14] S. Kawasaki, "SH-II A low power RISC microprocessor for consumer applications," in HOT Chips VI, Aug. 1994, pp. 79–103.

[15] A. Hasegawa et al., "SH-3: high code density, low power," IEEE Micro, 15(6), Dec. 1995, pp. 11–19.

[16] F. Arakawa et al., "SH4 RISC multimedia microprocessor," in HOT Chips IX Symposium Record, pp. 165–176, Aug. 1997.

[17] O. Nishii et al., "A 200 MHz 1.2 W 1.4GFLOPS microprocessor with graphic operation unit," in ISSCC Dig. Tech. Papers, Feb. 1998, pp. 288–289, 447.

[18] F. Arakawa et al., "SH4 RISC multimedia microprocessor," IEEE Micro, 18(2), March/April 1998, pp. 26–34.

[19] P. Biswas et al., "SH-5: the 64 bit SuperH architecture," IEEE Micro, 20(4), July/Aug. 2000, pp. 28–39.

[20] K. Uchiyama et al., "Embedded processor core with 64-bit architechture and its system-on-chip integration for digital consumer products," IEICE Transactions on Electronics, E84-C(2), Feb. 2001, pp. 139–149.

[21] F. Arakawa, "SH-5: a first 64-bit SuperH core with multimedia extension," in HOT Chips 13 Conference Record, Aug. 2001.

[22] F. Arakawa et al., "An embedded processor core for consumer appliances with 2.8GFLOPS and 36M polygons/s FPU," in ISSCC Digest of Technical Papers, vol. 1, Feb. 2004, pp. 334–335, 531.

[23] M. Ozawa et al., "Pipeline structure of SH-X core for achieving high performance and low power," in COOL Chips VII Proceedings, vol. I, pp. 239–254, April 2004.

[24] F. Arakawa et al., "An embedded processor core for consumer appliances with 2.8GFLOPS and 36M polygons/s FPU," IEICE Transactions on Fundamentals, E87-A(12), Dec. 2004, pp. 3068–3074.

[25] F. Arakawa et al., "An exact leading non-zero detector for a floating-point unit," IEICE Transactions on Electronics, E88-C(4), April 2005, pp. 570–575.

[26] F. Arakawa et al., "SH-X: an embedded processor core for consumer appliances," ACM SIGARCH Computer Architecture News, 33(3), June 2005, pp. 33–40.

[27] T. Kamei et al., "A resume-standby application processor for 3G cellular phones," in ISSCC Dig. Tech. Papers, Feb. 2004, pp. 336–337, 531.

[28] M. Ishikawa et al., "A resume-standby application processor for 3G cellular phones with low power clock distribution and on-chip memory activation control," in COOL Chips VII Proceedings, Vol. I, April 2004, pp. 329–351.

[29] M. Ishikawa et al., "A 4500 MIPS/W, 86 μA resume-standby, 11 μA ultra-standby application processor for 3G cellular phones," IEICE Transactions on Electronics, E88-C(4), April 2005, pp. 528–535.

[30] T. Yamada et al., "Low_power design of 90-nm SuperHTM processor core," in Proceedings of 2005 IEEE International Conference on Computer Design (ICCD), Oct. 2005, pp. 258–263.

[31] F. Arakawa et al., "SH-X2: an embedded processor core with 5.6 GFLOPS and 73M polygons/s FPU," in 7th Workshop on Media and Streaming Processors (MSP-7), Nov. 2005, pp. 22–28.

[32] T. Yamada et al., "Reducing consuming clock power optimization of a 90 nm embedded processor core," IEICE Transactions on Electronics, E89–C(3), March 2006, pp. 287–294.

[33] T. Kamei, "SH-X3: enhanced SuperH core for low-Power multi-processor systems," in Fall Microprocessor Forum 2006, Oct. 2006.

[34] F. Arakawa, "An embedded processor: is it ready for high-performance computing?" in IWIA 2007 Jan. 2007, pp. 101–109.

[35] Y. Yoshida et al., "A 4320 MIPS four-prcessor core SMP/AMP with individually managed clock frequency for low power consumption," in ISSCC Dig. Tech. Papers, Session 5.3, Feb. 2007.

[36] S. Shibahara et al., "SH-X3: flexible SuperH multi-core for high-performance and low-power embedded systems," in HOT CHIPS 19, Session 4, no 1, Aug. 2007.

[37] O. Nishii et al., "Design of a 90 nm 4-CPU 4320 MIPS SoC with individually managed frequency and 2.4 GB/s multi-master on-chip interconnect," in Proc. 2007 A-SSCC, Nov. 2007, pp. 18–21.

[38] M. Takada et al., "Performance and power evaluation of SH-X3 multi-core system," in Proc. 2007 A-SSCC, Nov. 2007, pp. 43–46.

[39] M. Ito et al., "An 8640 MIPS SoC with independent power-off control of 8 CPUs and 8 RAMs by an automatic parallelizing compiler," in ISSCC Dig. Tech. Papers, Session 4.5, Feb. 2008.

[40] Y. Yoshida et al., "An 8 CPU SoC with independent power-off control of CPUs and multicore software debug function," in COOL Chips XI Proceedings, Session IX, no. 1, April 2008.

[41] H.T. Hoang et al., "Design and performance evaluation of an 8-processor 8640 MIPS SoC with overhead reduction of interrupt handling in a multi-core system," in Proc. 2008 A-SSCC, Nov. 2008, pp. 193–196.

[42] Y. Yuyama et al., "A 45 nm 37.3GOPS/W heterogeneous multi-core SoC," in ISSCC Dig., Feb. 2010, pp. 100–101.

[43] T. Nito et al., "A 45 nm heterogeneous multi-core SoC supporting an over 32-bits physical address space for digital appliance," in COOL Chips XIII Proceedings, Session XI, no. 1, April 2010.

[44] F. Arakawa, "Low power multicore for embedded systems," in COMS Emerging Technology 2011, Session 5B, no. 1, June 2011.

[45] G. Hinton et al., "A 0.18-μm CMOS IA-32 processor with a 4-GHz integer execution unit," IEEE Journal of Solid-State Circuits, 36(11), Nov. 2001, pp. 1617–1627.

[46] S.C. Woo et al., "The SPLASH-2 programs: characterization and methodological considerations," in Proc. ISCA, 1995.

[47] M. Ito et al., ""Heterogeneous multiprocessor on a chip which enables 54x AAC-LC stereo encoding," in IEEE 2007 Symp. VLSI, June 2007, pp. 18–19.

[48] H. Shikano et al., "Heterogeneous multi-core architecture that enables 54x AAC-LC stereo encoding," IEEE Journal of Solid-State Circuits, 43(4), April 2008, pp. 902–910.

[49] T. Kurafuji et al., "A scalable massively parallel processor for real-time image processing," in IEEE Int. Solid-State Circuits Conf. Dig. Tech. Papers, Feb. 2010, pp. 334–335.

[50] K. Iwata et al., "256 mW 40 Mbps Full-HD H.264 high-Profile codec featuring a dual-macroblock pipeline architecture in 65 nm CMOS," IEEE Journal of Solid-State Circuits, 44(4), Apr. 2009, pp. 1184–1191.

[51] K. Iwata et al., "A 342 mW mobile application processor with full-HD multi-standard video codec and tile-based address-translation circuits," IEEE Journal of Solid-State Circuits, 45(1), Jan. 2010, pp. 59–68.

2 Special-Purpose Hardware for Computational Biology

SIDDHARTH SRINIVASAN

Computer simulations of proteins and other biomolecules have been the subject of significant research over the past few decades for a variety of reasons. These include improved understanding of the structure–function relationships within a protein; engineering of biomolecules to meet specific targets such as improved stability or binding specificity; and gaining a deeper insight into the interactions between proteins, and within proteins themselves, to better understand the mechanics of these biological entities.

Specifically, both molecular dynamics (MD) and Monte Carlo (MC) simulation techniques have been applied to study the conformational degrees of freedom of biomolecules. Biomolecules of interest for therapeutic purposes, such as antibodies and antibody derivatives, are generally very large in size, with $3N$ degrees of freedom, where N is the number of atoms in the protein. Both simulation techniques typically employ physics-based "force fields" to computationally estimate the interactions between the atoms and propagate motion in discrete intervals, which collectively describe one possible pathway of conformational change undertaken by that protein. While MD techniques rely on Newton's laws of physics to propagate motion and are a wholly deterministic algorithm, MC techniques rely on random displacements followed by a scoring function of some sort to determine the propagation of motion. In this chapter, we will study the implementation of both types of algorithms on specialized hardware architectures, intended to maximize their performance over what can be achieved using traditional central processing units (CPUs) and parallelization techniques.

Embedded Systems: Hardware, Design, and Implementation, First Edition.
Edited by Krzysztof Iniewski.
© 2013 John Wiley & Sons, Inc. Published 2013 by John Wiley & Sons, Inc.

2.1 MOLECULAR DYNAMICS SIMULATIONS ON GRAPHICS PROCESSING UNITS

MD simulations have been one of the most effective theoretical methods of exploring the conformational flexibility of biomolecules, by modeling the microscopic interactions between the individual atoms that make up a biomolecule, as well as the interactions between the biomolecule and its surrounding environment. The macroscopic properties of these biological systems are obtained from the microscopic information via statistical mechanics, which provide the rigorous mathematical models necessary to relate these properties with the positions and velocities of the constituent atoms. In MD, successive configurations of the system are generated from the current configuration using Newton's laws of motion. Each successive configuration depends on the current state of the system and the forces between the various particles. The trajectory of the system over time is obtained by solving the differential equations embodied in Newton's second law ($F = ma$), where the force F, acting on a particle in a particular direction, is proportional to its mass m and the acceleration a of the particle in that particular direction (Figure 2.1).

The force acting on a particle is thus dependent on the positions of every other particle in the system, and this results in a computational complexity of $O(N^2)$, where N is the number of particles in the system. At each time step, the positions of all the particles in the system are propagated by integrating these equations of motion, typically via finite difference methods. In such methods (known as *integration* methods), the total force on each particle at time t is calculated as the vector sum of its interactions with other particles. From the force, we obtain the acceleration of each particle using Newton's second law, which is then combined with the positions and velocities of the particles at time t to obtain the new positions and velocities at time $t + \delta t$. A

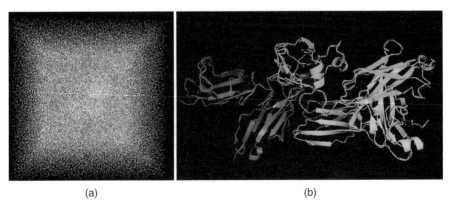

(a)   (b)

FIGURE 2.1. Snapshots of protein MD simulation. (a) Protein in periodic water box; (b) protein after 20 ns simulation.

standard integrator used in many MD engines is the *velocity Verlet* method [1], which allows for synchronized computations of velocities ($\mathbf{v}$), positions ($\mathbf{r}$), and accelerations ($\mathbf{a}$) at the same time and does not compromise precision:

$$\mathbf{r}(t + \delta t) = \mathbf{r}(t) + \delta t \mathbf{v}(t) + \frac{1}{2} \delta t^2 \mathbf{a}(t), \tag{2.1}$$

$$\mathbf{v}(t + \delta t) = \mathbf{v}(t) + \frac{1}{2} \delta t [\mathbf{a}(t) + \mathbf{a}(t + \delta t)]. \tag{2.2}$$

Thus, at each time step the relevant properties are computed, and the new positions and velocities are used to create a trajectory of the system over time.

2.1.1 Molecular Mechanics Force Fields

The energies and forces in an MD simulation are described by a *force field*, which specifies the functional form and parameters of the mathematical functions used to characterize the potential energy (PE) of a system of particles. Force field functions and parameter sets are derived from both experimental work and high-level quantum mechanical calculations. The standard force fields used for all-atom MD simulations typically consist of terms representing the bonded and nonbonded interactions between particles and are described in the following section.

2.1.1.1 Bond Term This term describes the harmonic stretching of a bond between a pair of connected particles. The potential is a harmonic potential of the form

$$V_{\text{bond}} = k(r - r_0)^2, \tag{2.3}$$

where r is the actual distance between the particles, r_0 is the equilibrium (or ideal) distance between them, and k is a force constant determining the strength of the bond.

2.1.1.2 Angle Term This term describes the harmonic stretching of an angle between three connected particles, and takes the form

$$V_{\text{angle}} = k(\theta - \theta_0)^2, \tag{2.4}$$

where θ is the angle between the particles, θ_0 is the equilibrium angle, and k is a force constant indicating the rigidity of the angle.

2.1.1.3 Torsion Term The torsion (or dihedral) term accounts for the angle between two planes described by four particles as described in Figure 2.2c, and ensures the planarity of atoms in a biomolecule. The backbone dihedral angles are particularly important since they define the *fold* of a protein backbone,

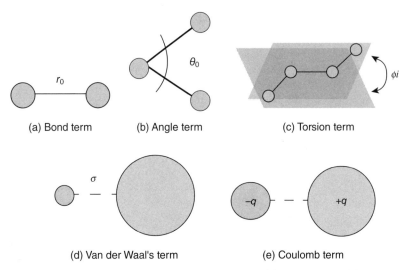

FIGURE 2.2. PE terms in standard MD force fields. (a) Bond term; (b) angle term; (c) torsion term; (d) Van der Waal's term; (e) Coulomb term.

which gives the protein its characteristic shape and folded structure. The allowable values for the backbone dihedrals for different protein amino acids can be described by the Ramachandran plot [2], and are important indicators of the physical stability of a protein structure. The dihedral terms are periodic, and are typically described using cosine terms as:

$$V_{\text{torsion}} = \sum \frac{V_n}{2}(1 + \cos(n\omega - \gamma)), \qquad (2.5)$$

where V_n are the amplitudes of the n different harmonic terms, ω are the phases, and γ is the torsion angle itself.

2.1.1.4 Van der Waal's Term   The Van der Waal's term describes the nonbonded interaction between pairs of atoms, and is a combination of the repulsive and attractive components of the interaction. The repulsive force arises at short distances and describes Pauli repulsion at short ranges due to overlapping electron orbitals. The attractive force, also referred to as the dispersion force, arises from fluctuations in the charge distribution in the electron clouds. This nonbonded interaction is most commonly modeled using the Lennard-Jones equation [1], comprising of the repulsive and attractive portions as a function of interatomic separation. The potential is of the form:

$$V_{\text{LJ}} = 4\varepsilon \left[\left(\frac{\sigma}{r} \right)^{12} - \left(\frac{\sigma}{r} \right)^{6} \right]. \qquad (2.6)$$

The value of the PE is zero at the equilibrium distance σ, and the potential is at its minimum at a separation of r_m, with a well depth ε.

2.1.1.5 Coulomb Term   The Coulomb term represents the electrostatic interaction between two charged particles, and is inversely proportional to the distance between the particles. It is described by:

$$V_{coulomb} = \frac{1}{4\pi\varepsilon_0}\frac{q_i q_j}{r},\tag{2.7}$$

where q_i and q_j are the charges of the two particles separated by a distance r. ε_0 is a constant representing the permittivity of free space.

Together, the Lennard-Jones and Coulomb terms represent the nonbonded interactions between all pairs of atoms in the biomolecular system to be simulated and account for the largest computational load in an MD simulation. They account for almost 80–90% of the total runtime of a typical MD simulation and are the primary target for different parallelization and spatial decomposition techniques intended to distribute the workload among different compute elements.

2.1.2 Graphics Processing Units for MD Simulations

MD simulations are extremely computationally demanding, since in the brute-force approach, each atom in the system interacts with every other atom via a set of interactions described by the force field, as described in Section 2.1.1. Even though several approximations are typically made during the computation of these interactions to reduce the computational cost, the algorithm complexity scales as $O(N^2)$ with the size of the system. On standard CPU-based supercomputers, the network latency is typically the limiting factor in the strong scaling* performance of MD simulations, since particle information needs to be continually updated as particles move during the simulation.

The majority of the computation in an MD simulation involves range-limited pairwise particle interactions, interactions between pairs of particles or grid points separated by less than some cutoff radius. In principle, both van der Waal's and electrostatic forces act between all pairs of particles in the simulated system. Van der Waal's forces fall off sufficiently quickly with distance that they are generally neglected for pairs of particles separated by more than some cutoff radius, usually chosen between 12 and 15 Å. In order to allow the energy profile to fall asymptotically to zero at the cutoff distance, rather than having an abrupt transition from a finite value to zero, a shifting function is sometimes applied to the potential.

---

* Strong scaling performance indicates how a computational problem scales with the number of processors for a fixed problem size.

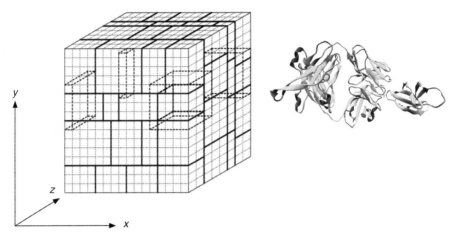

FIGURE 2.3. Spatial decomposition example for MD simulations.

In order to efficiently compute nonbonded interactions across multiple processing nodes, different *spatial decomposition* techniques are employed to communicate information between these processors, as shown in Figure 2.3. These algorithms partition particle data into subsets, and allocate different subsets to different processing nodes. Since particles move around during a simulation, it is necessary to share particle information between the various processors at each time step. The network latency thus becomes a limiting factor for the scaling performance of these simulations as the number of processors over which parallelization occurs is increased.

Graphics processing units (GPUs) have become increasingly attractive for high-performance computing (HPC) applications due to their highly data-parallel streaming architecture, massive multithreading capabilities, fast context switching, and high memory bandwidth. Since the size of the local data caches on GPUs is small, the implementation complexity increases dramatically when programming GPUs compared with traditional CPU-based architectures due to slow global memory access. In addition, modern programming interfaces like Compute Unified Device Architecture (CUDA) and OpenCL allow for easier programming methodologies, and have made GPU programming more accessible to application programmers.

2.1.2.1 GPU Architecture Case Study: NVIDIA Fermi GPUs are high-performance many-core processors capable of very high computation and data throughput. Once specially designed for computer graphics and difficult to program, today's GPUs are general-purpose parallel processors with support for accessible programming interfaces and industry-standard languages such as C and Python. Developers who port their applications to GPUs often achieve speedups of orders of magnitude versus optimized CPU implementations, depending upon the complexity and data access patterns of the underlying algorithm.

The first NVIDIA GPU based on the Fermi architecture [3], implemented with 3.0 billion transistors, features up to 512 CUDA cores. A high-level architectural block diagram is shown in Figure 2.4. A CUDA core executes a floating point or integer instruction per clock for each thread. The 512 CUDA cores are organized in 16 streaming multiprocessors (SM) of 32 cores each. The GPU has six 64-bit memory partitions, for a 384-bit memory interface, supporting up to a total of 6 GB of Graphics Double Data Rate, version 5 (GDDR5) dynamic random access memory (DRAM) memory. A host interface connects the GPU to the CPU via a peripheral component interconnect (PCI) Express bus. The GigaThread global scheduler distributes thread blocks to SM thread schedulers.

Each streaming multiprocessor features 32 CUDA processors. Each CUDA processor has a fully pipelined integer arithmetic logic unit (ALU) and floating point unit (FPU). The Fermi architecture implements the new Institute of

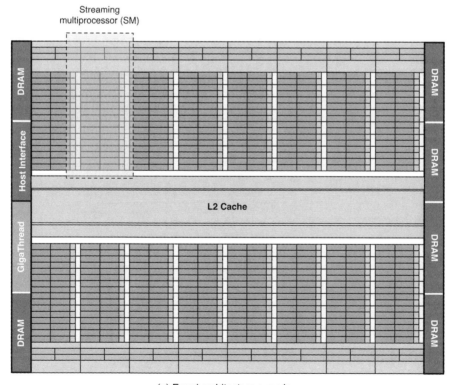

(a) Fermi architecture overview

FIGURE 2.4. Architecture of NVIDIA Fermi GPU. (a) Fermi architecture overview; (b) streaming multiprocessor.

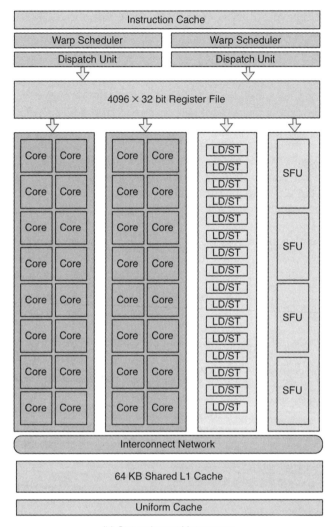

(b) Streaming multiprocessor

FIGURE 2.4. (*Continued*)

Electrical and Electronics Engineers (IEEE) 754-2008 floating-point standard, providing the fused multiply–add (FMA) instruction for both single- and double-precision arithmetic. This instruction improves over a multiply–add instruction by doing the multiplication and addition with a single final rounding step, with no loss of precision in the addition. This is more accurate than performing the operations separately, and is especially important in the computation of the Lennard-Jones and Coulomb potentials, with their millions of iterations over pairs of atoms at each MD time step.

Load and Store Units   Each streaming multiprocessor has 16 load and store units (LD/ST), allowing source and destination addresses to be calculated for 16 threads per clock cycle. Supporting units load and store the data at each address to cache or DRAM.

Special Function Units   Four special function units (SFUs) execute transcendental instructions such as sin, cosine, reciprocal, and square root, which are required for nonbonded interactions as well as long-range electrostatic interactions. Each SFU executes one instruction per thread, per clock cycle, while a warp executes over eight clock cycles. The SFU pipeline is decoupled from the dispatch unit, allowing the dispatch unit to issue instructions and data to other execution units while the SFU is occupied with a particular computational task.

Double-Precision Performance   Double-precision arithmetic is at the heart of HPC applications such as linear algebra, numerical simulation, and quantum chemistry. In MD simulations, accumulators for the nonbonded terms are typically implemented in double precision, as are the so-called constraint algorithms implemented to simulate the rigidity of certain bonds within the system. The Fermi architecture has been specifically designed to offer unprecedented performance in double precision, with up to 16 double-precision FMA operations performed per streaming multiprocessor, per clock cycle.

Configurable Shared Memory and Level 1 Cache   The on-chip shared memory greatly improves performance by allowing threads within a block to access the low-latency memory concurrently. Each streaming multiprocessor has 64 kB of on-chip memory that can be configured as 48 kB of shared memory with 16 kB of level 1 (L1) cache, or as 16 kB of shared memory with 48 kB of L1 cache.

2.1.2.2 *Force Computation on GPUs*   In traditional CPU-based MD simulations, pairwise computation of forces is typically accomplished using a neighbor list approach. Each particle in the system is associated with a list of its neighboring particles, where neighboring particles are only considered if they fall within a certain cutoff distance of the particle in question. This requires a force computation of the form

```
for p = 1 to total_particles:
     for n = 1 to neighbors(p):
          compute_forces(p, n)
```

Random memory access of the type `array[index[i]]` is well suited to traditional CPUs with large cache sizes that are optimized for random memory access. In contrast, GPUs have very small caches (on the order of tens of kilobytes), and therefore *data coalescing* is extremely critical to extract any performance advantage. Data coalescing involves aggregating information from atoms that are spatially close into a contiguous memory block and transferring that entire block to GPU memory as one transaction. This allows for force computations of the form

```
for p = 1 to len(array1)
    for n = 1 to len(array2)
        compute_forces(p, n)
```

This results in a data access pattern of `array[i]` rather than `array[index[i]]`, and can be more efficiently implemented on a GPU. Several approaches have been tried in the past to reduce indirect memory access patterns while computing nonbonded interactions. Friedrichs et al. [4] initially attempted porting implicit solvent simulations to GPUs using an N^2 approach, which scaled poorly with system size. Anderson et al. [5] used a voxel-based algorithm to generate neighbor lists, and reduced memory access times by taking advantage of the small texture cache on GPU architectures, in conjunction with spatial sorting of particles based on their positions. Another voxel-based method was used by Stone et al. [6] without the need for neighbor lists, and was incorporated into the MD package NAMD.

A typical algorithm to compute nonbonded interactions on GPUs breaks up particles into sets of 32. This number corresponds with the warp size (see Section 2.1.2.1) of NVIDIA GPUs, and maximizes the use of shared memory without the need for thread synchronization, since threads all run concurrently. A set of size greater than 32 would result in partial results being computed at each clock cycle, and would require the threads to be synchronized with each other and involve additional communication between compute cores. In this way the N^2 interactions between all the particles in a system are broken into $(N/32)^2$ tiles, with 32^2 interactions within each tile (Figure 2.5).

The simulation space is decomposed into equally sized tiles, and due to symmetry, only the upper diagonal tiles need to be computed. Each tile involves interactions between 64 atoms (32 in case of tiles on the diagonal), and to reduce the amount of data transferred from the GPU main memory, positions and velocities for these 64 atoms are loaded first. Therefore, for each 1024 interactions within a tile, information for only 64 particles needs to be transmitted, resulting in a data transfer-to-computation ratio of 1:16.

To further reduce the computation, the spatial decomposition algorithm first checks the tiles themselves to find those where the constituent particles may be too far apart to physically interact. Force computation is only done on those tiles which contain spatially relevant particles. In order to accomplish

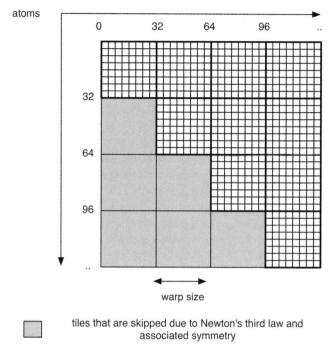

FIGURE 2.5. Spatial decomposition of particles per thread block.

this, similar algorithms are used as those for building neighbor lists, involving searches for particles within a certain cutoff using grid-based or distance-based algorithms. A bounding box is constructed for each of the two blocks in a tile. If the distance between these boxes is larger than the cutoff distance, then that tile is guaranteed to have noninteracting particles and is skipped over.

For the above algorithm to work correctly, atoms within a tile should be spatially close, otherwise the large bounding boxes would result in very few tiles being skipped. For a protein, this is accomplished by having contiguous particles within the polypeptide chain as a block, thus ensuring that spatial relevance is maintained. However, for the surrounding solvent, this is a larger issue, since multiple water molecules can be part of a block, and each is free to move independently of the other during the course of a simulation. This issue is addressed by periodically reordering water molecules to ensure spatial coherency. With such an implementation, an MD simulation on a single NVIDIA Fermi GPU is equivalent to a small CPU-based cluster with high-speed interconnects, demonstrating the advantages that GPU computing can bring to HPC applications if the appropriate programming complexity and optimizations are included.

2.2 SPECIAL-PURPOSE HARDWARE AND NETWORK TOPOLOGIES FOR MD SIMULATIONS

In this section we study the architecture of Anton [7, 8], a massively parallel special-purpose computer designed and built to accelerate MD simulations to the millisecond timescale. Such long timescales are impossible in traditional CPU-based supercomputers, and even on GPU-accelerated platforms the strong scaling performance of these implementations quickly becomes a bottleneck in simulating biologically relevant structures at these extreme timescales. Protein folding is one such application, where only recently have such folding events been observed on small systems using the massively distributed Folding@home project (http://folding.stanford.edu/). Anton was built with the express purpose of enabling scientists to study biochemical phenomena such as protein folding and other long timescale events that cannot currently be observed in laboratory settings or using current computational techniques.

The most dominant computation in a MD simulation is the computation of the pairwise nonbonded interactions between particles. Most MD engines use a range-limited pairwise interaction model, where particles within a certain cutoff distance of each other interact directly via a pairwise interaction model, and distant particles are handled using some long-range approximations. Anton uses an algorithm called k-Gaussian Split Ewald or k-GSE [9] to compute electrostatic interactions, where the long-range portion is computed efficiently in Fourier space. The charge on distant particles is interpolated onto a grid in Fourier space, and the force acting on particles is computed using the nearby grid points after an inverse Fourier transform. The hardware architecture is divided into the high-throughput subsystem for computing the nonbonded interactions, and the flexible subsystem to perform the rest of the tasks in an MD simulation.

2.2.1 High-Throughput Interaction Subsystem

The compute power behind Anton is delivered by the High-Throughput Interaction Subsystem (HTIS) that handles the pairwise nonbonded interactions, which are the most compute-intensive portions of MD calculations. An overview of the Anton architecture is shown in Figure 2.6. The rest of the computational tasks during a time step, namely the bonded interactions, charge spreading, and interpolation, as well as general application-specific integrated circuit (ASIC) bookkeeping, are done on the flexible subsystem. The HTIS comprises of 32 pairwise point interaction modules (PPIMs), each of which contains a pairwise point interaction pipeline (PPIP) that runs at 800 MHz and is capable of computing one interaction per clock cycle, for a theoretical peak rate of 25,600 interactions per microsecond per ASIC. The HTIS also provides specialized support for the Neutral Territory method, a parallelization scheme which significantly reduces inter-ASIC communication requirements relative to traditional parallelization methods [10]. Particles imported

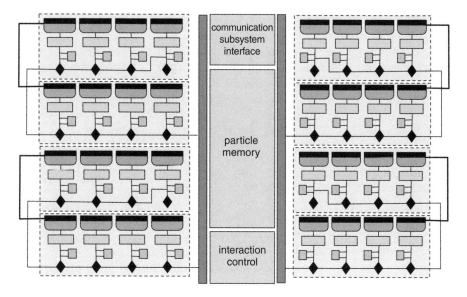

(a) HTIS architecture

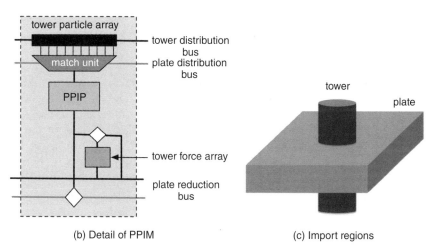

(b) Detail of PPIM   (c) Import regions

FIGURE 2.6. Anton: HTIS architecture. (a) HTIS architecture; (b) detail of PPIM; (c) import regions.

into each ASIC are divided into "tower" particles and "plate" particles as illustrated in Figure 2.6c, depending on the particles on which forces are being computed, and the particles they interact with. Sixteen particles at a time from the "tower" set are distributed to each of 32 PPIMs for a total of 512 particles. Then neighboring particles are streamed to each of these PPIMs to determine the interactions between the tower and neighboring particles. The selection of particles to stream is done using a custom distribution network known as the particle distribution network. In the range-limited approach, only particles that fall within the specified cutoff have their interactions evaluated, and this selection is done in parallel by eight match units per PPIM that identify interacting pairs in two clock cycles. Once the interaction between the "tower" and neighboring particles has been computed, two reductions take place: forces on all points of the tower are summed in the *tower force arrays*, and forces on the particles themselves are summed on the *force reduction network*.

2.2.1.1 Pairwise Point Interaction Modules
Each PPIM is comprised of subunits that are discussed in the succeeding sections.

Match Units The match units are responsible for eliminating interactions that do not fit within the parallelization framework, including particles that fall outside the pairwise range-limited cutoff. The match unit computes the square of the distance between particles using an approximate low-precision distance check, which uses 8×8 bit multipliers. This results in a die usage area of only 0.1 mm^2, or roughly 10% of the entire PPIM surface area. The match units also implement other algorithms required by the Neutral Territory method, to only compute interactions that fit the parallelization scheme. Specifically it avoids the double counting of interactions if the home boxes overlap with each other.

Pairwise Point Interaction Pipelines The PPIP is the compute core of each PPIM, and is an MD-specific ASIC comprising of 18 adders, 27 multipliers, 3 lookup tables, error detection logic, and numerous multiplexers. The input to each PPIP are two position vectors representing the positions of the two interacting particles, and a six-element vector comprising of the force field parameters for each particle (Van der Waal's parameters and point charges). The PPIP computes the square of the distance (according to Eqs. 2.6 and 2.7). Each PPIP is capable of producing one result every clock cycle and is fully pipelined, operating at about 800 MHz.

The PPIPs perform fixed-point arithmetic, which has many advantages over floating-point arithmetic in MD simulations. Since the functional forms of the computations performed by the PPIPs (see Section 2.1.1) is not always exactly the same, a significant amount of flexibility is incorporated into this unit. This is accomplished by using lookup tables, with 256 entries being used to interpolate values in between data points. The table entries are computed using piecewise cubic polynomials of variable length and implemented in floating-point precision, allowing for a high degree of accuracy along with the inherent flexibility of lookup tables themselves.

2.2.1.2 Particle Distribution Network
In order to efficiently compute pairwise interactions on the PPIMs, pipelining of input data is necessary

to hide the network latency costs associated with the distribution of particle data to the various compute units. To this end, the PPIMs are decomposed into four arrays of eight PPIMs each, and the assignment of "plate" and "tower" particles is broadcast to the PPIMs on a round-robin basis. The distribution network supports double buffering of tower points, and supports a storage datastructure called a *shadow tower*, which in effect acts like an input buffer for that PPIM. It contains particle information for the next set of computations, and is populated while the current set of interactions is being computed. This buffering allows for the communication overhead to be effectively hidden by the computations and enables a fully pipelined architecture.

The other advantage of buffering is to provide the PPIMs with a continuous stream of computation units even when there is a load imbalance between the interactions in different PPIM compute cycles. Frequently the number of matching pairs between iterations varies, since the number of matches could be zero or saturated. In this case, the pipelining provides a queue of about 20 interactions for the distribution bus to supply the compute units.

2.2.2 Hardware Description of the Flexible Subsystem

Aside from the computation of nonbonded interactions, various other less compute-intensive algorithms are employed in MD simulations. These include the bonded computations between atoms (bond, angle, and torsion forces), integration of motion according to Newton's laws, and computation of fast Fourier transforms (FFTs) and inverse FFTs required for the long-range nonbonded interactions. While the HTIS accelerates the computation of the most compute-intensive portions of the force computations, the bottleneck in the simulation shifts to the rest of the algorithms. An efficient hardware implementation of these algorithms is necessary to provide the extreme strong scaling performance of the simulation engine.

The flexible subsystem comprises of four processing "nodes," each of which comprises of a Tensilica core with associated memory caches, a remote access unit (RAU) for managing data flow, and two *geometry cores* (GCs). These components are connected by an interconnect called the *racetrack*, and communicate data with the rest of the system via a *ring interface unit* (Figure 2.7).

The various hardware components of the flexible subsystem and their functions are summarized below.

- Remote access unit
  - Comprises 128 message transfer descriptor (MTDs). Each MTD controls one data transfer or synchronization operation.
  - Contains a Tensilica compute core that coordinates data transfer and synchronization of data in a single MD time step using 128-bit vector operations.

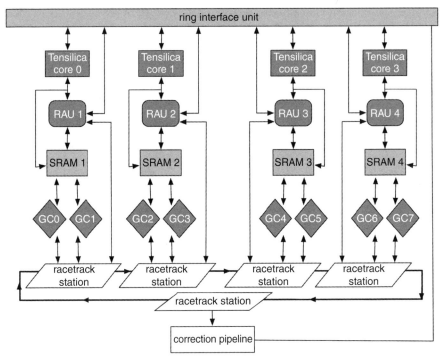

FIGURE 2.7. Anton: flexible subsystem architecture.

- Geometry cores
  - Single instruction, multiple data (SIMD) processor operating on four 32-bit values for a base datatype of a 128-bit vector.
  - Implements dot product and partial matrix determinants, and is used mainly for bonded interactions and constraint evaluations.
  - Contains caches for data and instructions, and manages the caches to ensure that no cache misses occur during a typical MD time step.
- Racetrack
  - Unidirectional 128-bit interconnect connecting the GCs, RAUs, and the other hardware components.
  - Contains deadlock-avoiding dual virtual channels.
  - Used by the Tensilica cores to broadcast atom information to the GCs, and in turn is used by the GCs to transmit force information to the force accumulators.
- Ring interface unit
  - Performs data transfers to the rest of the system outside the flexible subsystem.
  - Manages communication between the GCs/Tensilica cores and the DRAM in case of cache misses.

2.2.3 Performance and Conclusions

The specialized hardware used for speeding up nonbonded force computations using the high-throughput interconnect subsystem, as well as the accelerated computation of bonded forces, motion integration, and Fourier transforms on the flexible subsystem allow Anton to achieve a peak performance of almost two orders of magnitude over traditional general-purpose processors [7]. The computation of long-range electrostatic interactions is also accelerated on specialized hardware, which contributes significantly to the overall acceleration of the simulation. The ideal parallelization technique would ensure that each compute node in a distributed computing system would have information for only one atom, since this ensures the maximum possible scaling performance of such a system. Anton's parallelization methods ensure that the number of atoms per node is in the region of tens, ensuring almost linear scaling performance with the number of compute cores. Simulations on a millisecond timescale [11] are now possible, and can be applied to solve problems in different areas of computational biology and health sciences.

2.3 QUANTUM MC APPLICATIONS ON FIELD-PROGRAMMABLE GATE ARRAYS

In quantum mechanics, the Schrödinger equation describes how the quantum state of a physical system, comprising of a collection of bodies, varies with time. The solution to this equation is of great interest to researchers in a variety of fields, ranging from computational biology to quantum computing. However, for practical systems of interest, the number of degrees of freedom, or variables, is at least three times the total number of bodies in the system, which may be in the thousands or millions in some cases. Thus solving this equation is, for all practical purposes, nearly impossible even on today's modern parallel architectures. The Schrödinger equation relates the wave function (WF) Ψ of the quantum state with the Hamiltonian operator $\hat{H}$ via the following equation:

$$i\hbar\Psi = \hat{H}\Psi, \tag{2.8}$$

where i is the imaginary unit and $\hbar$ is the reduced Planck constant. This many-body WF or quantum state is typically approximated using a Hartree–Fock approximation, and most methods aim to compute the lowest energy state (also called the ground state) WF.

Quantum MC (QMC) methods such as diffusion MC (DMC) and variational MC (VMC) are useful tools to study these ground-state WFs and other properties of quantum systems [12]. VMC is an iterative algorithm, which refines an initial estimate of the WF $\psi(R)$ based on incremental Cartesian displacements of atoms and an acceptance criteria p. The algorithm can be defined as follows:

- Select a reference conformation $R(x, y, z)$ at random
- Compute the ground-state properties such as WF $\psi(R)$, energy at R
- Make a small displacement to one of the particles to obtain a new configuration $R'(x, y, z)$ and compute the ground-state properties $\psi(R')$
- Accept or reject the new configuration based on the acceptance criteria

$$p = \left| \frac{\psi(R')}{\psi(R)} \right|. \tag{2.9}$$

In the above algorithm, the computation of the WF and the energy is the most computationally demanding portion, and lends itself well to hardware implementation due to the nature of the WF itself. Typically these WF kernels are implemented using a large number of adders and multipliers, thereby being ideal candidates for specialized hardware implementations. In addition, the reconfigurable nature of field-programmable gate arrays (FPGAs) is also a desirable feature, since the particles in the system change chemical composition as the simulation advances, requiring the use of different interpolation parameters for different particles.

2.3.1   Energy Computation and WF Kernels

The interatomic PE as a function of distance is shown in Figure 2.8a. The functional form is similar to the nonbonded Lennard-Jones term in molecular simulations, and involves the computation of an expensive square-root term. Since a spline-based evaluation method is used, the square root can be effectively precomputed by building it into the supplied interpolation coefficients. In addition, the energy function is made up of two distinct regions, a repulsive region ($0 \leq \sigma^2 < r_{ij}^2$) and an attractive region ($r_{ij}^2 \geq \sigma^2$). In order to deal with the infinite attractive domain (Region II), an approximate logarithmic binning scheme is employed. The entire region is partitioned into bins such that the end points of each bin correspond with consecutive powers of two. This ensures that the size of each subregion increases stepwise with r_{ij}^2. Finally, each subregion is partitioned into several intervals of equal size.

A similar binning scheme is applied to the computation of the WF (see Figure 2.8b):

$$\psi = \prod T_1(r_i) \prod T_2(r_{ij}) \prod T_3(r_{ij}, r_{jk}, r_{ik}). \tag{2.10}$$

A two-level classification is also applied to the WF computation, where the cutoff σ is the maximum value of the WF, obtained by setting the first derivative of the function to zero. Thus the same hardware configurations can be used for both energy and WF computations.

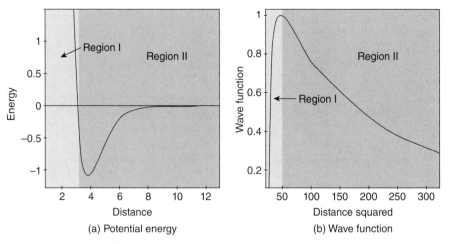

FIGURE 2.8. QMC components. (a) Potential energy; (b) wave function.

2.3.2 Hardware Architecture

Figure 2.9 shows a high-level overview of the FPGA architecture. The MC portion of the algorithm runs on the host processor, with the PE and WF kernels being implemented on independent FPGA units, labeled PE-FPGA and WF-FPGA, respectively. The main components of the FPGA platform are the memory substructure, the PE and WF computation engines, and the coefficient memory units. The computation engines themselves consist of a distance computation unit (DCU), which computes the square of the distance between particles, the energy computation kernel (FCU) that computes the intermediate potentials and WFs at each clock cycle, and an accumulator (ACU) which is responsible for the accumulation of these terms and transfer back to the host processor.

2.3.2.1 Binning Scheme Both the PE and the WF equations are transformed such that they can be written as a product of pairwise terms for easier computation, rather than a sum. This exponential transform also has the added property that the potential takes on values between 0 and 1, and the final inverse logarithm can then be performed on the host processor rather than the FPGA unit.

Potential Energy Transform

$$V_{\text{total}} = V_1 + V_2 = \sum V_1(r_{ij}) + \sum V_2(r_{ij}), \qquad (2.11)$$

$$V'(r_{ij}^2) = e^{-V_1(r_{ij})}. \qquad (2.12)$$

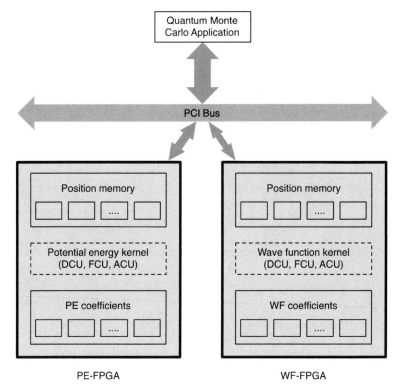

FIGURE 2.9. Block diagram of QMC FPGA.

A quadratic spline interpolation technique is used to generate the transformed potentials, where different schemes are used to look up the interpolation coefficients in each region due to the mathematical differences between them. Region I is composed of 256 bins, and are uniformly spaced at intervals between $r_{ij}^2 = 0$ and $r_{ij}^2 = \sigma^2$. For Region II, a logarithmic binning scheme is used, and a two-stage lookup technique is employed to find the interpolation coefficients. First, the subdivision within a particular bin, as described in Section 2.3.1, is selected, followed by the coefficients corresponding to that particular subdivision. The difference between the distance squared and the σ^2 is first used by the leading zero-count detector (LZCD) to compute the index of the subdivision, which is implemented using priority encoders. This is then used by the second stage, to determine the coefficients themselves.

2.3.3 PE and WF Computation Kernels

These hardware modules, illustrated in Figure 2.10, implement the actual energy and wavefront computation kernels themselves, and are nearly identical except for the accumulation of partial terms.

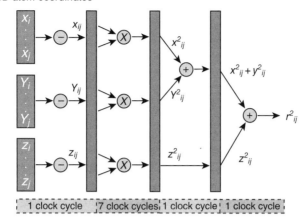

(a) Distance computation FPGA unit

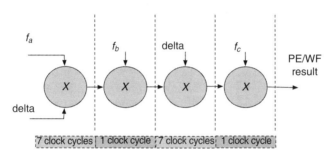

(b) PE/WF function computation unit

FIGURE 2.10. Computation kernel architecture for QMC application. (a) Distance computation FPGA unit; (b) PE/WF function computation unit. 3D, three dimensional.

2.3.3.1 *Distance Computation Unit (DCU)*

Every clock cycle, a pair of particle coordinates is read from the local on board memory using an address generator, which provides read addresses to the memory modules. These addresses are passed to the distance computation engine. With N particles, there are $N(N-1)/2$ distances, and the state machine repeats this address generation for all N particles. A 32-bit fixed-point representation is used, which provides a 52-bit squared distance value per clock cycle. The delta $r_{ij}^2 - \sigma^2$ is computed to decide whether a Region I or Region II computation is being performed.

2.3.3.2 *Calculate Function Unit and Accumulate Function Kernels*

Once the interpolation coefficients have been fetched using the binning scheme described in Section 2.3.2.1, these intermediate energy and WF values are accumulated as a running product for Region I and as a running sum in Region II. While the PE transformation requires both sum and product accumulators for the partial energy, the WF equation is not transformed, since it requires only the product accumulators. These results are then transferred to the host, which removes the scaling associated with the PE transform, and combines the results from Region I and Region II to reconstruct the full PE.

2.4   CONCLUSIONS AND FUTURE DIRECTIONS

In this chapter, we have studied the application of specialized hardware architectures to biomolecular simulations. While GPU computing has already found its way into mainstream HPC applications, the use of FPGAs and other programmable devices has lagged behind in finding wide acceptance, mainly due to the lack of software platforms and interfaces to make the devices more accessible to application programmers. With sophisticated programming platforms like CUDA and OpenCL, GPU computing is widely accessible to HPC programmers wishing to accelerate their applications without knowing the intricate details of the underlying architecture, and having to program at the lowest levels. Recently, significant developments have been made in the coprocessor architecture, which are hybrids between x86-based CPU and GPU architectures. Such architectures allow the same programming language capabilities as today's traditional high-end CPUs, and also incorporate the highly parallel streaming capabilities of current GPU architectures. The fact remains that traditional CPU-based clusters will be eventually replaced by some combination of hardware accelerators and the associated programming infrastructure, which have the potential to provide insight into a variety of HPC problems ranging from drug design to weather prediction models that are currently simply out of reach due to computational constraints.

REFERENCES

[1] A.R. Leach, Molecular Modelling: Principles and Applications, Prentice-Hall, Harlow, UK, 2001.

[2] G. Ramachandran, C. Ramakrishnan, V. Sasisekharan, "Stereochemistry of polypeptide chain configuration," Journal of Molecular Biology, 7, 1963, pp. 95–99.

[3] NVIDIA Corporation, NVIDIA Next Generation CUDA Compute Architecture: Fermi, 2009.

[4] M.S. Friedrichs, P. Eastman, V. Vaidyanathan, M. Houston, S. LeGrand, A.L. Beberg, D.L. Ensign, C.M. Bruns, V.S. Pande, "Accelerating molecular dynamics simulations on graphics processing units," Journal of Computational Chemistry, 30, 2009, pp. 864–872.

[5] J.A. Anderson, C.D. Lorenz, A. Travesset, "General purpose molecular dynamics simulations fully implemented on graphics processing units," Journal of Computational Physics, 227, 2008, pp. 5342–5359.

[6] J.E. Stone, J.C. Phillips, P.L. Freddolino, D.J. Hardy, L.G. Trabuco, K. Schulten, "Accelerating molecular modeling applications with graphics processors," Journal of Computational Chemistry, 28, 2007, pp. 2618–2640.

[7] R.H. Larson, J.K. Salmon, R.O. Dror, M.M. Deneroff, C. Young, J.P. Grossman, Y. Shan, J.L. Klepeis, D.E. Shaw, "High-throughput pairwise point interactions in Anton, a specialized Machine for molecular dynamics simulation," in Proc. 14th International Symposium on Computer Architecture (HPCA 2008), 2008, pp. 331–342.

[8] J.S. Kuskin, C. Young, J.P. Grossman, B. Batson, M.M. Deneroff, R.O. Dror, D.E. Shaw, "Incorporating flexibility in Anton, a specialized machine for molecular dynamics simulation," in High Performance Computer Architecture, 2008. HPCA 2008, 2008, pp. 343–354.

[9] Y. Shan, J.L. Klepeis, M.P. Eastwood, R.O. Dror, D.E. Shaw, "Gaussian Split Ewald: a fast Ewald mesh method for molecular simulation," The Journal of Chemical Physics, 122, 2005.

[10] D.E. Shaw, "A fast, scalable method for the parallel evaluation of distance-limited pairwise particle interactions," Journal of Computational Chemistry, 26, 2005, pp. 1318–1328.

[11] K. Lindorff-Larsen, S. Piana, R.O. Dror, D.E. Shaw, "How fast-folding proteins fold," Science, 334, 2011, pp. 517–520.

[12] A. Gothandaraman, G.D. Peterson, G.L. Warren, R.J. Hinde, R.J. Harrison, "FPGA acceleration of a Quantum Monte Carlo application," Parallel Computing, 34, 2008, pp. 278–291.

3   Embedded GPU Design

BYEONG-GYU NAM and HOI-JUN YOO

3.1   INTRODUCTION

As the mobile electronics advances, the real-time three-dimensional (3D) graphics has been widely adopted in mobile embedded systems such as smartphones and smart pads. The realization of 3D graphics on these power- and area-limited mobile devices has been a challenging issue because of the highly complex computations inherent in the 3D computer graphics. There have been studies on the hardware acceleration of 3D graphics pipeline to provide the required computing power within a limited power consumption [1,2]. However, these hardwired accelerators just provide limited graphics effects but cannot afford the programmability to adapt to advanced graphics algorithms. Recently, mobile 3D graphics standards like OpenGL ES [3] introduced the programmable graphics pipeline to accommodate a wide variety of advanced graphics effects on mobile devices. Therefore, high-performance 3D graphics processors consuming only limited amount of power are proposed to address these technology trends [4–7]. However, their design is optimized for fast matrix-vector multiplications for 3D geometry transformations but area efficiency is somewhat neglected.

In this chapter, a high-speed, low-power, and area-efficient 3D graphics processor, that is, a graphics processing unit (GPU), implementing a full 3D graphics pipeline is presented [8]. It demonstrates 141 Mvertices/s peak geometry processing performance with 968 k transistors only and 52.4 mW power consumption at 60 frames per second (fps). The geometry processor (GP) exploits the logarithmic number system (LNS) [9] for the high speed, low power, and area efficiency. The rendering engine (RE) also adopts the LNS

Embedded Systems: Hardware, Design, and Implementation, First Edition.
Edited by Krzysztof Iniewski.
© 2013 John Wiley & Sons, Inc. Published 2013 by John Wiley & Sons, Inc.

for all of the dividers in the rendering pipeline. For a further power reduction, the chip is divided into three power domains, each of which is controlled independently by dynamic voltage and frequency scaling (DVFS).

This chapter is organized as follows. Overall GPU architecture is discussed in Section 3.2, and each of the GPU modules are covered in Section 3.3. The chip-level power management methodology will be described in Section 3.4. The implementation and measurement results of the proposed GPU are presented in Section 3.5, and finally, we will summarize in Section 3.6.

3.2   SYSTEM ARCHITECTURE

Figure 3.1 illustrates the full 3D graphics pipeline accommodating application, geometry, and rendering stages. In the application stage, object-level algorithms like the artificial intelligence and physics simulation are performed for 3D gaming applications. The geometry stage processes various geometry transformation and lighting (TnL) operations at each vertex. The per-pixel operations like rasterization and texture image mapping are carried out in the rendering stage. Figure 3.2 shows the overall architecture of the proposed 3D graphics processor. It consists of an ARM10-compatible 32-bit reduced instruction set computer (RISC) processor, a 128-bit GP, an RE for each corresponding stage in Figure 3.1, and three power management units (PMUs) are included to control the power supply to each of the three modules. The RISC works as the host processor of this chip and runs application programs generating the geometry transformation matrices for 3D objects in a scene. The GP transfers the matrix from the matrix first in, first out (FIFO) to its constant memory (CMEM), and then performs a variety of geometry operations on the fetched vertices by running a vertex program from instruction memory (IMEM). The RE shades and texture-maps the pixels inside the triangle from the GP. The chip is divided into three different power domains: RISC, GP, and RE. Level shifters and synchronizers shown in Figure 3.2b are inserted between power domains to adjust signal levels and avoid the metastability of data being transferred. The 4 kB matrix FIFO and 64 bit index

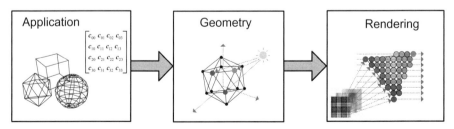

FIGURE 3.1. 3D graphics pipeline.

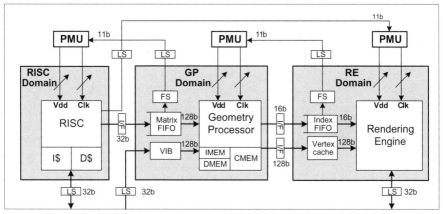

FIGURE 3.2. Proposed 3D graphics processor. (a) Overall architecture. (b) Interface circuit between power domains. DMEM, data memory; VIB, vertex input buffer; Clk, clock; FS, frequency selector.

FIFO are also used across the modules to maintain the pipeline throughput stable despite the delayed response of the PMUs. In these FIFOs, the Gray coding, where two consecutive values differ in only 1 bit, is used for encoding the read/write pointer for a reliable operation between different read/write clock domains. The domain interface FIFO is shown in Figure 3.3. As shown in this figure, the FIFO and reading block are assigned to the same voltage domain since the read clock between them can skew if the supply voltages to each of them are different from each other. The write clock to the FIFO is fixed at 200 MHz, the highest operating frequency at which the writing block can run, in order not to miss the data from the writing block even when it operates at its highest operating frequency. In the case of the *full* signal produced by the FIFO to be transmitted to the writing block, only the *full* signal occurring at the clock of the writing block needs to be transmitted. So the clock to the writing block is disabled upon the receipt of this *full* signal from the FIFO. Similarly, the clock to the reading block is also disabled on the receipt of the *empty* signal from the FIFO.

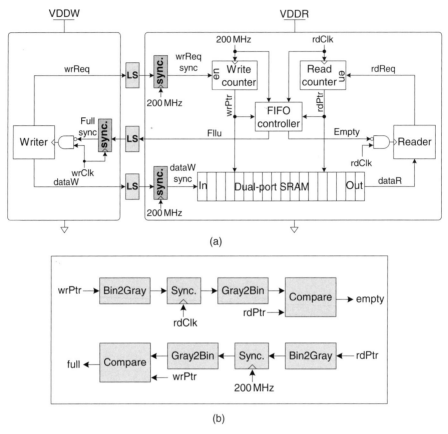

FIGURE 3.3. Domain interface FIFO. (a) FIFO organization. (b) FIFO controller. LS, level shifter; VDDW, VDD of writer; VDDR, VDD of reader.

3.3 GRAPHICS MODULES DESIGN

3.3.1 RISC Processor

The ARM10 compatible RISC processor is equipped with 4 kB instruction and 8 kB data caches. Its maximum operating frequency is 200 MHz. A 4 kB dual-port static random access memory (SRAM) with 128-bit word length is used to mitigate the discrepancies in the processing speed and data width between the RISC and the GP, as well as the response delay of the PMU. The memory interface of the RISC processor is optimized for the real-time 3D graphics applications so that the RISC can directly supply the matrix data to the GP through the matrix FIFO, bypassing the data cache.

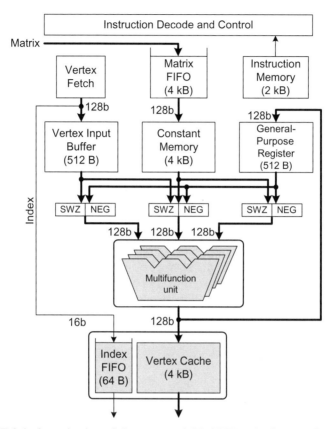

FIGURE 3.4. Organization of the proposed GP. SWZ, swizzling; NEG, negation.

3.3.2 Geometry Processor

The proposed GP is shown in Figure 3.4. The GP accommodates a multifunction unit that unifies the vector operations and elementary functions based on the LNS and the scheme introduced in Nam et al. [10]. Moreover, it unifies the matrix-vector multiplication (MAT) of 4×4 transformation matrix and four-element vertex vector with the vector operations and elementary functions in a single four-way arithmetic unit. Its fully pipelined architecture achieves six-cycle latency and half-cycle throughput for the matrix-vector multiplication; it produces a result every two cycles. The vertex cache is also equipped in the GP to reuse the vertices that already finished processing for other triangles. With the high throughput matrix-vector multiplication and the vertex cache, the GP achieves a peak performance of 141 Mvertices/s at 200 MHz operating frequency.

3.3.2.1 Geometry Transformation The geometry transformation in 3D graphics can be computed by the matrix-vector multiplication shown in Equation (3.1), where 16 multiplications and 12 additions are required to complete the operation:

$$
\begin{bmatrix} c_{00} & c_{01} & c_{02} & c_{03} \\ c_{10} & c_{11} & c_{12} & c_{13} \\ c_{20} & c_{21} & c_{22} & c_{23} \\ c_{30} & c_{31} & c_{32} & c_{33} \end{bmatrix} \begin{bmatrix} x_0 \\ x_1 \\ x_2 \\ x_3 \end{bmatrix} = \begin{bmatrix} c_{00} \\ c_{10} \\ c_{20} \\ c_{30} \end{bmatrix} x_0 + \begin{bmatrix} c_{01} \\ c_{11} \\ c_{21} \\ c_{31} \end{bmatrix} x_1 + \begin{bmatrix} c_{02} \\ c_{12} \\ c_{22} \\ c_{32} \end{bmatrix} x_2 + \begin{bmatrix} c_{03} \\ c_{13} \\ c_{23} \\ c_{33} \end{bmatrix} x_3. \tag{3.1}
$$

As the coefficients of the transformation matrix of Equation (3.1) for a given object does not change in a scene, they can be preconverted into the LNS before the processing starts. Therefore, only the vertex vector needs to be converted into the LNS at every processing cycle to perform the matrix-vector multiplication. Therefore, a complete matrix-vector multiplication is converted into Expression (3.2) that requires 4 logarithmic converters, 16 adders in the logarithmic domain, 16 antilogarithmic converters, and 12 floating-point adders for the summation of product terms to get the final results:

$$
2^{\left(\begin{bmatrix} \log_2 c_{00} \\ \log_2 c_{10} \\ \log_2 c_{20} \\ \log_2 c_{30} \end{bmatrix} + \log_2 x_0 \right)} + 2^{\left(\begin{bmatrix} \log_2 c_{01} \\ \log_2 c_{11} \\ \log_2 c_{21} \\ \log_2 c_{31} \end{bmatrix} + \log_2 x_1 \right)} + 2^{\left(\begin{bmatrix} \log_2 c_{02} \\ \log_2 c_{12} \\ \log_2 c_{22} \\ \log_2 c_{32} \end{bmatrix} + \log_2 x_2 \right)} + 2^{\left(\begin{bmatrix} \log_2 c_{03} \\ \log_2 c_{13} \\ \log_2 c_{23} \\ \log_2 c_{33} \end{bmatrix} + \log_2 x_3 \right)}. \tag{3.2}
$$

Expression (3.2) can also be computed in a two-phase manner described in Figure 3.5, requiring only eight adders in the logarithmic domain and eight antilogarithmic converters per phase. This scheme realizes the matrix-vector multiplication on the four-way arithmetic unit every two cycles.

3.3.2.2 Unified Multifunction Unit According to the scheme in Nam et al. [10], the vector operations require two logarithmic converters per vector lane, and the elementary functions like trigonometric functions and power functions require multipliers in the logarithmic domain. In addition, eight

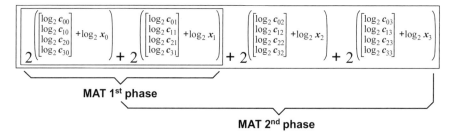

FIGURE 3.5. Two-phase matrix-vector multiplication.

antilogarithmic converters are required for matrix-vector multiplication as explained above. Therefore, a Booth multiplier is prepared to implement the multiplications required for the elementary functions and is made as a programmable Booth multiplier (PMUL) to operate as the logarithmic and antilogarithmic converters as well. It works as a logarithmic converter for processing vector operations by just adding a 64-byte lookup table (LUT) and an antilogarithmic converter for the matrix-vector multiplication by adding a 56-byte LUT. This programmable scheme is enabled by the similar structure shared between the 32×6-bit Booth multiplier and the logarithmic and antilogarithmic converters. They are sharing the adder tree consisting of carry-save adders (CSAs) and a carry propagation adder (CPA) to implement the partial product reduction in a Booth multiplier or the linear interpolation for logarithmic and antilogarithmic converters introduced in Nam et al. [10]. This PMUL is described in Figure 3.6 and is located at the E2 stage of the unified multifunction unit shown in Figure 3.7.

Therefore, the required eight antilogarithmic converters for the matrix-vector multiplication are obtained by programming the PMUL into four antilogarithmic converters together with the four antilogarithmic converters that exist in the E3 stage. The four CPAs in the E1 stage and four CPAs in the E3 stage make the required eight CPAs to implement the additions in logarithmic domain. The four multiplication results from the antilogarithmic converters in the E2 stage and the other four from the E3 stage are added through the floating-point (FLP) adders in the E4 stage to get the first-phase result of the matrix-vector multiplication. With the same process being repeated and accumulating with the first-phase result, the final result is obtained for the matrix-vector multiplication in the second phase. This results in a half-cycle throughput of the matrix-vector multiplication that has traditionally been a quarter-cycle throughput, that is, producing results every four cycles.

3.3.2.3 *Vertex Cache* In this GP, a TnL vertex cache is provided to reuse the vertices processed for other triangles without running the TnL code again. The vertex cache consists of the cache controller and the physical entry memory as shown in Figure 3.8. The vertex cache is implemented by just adding the cache replacement scheme to the existing domain interface FIFO between the GP and the RE. Therefore, the vertex cache is implemented with almost no overhead. The 4 kB physical entry memory contains 16 entries of vertex output buffer. The vertex cache controller described in Figure 3.8 keeps the 16 counters to count the indices pointing to each cache entry. If an incoming index is found in the cache, the index is pushed into the index FIFO and the counter increments. The counter decrements when the RE pops the corresponding index. The cache entry is replaced when the counter is set to zero. This 16-entry vertex cache shows a 58% hit rate and results in a single-cycle TnL for the vertices found in the vertex cache. The half-cycle throughput MAT and this vertex cache achieve the peak geometry processing performance of 141 Mvertices/s at 200 MHz.

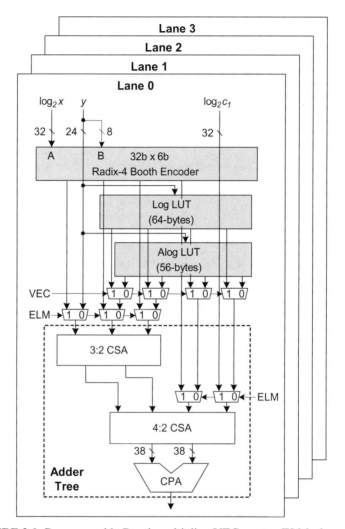

FIGURE 3.6. Programmable Booth multiplier. VEC, vector; ELM, elementary.

3.3.3 Rendering Engine

Figure 3.9 shows the block diagram of the RE. It consists of a triangle setup engine (TSE), rasterizer, and pixel pipeline. The RE varies its operation cycles according to the input triangle size. The pixel pipeline is designed to operate with single-cycle throughput to maximize the rendering throughput at a low clock frequency. The TSE and rasterizer are implemented as multicycle operations by folding the internal arithmetic units three times because a single-cycle implementation of these blocks would take as large as about 40% of the size of the entire RE. Note that this multicycle design approach does not form a

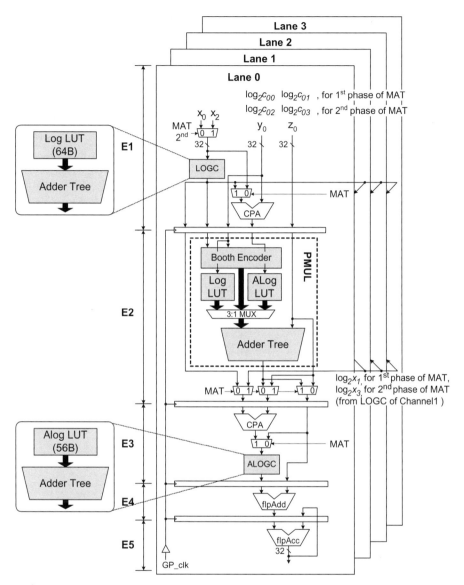

FIGURE 3.7. Unified multifunction unit. LOGC, logarithmic converter; ALOGC, antilogarithmic converter; MUX, multiplexer.

bottleneck in the pipeline since the geometry processing stage usually takes tens of cycles to run the TnL code for a triangle and each triangle corresponds to a number of pixels processed in the rendering stage.

In this work, all the divisions required in the RE are implemented based on the LNS for a power- and area-efficient design. The TSE computes 11-way vector interpolation to set up 11 components in parameters for a triangle

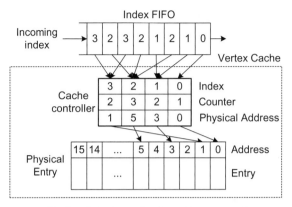

FIGURE 3.8. Organization of vertex cache.

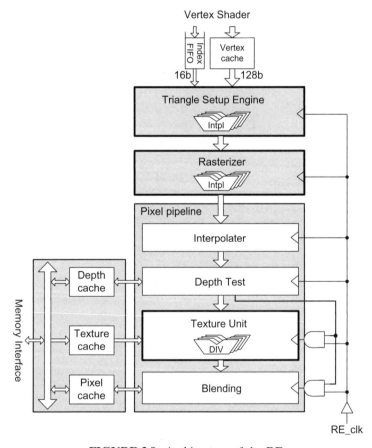

FIGURE 3.9. Architecture of the RE.

rasterization such as screen coordinate, color, and texture coordinates, that is, $(x, z, w, r, g, b, a, s, t, u, v)$. The TSE is implemented as a threefolded four-way single instruction, multiple data (SIMD) unit and interpolates 11-way vector in three cycles. The sequence of division and multiplication required for an interpolation are implemented as a sequence of subtraction and addition in the logarithmic domain as shown in Equation (3.3):

$$p_0 + (\Delta p / \Delta y) \times y = p_0 + 2^{(\log_2 \Delta p - \log_2 \Delta y + \log_2 y)}. \tag{3.3}$$

The implementation of this operation is depicted in Figure 3.10. The same scheme is applied to setting up horizontal scanline parameters as well.

In the pixel pipeline, per-pixel rendering operations like rasterization, depth comparison, texture image mapping, and pixel blending are carried out. The pixel pipeline takes an early depth test to gate off the clock to the pipeline in processing pixels failing the depth test. This leads to power saving and memory bandwidth reduction by preventing the texture image mapping and blending operations from being applied to the invisible pixels unnecessarily. In a texture unit, a four-way SIMD divider, $(s, t, u, v)/w$, is required for the perspective correct texture address generation and is also implemented using the logarithmic arithmetic as expressed in Equation (3.4), resulting in area and power reductions:

$$(s, t, u, v)/w = 2^{(\log_2 s, \log_2 t, \log_2 u, \log_2 v) - \log_2 w}. \tag{3.4}$$

The implementation for this is shown in Figure 3.11. The texture unit also implements bilinear MIPMAP filtering rather than the trilinear MIPMAP algorithm for a cost-effective and power-efficient implementation of the texture-mapping operation [11].

3.4   SYSTEM POWER MANAGEMENT

The system power efficiency can be improved by scaling the supply voltage since the power consumption is quadratically proportional to the supply voltage [12]. Therefore, the dynamic voltage scaling (DVS) or DVFS schemes can drastically reduce system power consumption, and these are adopted for the battery-powered handheld systems [13–16].

3.4.1   Multiple Power-Domain Management

In this work, triple-power domains with independent frequencies and voltages are continuously tuned by tracking the workload on each module. This chip uses two levels of hierarchical power management scheme, that is, interframe and intraframe. Since 3D graphics scenes are displayed at a given frame rate, the RE should draw a finite number of pixels within the given period for a

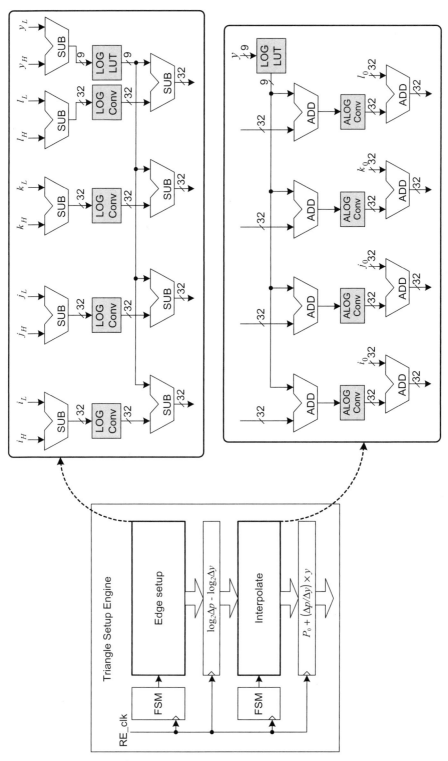

FIGURE 3.10. Four-way SIMD interpolator. FSM, finite state machine.

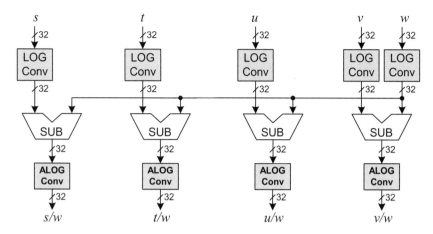

FIGURE 3.11. Four-way SIMD divider.

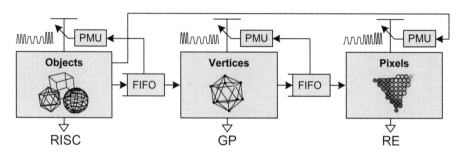

FIGURE 3.12. Multiple-domain power management scheme.

single frame. Therefore, at the interframe level, the target frequency and supply voltage for the RE is determined according to the target frame rate. In every completion of a scene, the software library running on the RISC measures the time elapsed for a scene and adjusts the frequency and supply voltage adaptively for processing the next scene.

Since the objects in a scene are composed of a number of polygons which are again composed of a number of pixels, the workloads on the RISC, GP, and RE that operate on the object, triangle, and pixel, respectively, should be completely different. Therefore, at the intraframe level, the RISC, GP, and RE are divided into different power domains from each other and each of the frequencies and voltages are controlled independently according to the workload. Figure 3.12 illustrates the proposed power management scheme.

The power management loop for each domain is described in Figure 3.13. The workloads on the RISC and GP are obtained by measuring the occupation levels at the 16-entry matrix FIFO and 32-entry index FIFO, respectively. Each occupation level is compared with the reference level to produce the *Err* value,

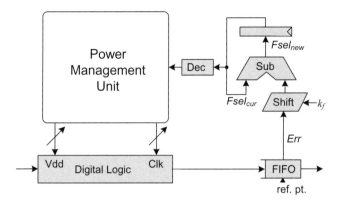

FIGURE 3.13. Power management loop. Dec, decoder.

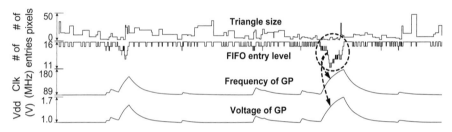

FIGURE 3.14. DVFS operation diagram.

which is shifted by the gain, k_f. New target frequency, $Fsel_{new}$, is calculated from the frequency selection logic by subtracting the shifted Err from $Fsel_{cur}$, as shown in Figure 3.13. In Figure 3.14, the droop of the FIFO entry level, that is, the high workload due to small triangles, increases the frequency and supply voltage of the GP to speed up its operation until the FIFO entry level restores to the reference point.

3.4.2 Power Management Unit

This chip includes three PMUs for clock and supply voltage generation of three power domains. The PMU is based on a phase-locked loop (PLL) structure and produces new target frequencies according to the given frequency information from the frequency selection logic to its clock divider. Figure 3.15 shows the circuit diagram of the PMU. It merges a linear regulator in its PLL loop to adjust the scaling of clock frequency and supply voltage correspondingly. Here, two separate regulators are accommodated for the digital logic and the voltage-controlled oscillator (VCO), respectively, to isolate the PMU loop from the digitally switching noise. In the PMU loop, the new target frequency given to the divider (DIV) is converted into a new control voltage

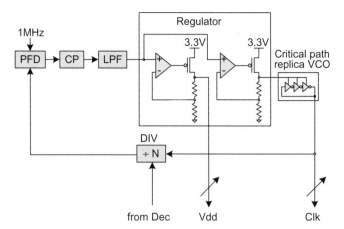

FIGURE 3.15. Proposed PMU.

through a phase frequency detector (PFD), a charge pump (CP), and a low-pass filter (LPF), and finally, the control voltage is given to the voltage regulator. The voltage regulator generates a new target supply voltage according to the given control voltage and the VCO generates a new clock frequency accordingly. The ring oscillator used for the VCO models the critical path of the digital logic in target power domain. The delay of the ring oscillator is made to be programmable so that the three PMUs can share the same design and be deployed at three different power domains with a proper programming of the ring oscillator delay to accurately model the actual critical path delay in each power domain.

The PMUs for the RISC and GP range from 89 to 200 MHz with 1 MHz step, or from 1.0 to 1.8 V with 7.2 mV step. The PMU for the RE changes from 22 to 50 MHz with the 250 kHz step with the same voltage characteristics with others. Since all of the blocks are designed with fully static circuits, the blocks can keep running during the frequency and voltage change. Figure 3.16 shows the measured DVFS waveforms of the PMU with a regulation speed of 9 mV/1 μs.

3.5   IMPLEMENTATION RESULTS

3.5.1   Chip Implementation

The proposed GPU is fabricated using 0.18 μm six-metal complementary metal–oxide–semiconductor (CMOS) technology. The core size is 17.2 mm^2 and has 1.57 M transistors and 29 kB SRAM. The graphics performance of this chip is 141 Mvertices/s for peak geometry transformations and 50 Mpixels/s for rendering operations, including texture image mapping. It integrates an

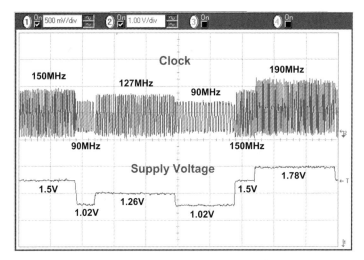

FIGURE 3.16. Measurement waveform of the PMU.

RISC, a GP, and an RE into a small area of 17.2 mm² and achieves an outstanding performance of 141 Mvertices/s by using the LNS in implementing the graphics pipeline. With the multiple power-domain management scheme, this chip consumes 52.4 mW when the scenes are drawn at 60 fps, which is a 51% reduction from the previous work [7]. Each PMU takes 0.45 mm² and consumes less than 5.1 mW in this scheme. The maximum power consumption of this chip is 153 mW when all components are operating at their full speeds, that is, 200 MHz and 1.8 V for both the RISC and GP, and 50 MHz and 1.8 V for RE. The area taken by each component is shown in Figure 3.17 and the power consumption according to different requirements on frame rate is shown in Figure 3.18. Table 3.1 summarizes key features of the chip and the chip micrograph is shown in Figure 3.19.

3.5.2 Comparisons

The comparison with other GPUs is listed in Table 3.2. The work in Yu et al. [7] shows similar processing speed and power consumption to ours even though it only integrates the GP, excluding the RE and application stage processor. As shown in this table, our chip shows the highest geometry processing speed and lowest area and power consumption among the compared works.

The individual graphics processing modules, GP and RE, are also compared separately with others using a proposed figure of merits (FoMs). Since the area and the power consumption as well as the performance are also important factors for the handheld devices, the FoM for these systems should take these factors into account. Therefore, we used the FoM of normalized performance

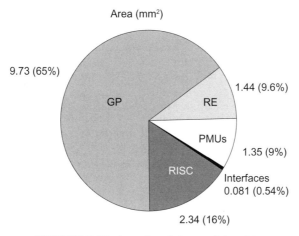

FIGURE 3.17. Area breakdown of the chip.

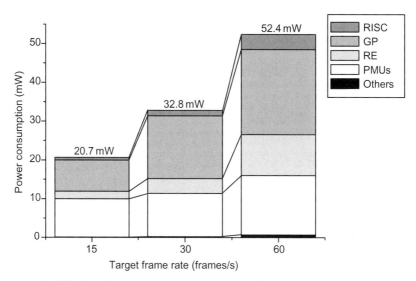

FIGURE 3.18. Power consumption according to target frame rate.

by the area and power consumption for the comparison. The FoMs used for the GP are expressed in Equation (3.5):

$$\text{FoM}_{GP} = \frac{\text{Performance (Kvertices/s)}}{\text{Power (mW)} \times \text{Area (mm}^2)}, \tag{3.5}$$

and those for the RE are expressed in Equation (3.6):

TABLE 3.1. Chip Characteristics

Process technology	TSMC 0.18 μm one-poly six-metal CMOS technology
Die size	Core: 17.2 mm^2 (9.7 mm^2 for GP)
	Chip: 25 mm^2
Power supply	1.0–1.8 V for core, 3.3 V for input/output (I/O)
Operating frequency	RISC: 89–200 MHz
	GP: 89–200 MHz
	RE: 22–50 MHz
Transistor counts	1.6 M Logic (968 K for GP)
	29 kB SRAM
Power consumption	52.4 mW at 60 fps with Quarter Video Graphics Display (QVGA)
	153 mW at full speed (86.8 mW for GP)
Processing speed	RISC	200 million instructions per second (MIPS)
	GP	141 Mvertices/s (geometry transformation)
	RE	50 Mpixels/s, 200 Mtexels/s (bilinear MIPMAP texture filtering)
Features	Standards	OpenGL ES 2.0
		Shader Model 3.0 (without texture lookup instruction)
	Power management	Multiple DVFS power domains
		• Three power domains with DVFS
		• One fixed power domain for PMU

$$\text{FoM}_{\text{RE}} = \frac{\text{Performance (Kpixels/s)}}{\text{Power (mW)} \times \text{Area (mm}^2)}. \qquad (3.6)$$

Based on these FoMs, the proposed GP and RE show 167.42 Kvertices/s/mW/ mm^2 and 1.98 Mpixels/s/mW/mm^2, respectively. Figure 3.20 compares the FoMs of GP and RE with those of other chips, of which performance, power, and area information are listed in Tables 3.3 and 3.4. According to the proposed FoMs, our proposed GPU shows maximum 3.5 times and 4.62 times improvements for GP and RE, respectively.

3.6 CONCLUSION

A power- and area-efficient embedded GPU is proposed for mobile multimedia systems. The GPU integrates a full 3D graphics pipeline with an ARM10 compatible RISC processor, a GP, an RE, and three PMUs. The GP includes a four-way, 32-bit floating-point arithmetic unit, which adopts the LNS at its arithmetic core and performs the 3D geometry transformation in every two

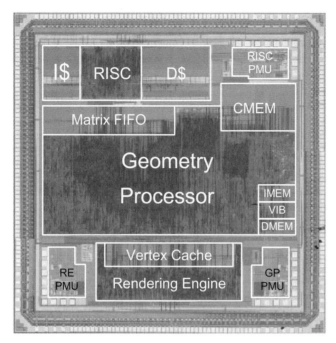

FIGURE 3.19. Chip micrograph.

TABLE 3.2. Comparison of Handheld GPUs

	Performance		Power Consumption (mW)			
Reference	In vertices/s	In pixels/s	At Full Speed	At 60 fps	Area (Transistors)	Functions
Arakawa et al. [4]	36 M	N.A.	250	N.A.	N.A.	GP only
Kim et al. [5]	33 M	166 M	407	N.A.	7.0 M	GP + RE
Sohn et al. [6]	50 M	50 M	155	N.A.	2.0 M	GP + RE
Yu et al. [7]	120 M	N.A	157	106	1.5 M	GP only
This work	141 M	50 M	153	52.4	1.57 M	GP + RE

N.A., not applicable.

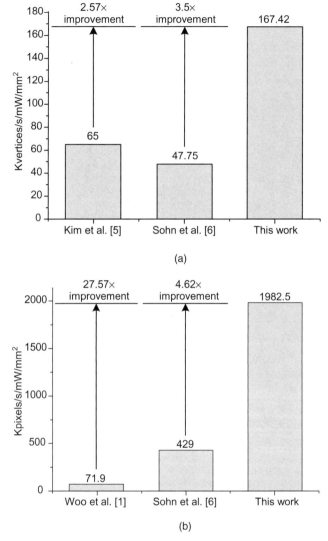

FIGURE 3.20. Comparison results. (a) Comparison of FoM for GPs. (b) Comparison of FoM for RE.

TABLE 3.3. Comparison of GPs

Reference	Performance (Mvertices/s)	Power consumption (mW)	Area (mm²)	Technology (μm)
Sohn et al. [6]	50	155	10.2	0.18 Logic
Yu et al. [7]	120	157	16	0.18 Logic
This work	141	86.8	9.7	0.18 Logic

TABLE 3.4. Comparison of REs

Reference	Performance (Mpixels/s)	Power consumption (mW)	Area (mm²)	Technology (µm)
Woo et al. [1]	66	58.1	15.8	0.16 DRAM
Yu et al. [7]	50	25.5	4.57	0.18 Logic
This work	50	19.4	1.3	0.18 Logic

cycles. This high-speed geometry transformation and the 58% hit-rate vertex cache achieve a peak geometry processing rate of 141 Mvertices/s at 200 MHz clock frequency. A test chip is fabricated using 0.18 µm six-metal CMOS technology. It contains 1.57 M transistors and 29 kB SRAM in a 17.2 mm² area. The chip is divided into three independent power domains of RISC, GP, and RE, and each domain adopts the DVFS for 52.4 mW power consumption at 60 fps. The maximum speeds for each power domain are 200 MHz for the RISC and GP, and 50 MHz for the RE, and the associated total power consumption is 153 mW. According to the proposed FoMs, this work is 3.5 times and 4.62 times better in geometry and rendering operations than previous works.

REFERENCES

[1] R. Woo, S. Choi, J.-H. Sohn, et al., "A 210 mW graphics LSI implementing full 3D pipeline with 264Mtexels/s texturing for mobile multimedia applications," in IEEE Int. Solid-State Circuits Conf. Dig. Tech. Papers, Feb. 2003, pp. 44–45.

[2] M. Imai, T. Nagasaki, J. Sakamoto, et al., "A 109.5 mW 1.2 V 600 M texels/s 3-D graphics engine," in IEEE Int. Solid-State Circuits Conf. Dig. Tech. Papers, Feb. 2004, pp. 332–333.

[3] Khronos Group, OpenGL-ES 2.0. Available: http://www.khronos.org

[4] F. Arakawa, T. Yoshinaga, T. Hayashi, et al., "An embedded processor core for consumer appliances with 2.8GFLOPS and 36M polygons/s FPU," in IEEE Int. Solid-State Circuits Conf. Dig. Tech. Papers, Feb. 2004, pp. 334–335.

[5] D. Kim, K. Chung, C.-H. Yu, et al., "An SoC with 1.3 Gtexels/s 3-D graphics full pipeline for consumer applications," in IEEE Int. Solid-State Circuits Conf. Dig. Tech. Papers, Feb. 2005, pp. 190–191.

[6] J.-H. Sohn, J.-H. Woo, M.-W. Lee, et al., "A 50Mvertices/s graphics processor with fixed-point programmable vertex shader for mobile applications," in IEEE Int. Solid-State Circuits Conf. Dig. Tech. Papers, Feb. 2005, pp. 192–193.

[7] C.-H. Yu, K. Chung, D. Kim, et al., "A 120Mvertices/s multi-threaded VLIW vertex processor for mobile multimedia applications," in IEEE Int. Solid-State Circuits Conf. Dig. Tech. Papers, Feb. 2006, pp. 408–409.

[8] B.-G. Nam, J. Lee, K. Kim, et al., "A 52.4 mW 3D graphics processor with 141Mvertices/s vertex shader and 3 power domains of dynamic voltage and frequency scaling," in IEEE Int. Solid-State Circuits Conf. Dig. Tech. Papers, Feb. 2007, pp. 278–279.

[9] J.N. Mitchell Jr., "Computer multiplication and division using binary logarithms," IRE Transactions on Electronic Computers, 11, Aug. 1962, pp. 512–517.

[10] B.-G. Nam, H. Kim, H.-J. Yoo, "Unified computer arithmetic for handheld GPUs," in K. Iniewski, ed., Chapter 27 in Circuits at the Nanoscale: Communications, Imaging, and Sensing, CRC press, Boca Raton, FL, Sept. 2008.

[11] B.-G. Nam, M.-W. Lee, H.-J. Yoo, "Development of a 3-D graphics rendering engine with lighting acceleration for handheld multimedia systems," IEEE Transactions on Consumer Electronics, 51(3), Aug. 2005, pp. 1020–1027.

[12] A. Chandrakasan, S. Sheng, R.W. Bordersen, "Low-power CMOS digital design," IEEE Journal of Solid-State Circuits, 27(4), Apr. 1992, pp. 473–484.

[13] T. Kuroda, K. Suzuki, S. Mita, et al., "Variable supply-voltage scheme for low-power high-speed CMOS digital design," IEEE Journal of Solid-State Circuits, 33(3), Mar. 1998, pp. 454–462.

[14] T.D. Burd, T.A. Pering, A.J. Stratakos, et al., "A dynamic voltage scaled microprocessor system," IEEE Journal of Solid-State Circuits, 35(11), Nov. 2000, pp. 1571–1580.

[15] K.J. Nowka, G.D. Carpenter, E.W. MacDonald, et al., "A 32-bit PowerPC system-on-a-chip with support for dynamic voltage scaling and dynamic frequency scaling," IEEE Journal of Solid-State Circuits, 37(11), Nov. 2002, pp. 1441–1447.

[16] S. Akui, K. Seno, M. Nakai, et al., "Dynamic voltage and frequency management for a low-power embedded microprocessor," IEEE Journal of Solid-State Circuits, 40(1), Jan. 2005, pp. 28–35.

4 Low-Cost VLSI Architecture for Random Block-Based Access of Pixels in Modern Image Sensors

TAREQ HASAN KHAN and KHAN WAHID

4.1 INTRODUCTION

Commercial high-speed image sensors [1–6] send pixels in a raster-scan fashion, and at a much higher rate compared with the speed of conventional low-cost microcontroller units (MCUs), such as Alf and Vegard's Risc (AVR) [7], Peripheral Interface Controller (PIC) [8], and so on. As a result, random access of image pixels and their real-time processing is not possible. Besides, inside the image sensors, no buffering is provided. Hence, given the limited size of internal memory in these microcontrollers, storing a complete high-resolution image frame and accessing pixels randomly of the same frame are also an issue. We have recently offered a low-cost solution to the abovementioned problems by introducing a bridge hardware, referred to as iBRIDGE, that bridges the speed gap between high-speed image sensors and low-speed microcontrollers (or image processors) [9]. The iBRIDGE offers several features such as random access of image pixels for any digital video port (DVP)-based image sensors, memory buffering, on-chip clock generator, built-in inter-integrated circuit (I2C) protocol [10], power management, and so on.

However, the first iBRIDGE design suffers from a few shortcomings, such as the large number of inputs/outputs (I/Os); as a result, the microcontroller also needs a large number of I/Os for interfacing. Although the first iBRIDGE allows random access, the block-based (BB) pixel access is inconvenient as the column and row position of each pixel in the block need to be provided each time during a pixel access. In this chapter, we present a second version of the bridge design, known as the iBRIDGE-BB, with fewer multiplexed I/O pins and that allows random access of BB pixels (i.e., 4×4, 8×8, 16×16, or

Embedded Systems: Hardware, Design, and Implementation, First Edition.
Edited by Krzysztof Iniewski.
© 2013 John Wiley & Sons, Inc. Published 2013 by John Wiley & Sons, Inc.

1	2	3	4	5	6	7	8	9	10	11	12
13	14	15	16	17	18	19	20	21	22	23	24
25	26	27	28	29	30	31	32	33	34	35	36
37	38	39	40	41	42	43	44	45	46	47	48

(a)

1	2	3	4	17	18	19	20	33	34	35	36
5	6	7	8	21	22	23	24	37	38	39	40
9	10	11	12	25	26	27	28	41	42	43	44
13	14	15	16	29	30	31	32	45	46	47	48

(b)

FIGURE 4.1. Pixel sequence in (a) raster scan (typical image sensor) and (b) for 4×4 block (in transform coding).

32×32) in both vertical and horizontal sequences efficiently. In Figure 4.1a, the pixel sequence coming from a commercial image sensor in raster-scan fashion is shown. However, when a 4×4 image block needs to be accessed, the required pixel sequence is different and is shown in Figure 4.1b. Using the proposed bridge, a block of pixels can be accessed without providing the column and row positions of each pixel in the block, rather the addressing is handled by the iBRIDGE-BB automatically. BB access is particularly useful in applications [11–15] that use transform coding or BB compression, such as in modern video codecs, video surveillance systems, real-time video tracking, robotics vision, and so on. Along with BB random access, single pixel-based random access feature is also supported in iBRIDGE-BB.

There are some commercial image sensors, such as MT9V011 from Aptina (San Jose, CA) [3] and OVM7690 from OmniVision (Santa Clara, CA) [2], that support partial access of image segments known as "windowing." By configuring the control resisters, the top-left and bottom-right corners of the desired area can be specified. The image sensor then captures and sends an image of the specified rectangle. However, it is not possible to access (and capture) other segments of the same frame with this feature, which is required in several image-coding applications such as transform coding.

4.2 THE DVP INTERFACE

Most leading commercial complementary metal–oxide–semiconductor (CMOS) image sensors, both standard definition (SD) and high definition (HD), send image data using a common standard interface, known as the DVP interface. The common I/O pins of a typical CMOS image sensor are shown at the left side of Figure 4.2. The *VD* (or *VSYNC*) and *HD* (or *HSYNC*) pins indicate the *end of frame* and *end of row*, respectively. Pixel data bytes are

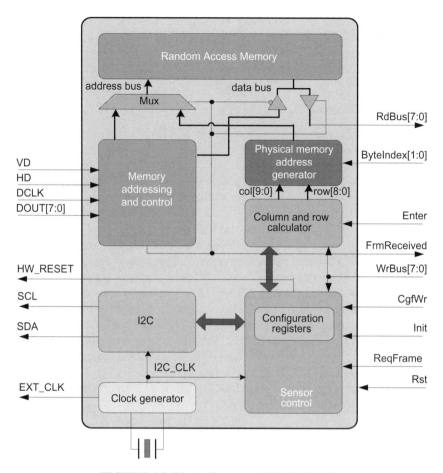

FIGURE 4.2 Block diagram of iBRIDGE-BB.

available for sampling at the *DOUT[7:0]* bus at the positive edge of the *DCLK* signal. The *EXTCLK* is the clock input for the image sensor. The frequency of *DCLK* is equal to or one-half of the frequency of *EXTCLK*, depending on the configuration of the image sensor. The initialization and configuration of the image sensor is done by the two-wire (*SCL* and *SDA*) I2C protocol. In the context of image sensor, it is often called the *Serial Camera Control Bus* (SCCB) interface [2]. The frame size, color, sleep mode, and wake-up mode can be controlled by sending I2C commands to the image sensor.

4.3 THE IBRIDGE-BB ARCHITECTURE

The proposed bridge (referred to as iBRIDGE-BB) is placed in between the image sensor and the image processor or the microcontroller. Figure 4.2 shows the interface pins of the iBRIDGE-BB and its internal blocks. The pins on the left-hand side (which is the DVP interface) are connected with an image sensor, while those on the right-hand side are connected to the image processor or the I/O ports of the microcontroller.

4.3.1 Configuring the iBRIDGE-BB

The operation starts by first configuring the iBRIDGE-BB's internal registers with a set of predefined addresses and data so that it can properly communicate with the image sensor. Image sensors of different manufactures have a different *device ID* (or *slave address*), which is used for communication using the I2C protocol [10]. The image sensors also have internal control registers used to configure the functionality, such as frame size, color mode, sleep mode, and so on. These registers are read and written by the I2C protocol. The internal register mapping is different for different manufacturers, so the iBRIDGE-BB needs to be configured with the proper configuration mapping of the image sensor (found on the datasheet). Table 4.1 shows 17 of such configuration registers implemented inside the iBRIDGE-BB. The table may be extended to accommodate additional special features. In order to write to a register of the iBRIDGE-BB, the register address is first placed on *WrBus* bus, and then a pulse is given at the *CfgWr* pin. After that, the data that need to be written in that address are placed on the *WrBus*, and another pulse is given at the *CfgWr* pin. Thus, the iBRIDGE-BB can be configured for any DVP-compatible image sensor, which makes it universal. After the iBRIDGE-BB is configured once, it can be used to capture image frame.

The iBRIDGE-BB can be configured either to random *BB access mode* or to random single *pixel-based access mode* using the register address 10h. In BB access mode, the block size (i.e., 4×4, 8×8, 16×16, or 32×32) and the pixel access sequence can also be configured.

4.3.2 Operation of the iBRIDGE-BB

After configuration, the image-capturing process can be started by giving a pulse at the *Init* pin. The image sensor will be configured with the frame size and color mode according to the information provided in the configuration registers, and then image data will begin to store in the bridge's memory sequentially. During the image-capturing process, *RdBus* goes to high impedance state. As soon as the image-capturing process is completed, the *Frm-Received* pin goes from low to high. At this time, the image sensor is taken to sleep mode to save power. The *WrBus*, *Enter*, *ByteIndex*, and *RdBus* pins are then used by the image processor to access any pixel or a block of pixels at any desired speed and in a random access fashion.

To access a pixel in pixel-based random access mode, the pixel column is placed on the *WrBus* and a pulse is given at the *Enter* pin. The pixel row is then placed on the *WrBus* and another pulse is given at the *Enter* pin. Then,

TABLE 4.1. Configuration Register Mapping

Register Address [7:0]	Register Data [7:0]	Description
00h	Device ID	The unique device ID of an image sensor. It is used for I2C communication.
01h	Total command	Total number of commands that need to be sent at the time of configuring the image sensor. Maximum value is 4.
02h	Sleep reg. adr.	The address of the register to control sleep mode.
03h	Sleep reg. data	Data that need to be written in the sleep register to switch the sensor to sleep mode.
04h	Wake reg. adr.	The address of the register to control wake-up mode.
05h	Wake reg. data	Data that need to be written in the wake-up register to switch the sensor to wake-up mode.
06h	Command_1 reg. adr.	The address of the register for command_1
07h	Command _1 reg. data	The data for the command_1 register
08h	Command _2 reg. adr.	The address of the register for command_2
09h	Command _2 reg. data	The data for the command_2 register
0Ah	Command _3 reg. adr.	The address of the register for command_3
0Bh	Command _3 reg. data	The data for the command_3 register
0Ch	Command _4 reg. adr.	The address of the register for command_4
0Dh	Command _4 reg. data	The data for the command_4 register
0Eh	ImageWidth/4	The image width divided by 4
0Fh	Bytes_per_pixel	Number of bytes used per pixel
10h	Bit [7:4] unused Bit [3] random access mode 0: pixel based 1: block based Bit [2] pixel access sequence in block 0: vertical 1: horizontal Bit [1:0] block size 00: 4×4 01: 8×8 10: 16×16 11: 32×32	Configure random access mode, pixel access sequence in block (i.e., pixels come horizontally from left to right from the top row *or* vertically from top to bottom from the left row), and block size.

the pixel data arrive at the *RdBus* and can be sampled by the image processor. If the pixel has more than 1 byte (such as in 3 bytes in RGB888, 2 bytes in RGB565), then the consecutive bytes can be accessed by proving the offset address in the *ByteIndex[1:0]* bus. For instance, placing 00_2 on the *ByteIndex* will cause the first byte of the pixel to appear on *RdBus*, placing 01_2 will give the second byte, and so on.

To access a block of pixels in BB random access mode, the top-left column and row positions of the block are placed on the *WrBus* and pulses are given at *Enter* pin according to the same method in pixel-based random access mode. Then, the first pixel data of the block arrives at the *RdBus* and can be sampled by the image processor. The other pixels of the block can be accessed by simply providing pulses at the *Enter* pin; the physical addressing of those pixels in the memory are handled by the iBRIDGE-BB automatically. For example, if the iBRIDGE-BB was configured for an 8×8 block, then the remaining pixels of the block can be accessed by proving 63 clock pulses at the *Enter* pin. The direction of pixel access in the block (i.e., pixels come horizontally from left to right from the top row, or vertically from top to bottom from the left row), as shown in Figure 4.3, can also be set during configuration of the iBRIDGE-BB. If the pixel has more than 1 byte, then the consecutive bytes can be accessed by proving the offset address in the *ByteIndex[1:0]* bus.

After the desired pixel values are accessed, the process of capturing the next frame with the same configuration can be repeated by giving a pulse at the *ReqFrame* pin. As soon as the image-capturing process is completed, the *FrmReceived* pin goes from low to high and then the pixels can be accesses as discussed above.

4.3.3   Description of Internal Blocks

4.3.3.1   Sensor Control   This block takes input data for the *configuration registers*, and then configures and controls different modes of the image sensor. After the *Init* signal is received, it generates the *HW_RESET* signal for the image sensor and then waits for the 2000 *EXT_CLK cycle*, which is required for the image sensor to accept I2C commands for the first time. After the wait period, it sends commands to the image sensor using the *I2C block* to configure it to the required frame size and color. The command frames are made by taking the information provided in the configuration register as shown in Table 4.1. After a *wake up* command is sent, the image sensor starts to produce image data. After a complete frame is received in the bridge's memory, the controller sends a *sleep mode* command to the sensor to reduce the power consumption. When a *ReqFrame* signal is received, it sends the *wake up* command and the next image data starts to store in the memory. The finite state machine (FSM) implemented in this block is shown in Figure 4.4.

4.3.3.2   I2C   This block is used to generate the *I2C protocol* bits in *single master mode* [10]. This protocol allows communication of data between I2C

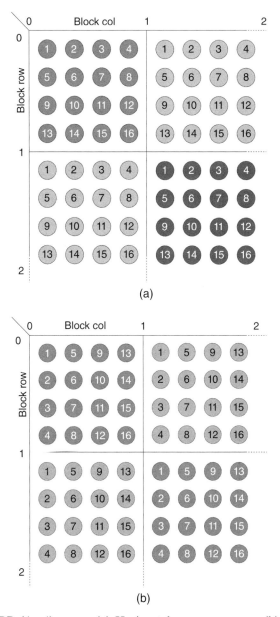

FIGURE 4.3. BB (4×4) access. (a) Horizontal access sequence; (b) vertical access sequence.

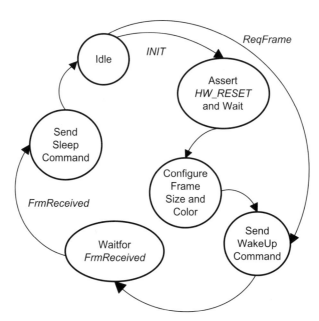

FIGURE 4.4. The FSM in the sensor control block.

devices over two wires. It sends information serially using one line for data (*SDA*) and one for clock (*SCL*). For our application, the iBRIDGE-BB acts as *master* and the image sensor acts as the *slave device*. Only the required subset of the I2C protocol is implemented to reduce the overall logic usage.

4.3.3.3 Memory Addressing and Control   This block, as shown in Figure 4.5, manages the data pins for the image sensor interface and generates address and control signals for the *random access memory* (RAM) block of the iBRIDGE-BB. It implements an *up counter* and is connected with the address bus of the memory. The *DOUT [7:0]* is directly connected with the data bus of the memory. When *VD* and *HD* are both high, valid image data comes at the *DOUT [7:0]* bus. In the valid data state, at each negative edge event of *DCLK*, the address up counter is incremented. At each positive edge event of *DCLK*, *WR* signal for the memory is asserted. After a complete frame is received, the address up counter is cleared and the *FrmReceived* signal is asserted high.

4.3.3.4 Random Access Memory (RAM)   A single-port RAM block is used to store a frame. Depending upon the application's requirement, a different memory size can be chosen. The memory requirements for different image sizes are shown in Table 4.2. In the iBRIDGE-BB, one multiplexer (Mux) for address bus and two tri-state buffers for data bus are used for proper writing in and reading from the memory as shown in Figure 4.2.

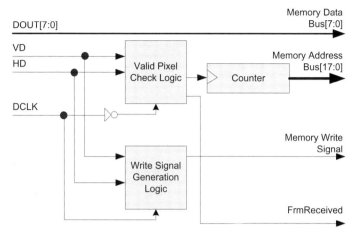

FIGURE 4.5. Memory addressing and control block.

TABLE 4.2. Memory Requirements for Different Image Sizes

Image Size (W × H) in Pixels	Memory Requirement (bytes)	
	Gray (8 bits/pixel)	RGB565/YUV422 (16 bits/pixel)
subQCIF (128 × 96)	12,288	24,576
QQVGA (160 × 120)	19,200	38,400
QVGA (320 × 240)	76,800	153,600
VGA (640 × 480)	307,200	614,400

subQCIF, subquarter Common Intermediate Format; QQVGA, Quarter QVGA; VGA, Video Graphics Array.

4.3.3.5 Column and Row Calculator This module implements the logics required to support both BB and pixel-based access mode. In BB access mode, this module first takes the *block column* and *block row* as input. Then, at each *Enter* pins positive-edge event, the *pixel column (col[9:0])* and *pixel row (row[8:0])* positions are generated for accessing the pixels as a block according to the information provided in the configuration register (address: 10h). The FSM implemented inside this module is shown in Figure 4.6a. The pseudocode for generating the next pixels column and row addresses for an 8 × 8 block in horizontal sequence is shown in Figure 4.6b. In pixel-based access mode, this module takes the *pixel column* and *pixel row* as input and then directly writes them on the *col* and *row* buses namely.

4.3.3.6 Physical Memory Address Generator This block takes the *col[9:0]*, *row[8:0]*, and *ByteIndex[1:0]* as inputs and generates the physical memory

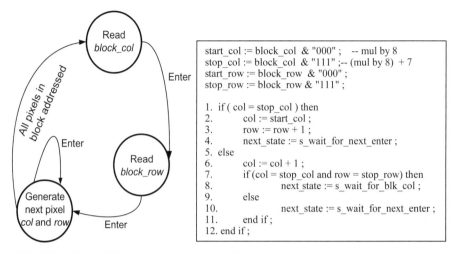

FIGURE 4.6. (a) FSM of BB pixel access; (b) pseudocode for generating next pixels column and row.

address from column and row position of the frame. To access a pixel value at column C, where $(0 \leq C \leq W - 1)$, and at row R where $(0 \leq R \leq H - 1)$, the physical memory address is calculated using Equation (4.1). Here, W is the image width and H is the image height. The image width W and *Bytes_per_pixel* are taken from the configuration register as shown in Table 4.1. If the *Bytes_per_pixel* is more than one, then the other consecutive bytes can be accessed by placing the offset on *ByteIndex* bus:

$$Adr = Bytes\_per\_pixel \times C + (Bytes\_per\_pixel \times W \times R) + ByteIndex \quad (4.1)$$

4.3.3.7 Clock Generator   This block generates the clock signal at the *EXT_CLK* pin, which must be fed to the image sensor. An external crystal needs to be connected with the iBRIDGE-BB; the frequency of the crystal should be the required *EXT_CLK* frequency of the image sensor. A parallel resonant crystal oscillator can be implemented to generate the clock [16]. An 800 KHz clock signal, called the *I2C_CLK*, is also required for operating the *I2C* and the *sensor control* blocks. This clock signal can be generated by dividing the *EXT_CLK* using a *mod* n *counter*. The *I2C* block generates a clock at *SCL* pin having half of the *I2C_CLK* frequency.

4.4 HARDWARE IMPLEMENTATION

4.4.1 Verification in Field-Programmable Gate Array

The proposed iBRIDGE-BB has been modeled in Very High Speed Integrated Circuit (VHSIC) hardware description language (VHDL) and then synthesized in Altera (San Jose, CA) DE2 board's field-programmable gate array (FPGA) [17] for verification. The internal modules of the iBRIDGE-BB, except the RAM, have been synthesized onto the Cyclone II FPGA. It occupies 584 logic elements (LE), 368 registers, and 6 embedded 9-bit multiplier elements. The iBRIDGE-BB's RAM module is connected with the 512 kB SRAM of the DE2 board. The on-board clock generator is used as the clock input for the bridge. The maximum frequency of $DCLK$ for the Altera Cyclone II implementation is 238.89 MHz. The other I/O pins of iBRIDGE-BB are assigned with different general-purpose input/output (GPIO) ports of the DE2 board. A commercial image sensor (TCM8230MD) from Toshiba (Minato, Tokyo, Japan) and a commercial microcontroller (ATmega644) from Atmel (San Jose, CA) have been used as the image sensor and the image processor or microcontroller, respectively. The microcontroller is then connected to a personal computer (PC) using a *Universal Serial Bus (USB) to RS232* port converter. A software is written in MS Visual Basic to display the captured images. The block diagram of the experimental setup and the photograph of the actual setup are shown in Figure 4.7.

Using the software in PC, a user can randomly enter the column and row positions of the captured image and then the software sends the position to the microcontroller through the PC's COM port. The microcontroller

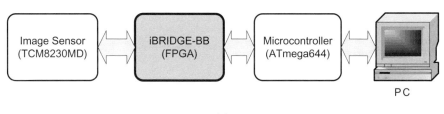

(a)

(b)

FIGURE 4.7. Experimental setup for verification. (a) Block diagram; (b) photograph.

FIGURE 4.8. (a) Full image; (b) BB random access (16×16); (c) BB random access (32×32); (d) pixel-based random access.

writes the column and row values in the iBRIDGE-BB, and then reads the corresponding block of pixel data (in BB access mode) or a pixel data (in pixel-based access mode) and sends them to the PC for plotting. In Figure 4.8, several randomly accessed blocks and single pixels captured by the microcontroller using the proposed iBRIDGE-BB are shown.

4.4.2 Application in Image Compression

The iBRIDGE-BB can be effectively used in low-cost embedded system application where image pixels need to be accessed in BB fashion for compression. If we try to make a real-time image processing (i.e., compression, pattern recognition, etc.) embedded system with a low-cost MCU and a commercial image sensor without using the iBRIDGE-BB, then the design will face the following problems:

- High-speed image sensors produce data at such a high rate that it cannot be processed in real time. As a consequence, most high-speed image sensors are difficult to use in low-power and low-speed embedded systems. For example, a Toshiba image sensor (TCM8230MD) produces data at the speed of 6–12 MHz [1]. Most microcontrollers, on the other hand, operate at a lower speed. For example, Atmel's ATmega644 can operate at a maximum speed of 8 MHz without external crystal [7]. Therefore, real-time processing of the pixel values is difficult.

- No buffering is provided inside the image sensors. Most microcontrollers have limited internal memory and may not be able to store a complete frame unless external memory is provided. For an example, the ATmega644 has 4 kB of internal memory. However, to store a Quarter Video Graphics Array (QVGA) (i.e., 320 × 240) grayscale image (1 byte/pixel), 76.8 kB memory is required.

- Several image and video compression algorithms such as JPEG, MPEG, VC1, H.264, and so on, need to access 8 × 8 or 4 × 4 pixel blocks [18]. However, commercial CMOS image sensors [1–6] send image data in a row-by-row (raster-scan) fashion and they do not have buffer memory; as a result, the data cannot be accessed in BB fashion efficiently. To access the pixels as a block, the image needs to be stored in the memory first and then the addressing of the pixels in the block needs to be calculated each time inside the MCU.

It should be noted that the commercial high-speed image sensors may be interfaced with more advanced MCUs (such as AT91CAP7E and AT91SAM7S512 from Atmel [19]). However, these microcontrollers contain many additional features (such as six-layer advanced high-speed bus [AHB], peripheral direct memory access [DMA] controller, USB 2.0 full speed device, configurable FPGA Interface, etc.) that may not be required for simple imaging applications. Besides, programming such microcontrollers and implementing the required protocols increase the design cycle time. The purpose of the proposed bridge hardware is to provide a compact, ready-made, and easy-to-use solution that enables interfacing of commercial general-purpose image sensors with simple microcontrollers that are low cost and easy to program (such as AVR, PIC etc.). Thus, the bridge hardware helps to shorten the design/development cycle time and facilitates rapid system-level prototyping.

To demonstrate the effectiveness of iBRIDGE-BB in real-time image compression application, we have used the hardware setup shown in Figure 4.7 to design a simple video surveillance system which includes a commercial image sensor, iBRIDGE-BB, and a low-cost MCU. The iBRIDGE-BB is configured to get color images from the image sensor of resolution 320 × 240 in YUV format. It is also configured to get pixels in 8 × 8 blocks in horizontal access sequence. In the microcontroller, a classical discrete cosine transform (DCT)-based compression algorithm is implemented for compressing the luminance (Y) component. Using the forward DCT as shown in Equation (4.2), the 64 pixel values p_{xy} are transformed to get the transform coefficients, S_{uv}:

$$S_{uv} = \frac{1}{4}C(u)C(v)\sum_{x=0}^{7}\sum_{y=0}^{7} p_{xy} \cos\frac{(2x+1)u\pi}{16}\cos\frac{(2y+1)v\pi}{16}, \quad 0 \le u, v \le 7, \quad (4.2)$$

where $C(0) = 1/\sqrt{2}$ and $C(i) = 1$ for $1 \le i \le 7$.

The interesting feature of DCT, which makes it widely useful in data compression, is that it takes correlated input data and concentrates its energy in a few transform coefficients. If the input data are correlated, such as in image pixels, then the first few top-left coefficients contains the most important (low-frequency) information and the bottom-right coefficients contains the less important (high-frequency) information [18]. The microcontroller computes the 15 top-left transform coefficients and sends them to the PC using 16 bytes (2 bytes for direct current [DC] and 1 byte for each alternating current [AC] coefficient in signed integer format).

The YUV color space has the advantage that the chrominance components (U and V) can be subsampled [20] without losing critical image quality. The microcontroller computes the average value of the U and V components in an 8×8 block and sends them to the PC using 2 bytes in unsigned integer format. Thus, for one 8×8 block of pixels (each containing 24 bits/pixel), the microcontroller only sends 18 bytes. In this way, the compression ratio is achieved to 90.63%.

In the PC side, the inverse DCT transformation is computed and the images are reconstructed. Some sample reconstructed images are shown in Figure 4.9.

Thus, we see that using the proposed iBRIDGE-BB, a low-cost MCU which has limited internal memory and low speed, can capture images using a high-speed, raster scan-based image sensor and can implement BB compression algorithm efficiently.

(a)   (b)

FIGURE 4.9. Reconstructed QVGA images of the video surveillance system (compression ratio 90.63%). (a) Building entrance; (b) hallway.

TABLE 4.3. Synthesis Results in ASIC

Technology	0.18 μm CMOS
Die dimension (W × H)	1.27 × 1.27 mm
Core dimension (W × H)	0.27 × 0.26 mm
Number of cells	1749
Number of gates	5733
Maximum DCLK frequency	316 MHz
Core power consumption	0.84 mW/MHz at 3.0 V

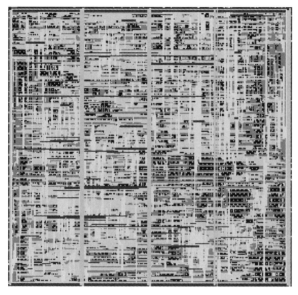

FIGURE 4.10. iBRIDGE-BB core.

4.4.3   Application-Specific Integrated Circuit (ASIC) Synthesis and Performance Analysis

The iBRIDGE-BB is implemented using Artisan 0.18 μm CMOS technology. The synthesis results are shown in Table 4.3. The chip layout (without the memory block) is shown in Figure 4.10. The design consumes 8.4 mW of power when running at 10 MHz with 3.0 V supply.

In order to show the significance of the proposed iBRIDGE-BB, we present comparisons with other work in Table 4.4. Here, we compare the performance of the iBRIDGE-BB with some random access image sensors; however, the image pixel arrays are dedicated with fixed resolution. While comparing with other sensors, we need to remember that the iBRIDGE-BB does not contain any dedicated image sensor; rather, it facilitates the interfacing of the image sensor with the image processor and enables random access in both

TABLE 4.4. Hardware Comparisons with Other Sensors

Design	Technology (μm)	Pixel Array (Resolution)	Random Access? Pixel	Random Access? Block	Size	Chip Area (mm²)	Data Rate	Power (mW)
Pecht et al. [21]	3.0	80 × 80	Yes	No	7.9 × 9.2	72.68	–	–
Scheffer et al. [22]	0.5	2048 × 2048	Yes	No	16.3 × 16.5	265.69	–	<100
Decker et al. [23]	0.8	256 × 256	Yes	No	–	–	390 fps	52
Dierickx et al. [24]	0.5	2048 × 2048	Yes	No	16 × 16	256	8 fps	–
Proposed iBRIDGE-BB (without sensor)	0.18	Any size	Yes	Yes	1.27 × 1.27	1.61	316 megasamples/s	8.4 at 3 V
iBRIDGE-BB with sensor (OV2710) [2]	0.18	Full HD 1920 × 1080	Yes	Yes	–	45.39	30 fps	358.4 at 3 V

"–", not reported; fps, frames per second.

BB and single pixel-based mode. In that sense, the proposed iBRIDGE-BB can be connected to any general-purpose DVP-based image sensors of any resolutions—this is a key advantage. As an example, in Table 4.4, we present the performance when the iBRIDGE-BB is interfaced with an advanced OmniVision HD image sensor (OV2710). With such a setup, the performance of the iBRIDGE-BB is noticeably better compared with all the sensors in terms of pixel array, BB access, silicon area, data rate, and power consumption. Note that the die area (i.e., core area plus the I/O pads) is used for the iBRIDGE-BB in these tables.

In Table 4.5, we present the performance of the iBRIDGE-BB when interfaced with both SD (TCM8230MD) and HD (OV2710) image sensors. It can be seen that, with a very little increase in hardware (i.e., 1.61 mm²) and power consumption (i.e., 8.4 mW), any DVP-compatible commercial image sensor can be converted to a high-speed randomly accessible image sensor.

4.5 CONCLUSION

In this chapter, the design of a bridge architecture, named as iBRIDGE-BB, is proposed, which enables low-cost MCUs to do real-time image processing efficiently. The proposed bridge overcomes the speed gap between commercially available CMOS image sensors and microcontrollers. By using the proposed bridge, a slow and low power microcontroller (or image processor) with less memory capacity can communicate with the image sensor to capture images of large size. The pixel data can also be accessed efficiently in random BB or single pixel-based mode through a parallel memory access interface at a desired speed. BB access (8×8 or 4×4) is particularly useful in compression application such as JPEG, MPEG, VC1, H.264, and so on. The design is also power efficient, as the iBRIDGE-BB forces the image sensor to sleep mode after capturing an image frame to the memory. The design is implemented using CMOS 0.18 μm Artisan library cells. The synthesis results show that the iBRIDGE-BB supports a data rate up to 316 MHz and suitable for rapid prototyping in different high-speed and low-power embedded system applications such as modern video codecs, video surveillance systems, real-time video tracking, robotics vision, and so on.

ACKNOWLEDGMENTS

The authors would like to acknowledge the Natural Science and Engineering Research Council of Canada (NSERC) for its support to this research work. The authors are also indebted to the Canadian Microelectronics Corporation (CMC) for providing the hardware and software infrastructure used in the development of this design.

TABLE 4.5. Performance Advantage of the iBRIDGE-BB with Commercial Image Sensors

Design		Pixel array	Area (mm2)	Data rate (fps)	Power (mW)	Random access?	
						Pixel	Block
Toshiba SD (TCM8230MD) [1]	Without iBRIDGE-BB	VGA 640 × 480	36.00	30	60.0	No	No
	With iBRIDGE-BB	VGA 640 × 480	37.61	30	68.4	Yes	Yes
OmniVision HD sensor (OV2710) [2]	Without iBRIDGE-BB	Full HD 1920 × 1080	43.78	30	350.0	No	No
	With iBRIDGE-BB	Full HD 1920 × 1080	45.39	30	358.4	Yes	Yes

REFERENCES

[1] Toshiba TCM8230MD Image Sensor (2012). [Online]. Available: http://www.sparkfun.com

[2] OmniVisoin OVM7690 CameraCube (2012). [Online]. Available: http://www.ovt.com

[3] Aptina MT9V011 Image Sensor (2012). [Online]. Available: http://www.aptina.com

[4] National LM9618 Image Sensor (2012). [Online]. Available: http://www.datasheetcatalog.org

[5] Kodak KAC-9630 Image Sensor (2012). [Online]. Available: http://www.kodak.com

[6] Pixelplus PO6030 Image Sensor (2012). [Online]. Available: http://www.pixelplus.com

[7] Atmel 8 bit AVR (2012). [Online]. Available: http://www.atmel.com

[8] PIC 8 bit Microcontrolller (2012). [Online]. Available: http://www.microchip.com

[9] T.H. Khan, K. Wahid, "A DVP-based bridge architecture to randomly access pixels of high speed image sensors," EURASIP Journal on Embedded Systems, Hindawi Publishing Corporation, vol. 2011, 2011. Article ID 270908.

[10] The I2C Bus Specification (2012). [Online]. Available: http://www.nxp.com

[11] T. Silva, C. Diniz, J. Vortmann, L. Agostini, A. Susin, S. Bampi, "A pipelined 8×8 2-D forward DCT hardware architecture for H.264/AVC high profile encoder," Advances in Image and Video Technology, LNCS, 4872, 2007, pp. 5–15.

[12] L. Pillai, "Video compression using DCT," *Xilinx Application Note: Virtex-II Series,* San Jose, CA, 2002.

[13] K. Wahid, S. Ko, D. Teng, "Efficient hardware implementation of an image compressor for wireless capsule endoscopy applications," in Proc. Int. Joint Conference on Neural Networks, 2008, pp. 2761–2765.

[14] P. Turcza, M. Duplaga, "Low-power image compression for wireless capsule endoscopy," in Proc. IEEE International Workshop on Imaging Systems and Techniques, 2007, pp. 1–4.

[15] M. Lin, L. Dung, P. Weng, "An ultra-low-power image compressor for capsule endoscope," BioMedical Engineering OnLine, 5(14), 2006, pp. 1–8.

[16] R. Wagner, UART Crystal Oscillator Design Guide, Data Communications Application Note. (Mar 2000) [Online]. Available: http://www.exar.com

[17] Altera DE2 Board (2012). [Online]. Available: http://www.altera.com

[18] D. Salomon, *Data Compression: The Complete Reference*, 3rd ed., Springer-Verlag, New York, 2004.

[19] J. Uthus, O. Strom, "MCU architectures for compute-intensive embedded applications," *Atmel White Paper*, 2005.

[20] G. Chan, "Towards better chroma subsampling," SMPTE Journal, 117, 2008, pp. 39–45. Issue: 2008 05/06 May/June

[21] O.Y. Pecht, R. Ginosar, Y.S. Diamand, "A random access photodiode array for intelligent image capture," IEEE Transactions on Electron Devices, 38(8), 1991, pp. 1172–1180.

[22] D. Scheffer, B. Dierickx, G. Meynants, "Random addressable 2048×2048 active pixel image sensor," IEEE Transactions on Electron Devices, 44(10), 1997, pp. 1716–1720.

[23] S. Decker, R.D. McGrath, K. Brehmer, C.G. Sodini, "A 256 × 256 CMOS imaging array with wide dynamic range pixels and column-parallel digital output," IEEE Journal of Solid-State Circuits, 33(12), 1998, pp. 2081–2091.

[24] B. Dierickx, D. Scheffer, G. Meynants, W. Ogiers, J. Vlummens, "Random addressable active pixel image sensors," in Proc. of SPIE, vol. 2950, Berlin, 1996, pp. 2–7.

5 Embedded Computing Systems on FPGAs

LESLEY SHANNON

Field-programmable gate arrays (FPGAs) provide designers with a programmable hardware substrate consisting of configurable logic blocks (CLBs) and a configurable routing fabric. This design flexibility comes at the cost of increased power, area, and latency for a given implemented on an FPGA as opposed to an application-specific integrated circuit (ASIC). However, given the rising cost of fabricating an ASIC, FPGAs are used for an increasing number of low- and medium-volume applications. Furthermore, the increasing design resources available on each new generation of FPGA devices due to Moore's law allows designers to implement circuits of increasing complexity on a single FPGA. Since the late 1990s, FPGAs have been used to implement not only glue logic or hardware accelerators (e.g., fast Fourier transforms [FFT]), but complete systems-on-chip (SoCs) [1, 2]. The ability to combine one or more processors plus the required custom hardware accelerators for an embedded computing system on a single die, without having to pay the huge fabrication costs of custom integrated circuits (ICs), makes FPGAs an extremely attractive option for embedded computing systems.

This chapter starts by introducing the fundamental concepts common to commercial FPGA architectures and the different configuration technologies used by different vendors. It is followed by a discussion of the software support tools and operating system (OS) choices available for system design on FPGAs. The chapter concludes with a summary of the benefits and challenges inherent to using an FPGA as the embedded computing system platform as well as future directions of research and commercial infrastructure development that may benefit the embedded systems community.

Embedded Systems: Hardware, Design, and Implementation, First Edition.
Edited by Krzysztof Iniewski.
© 2013 John Wiley & Sons, Inc. Published 2013 by John Wiley & Sons, Inc.

5.1 FPGA ARCHITECTURE

Figure 5.1 illustrates the general structure of the configurable fabric of today's FPGAs. Unlike the initial generations of FPGAs, which had a relatively homogenous fabric comprising a two-dimensional array of CLB, modern FPGAs are generally heterogeneous [3]. As shown in Figure 5.2, typically, a

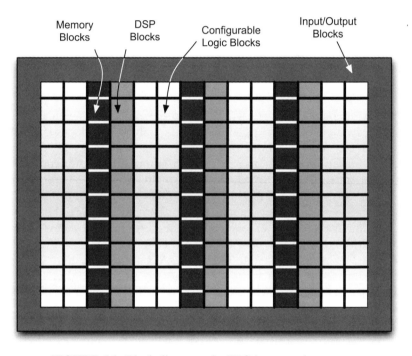

FIGURE 5.1. Block diagram of a FPGAs general structure.

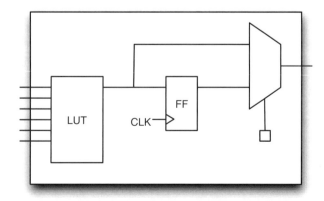

FIGURE 5.2. Block diagram of a typical CLB. CLK, clock; FF, flip-flop.

CLB combines a multi-input lookup table (LUT) with a flip-flop. LUTs are typically referred to as "n-LUTs," where n refers to the number of inputs to the LUT and can implement any combinational equation with up to n variables. The values stored in the LUT are determined by the circuit being implemented and are part of the programming "bitstream." The flip-flop enables the implementation of sequential circuits (e.g., state machines) and by multiplexing this output with the combinational, as shown in Figure 5.2, a configuration bit can be used to select either the combinational or registered output. If the combinational output is selected, this can be routed to a second CLB, enabling the FPGA to implement a combinational equation with greater than n inputs.

The thick black lines in Figure 5.1 represent the programmable routing fabric of the FPGA. Each routing channel, as represented by a single black line in the grid, comprises multiple routing tracks with different length wire segments. Wire segments are connected together using additional configuration bits. At each intersection of the routing channels is a switch block that allows vertical and horizontal tracks to be connected via configuration bits. A complete programming bitstream is generated for a device as the output of the synthesis tools and initializes all of the configuration bits in the device to the appropriate values needed to implement the user's circuits.

Besides the configurable routing and CLBs, another feature that has remained standard is that they provide a large number of programmable user inputs/outputs (I/Os) that can be used to transport data on and off chip. Furthermore, each I/O can be configured with numerous single-ended and differential signal standards. These two characteristics can greatly facilitate the fabrication of a printed circuit board with numerous peripherals that may need to be monitored and controlled by the FPGA in embedded applications.

Along with these more traditional components of an FPGA architecture, modern FPGAs generally contain embedded intellectual property (IP) blocks that are used to accelerate commonly used operations. The two most common IP cores are embedded block random access memories (BRAMs) and some form of digital signaling processing (DSP) block (e.g., a multiplier, a multiply-accumulator, etc). As memory is a key component of practically every computing system, BRAMs are the more common of the two IP cores. With these resources in the hardware fabric, an FPGA can be used to implement practically any system design (assuming sufficient resources); in fact, today's largest commercial devices can implement complex SoCs, ranging from multiprocessor SoCs (MPSoCs) [1, 4] to networks-on-chip [5, 6].

5.2   FPGA CONFIGURATION TECHNOLOGY

This section describes the two main technologies used to enable configuration of the programmable fabric. Specifically, static RAM (SRAM) and flash-based technologies are discussed, highlighting the opportunities and challenges for both technologies in embedded computing systems.

5.2.1 Traditional SRAM-Based FPGAs

The two largest FPGA vendors, Altera [7] and Xilinx [8], also sell the largest devices on the market. As such, both FPGA vendors provide lightweight soft processors that can be mapped to their devices; Altera has the NIOS [9] and Xilinx the MicroBlaze [10]. Furthermore, they both support the integration of custom accelerators into their processors. The NIOS has a more tightly integrated model, where the custom operations are integrated into the instruction pipeline. For operations with low latency, such as a multiply-accumulate function, this tight integration provides better performance. Conversely, the Micro-Blaze follows a more loosely coupled model; the processor includes additional unidirectional read and write ports, called Fast Simplex Links (FSLs), to which the processor can offload or upload data for independent processing. For operations with a longer latency, such as an FFT, this independence ensures that the MicroBlazes pipeline need not stall to wait for the completion of the operation.

Both Altera and Xilinx use SRAM technology to store the LUT configurations as well as to turn on the switches that connect wire segments in the connection blocks, routing channels, and switch blocks via SRAM bits. As such, when the device is powered down, its configuration "bitstream" is lost and the device must be reconfigured again on power up. Obviously, this means that a device can be used for more than one application during its lifetime, as there is no requirement to restore the previous bitstream on power up. However, a more interesting feature of SRAM-based FPGAs is that it is possible to leave a portion of the device statically configured, while the remainder of the device is reconfigured. This is known as dynamic partial reconfiguration (DPR) [11]. Although Xilinx has been supporting DPR for more than a decade, it is only with the latest generation of Stratix devices (Stratix V) that Altera will also be supporting DPR. This suggests that while reconfigurable computing has been fascinating researchers since the 1990s, it may take on more significance commercially in the future.

5.2.1.1 DPR   From a user's perspective, DPR means that a portion of their design may be allowed to continue to operate while another portion of the design is "swapped out" in favor of new functionality. For example, an embedded computing may be divided into a static component (e.g., the processors running software) that must remain active and available throughout the application's lifetime and a set of modules (e.g., the hardware accelerators) that are only required on occasion and may be allowed to share the same region of the fabric temporally as they are required. Computing systems that reconfigure their underlying architecture at runtime to process data are commonly referred to as *reconfigurable computing* architectures [12].

Reconfigurable computing architectures range from those with a reconfigurable pipeline [13–16] to those with a more traditional architectural; specifically, a traditional processor coupled with a reconfigurable fabric [17–19]. This coupling may be tightly integrated, such that custom functionality mapped to the fabric becomes a type of custom instruction, as with the NIOS processor. Alternatively, the accelerators may be more loosely integrated, such that the custom operations act as a semi-independent coprocessor, as with the MicroBlaze.

Since the FSLs are basically first in, first outs (FIFOs), it would be possible to connect these links to a dynamically reconfigurable region where users can swap in different hardware accelerators for their application as desired. However, this option has not been pursued by researchers to date. Instead, there has been significant focus on the challenges of practically integrating DPR into a computing system. Specifically, how will tasks be "scheduled" (i.e., mapped onto the hardware fabric) at runtime [20]? Another concern is moving data in the system between the processor and custom accelerators without negatively impacting the system performance [21] and ensuring that the system memory is coherent and protected, particularly in the case of multithreaded programs [22].

Along with the system-level challenges of using DPR in a system, there are other concerns. Although DPR allows designers to reduce the footprint of their design (by having modules temporally share a region of the fabric), thereby reducing a design's power consumption and potentially its required device size, reconfiguration comes at a cost. Specifically, while these devices are capable of operating in the hundreds of megahertz, reconfiguration times are on the order of milliseconds. Thus, reconfiguration requires numerous clock cycles, where the time is directly proportional to the size of the region to be reconfigured. Therefore, the question becomes, is the cost of the latency introduced by having to reconfigure the fabric worth it for an application? Instead, the designer could simply choose to use a larger device. Another concern when trying to incorporate DPR into a design is that the design process itself is nontrivial and, thus, increases the design cost. While Xilinx has improved the facility with which someone may create a design that includes DPR, this is still a nontrivial task, requiring considerable knowledge about FPGAs. As Altera's Stratix V devices are not readily available at this time, their new flow has not been as thoroughly evaluated for ease of use. It is hoped, however, that the tool flow for DPR will continue to improve now that it is supported by both major vendors; as such, researchers tend to focus much of their efforts on the previous issues discussed in this section.

5.2.1.2   Challenges for SRAM-Based Embedded Computing System   The two greatest challenges for SRAM-based computing systems are power and reliability. Since the device's configuration relies on SRAM, single event upsets (SEUs) can occur if an ionizing charge interacts with the SRAM bit, causing it to change states. Generally, this is only a concern for designs operating at higher altitudes or in outer space; however, researchers have investigated numerous techniques looked to provide fault tolerance for FPGA designs [23, 24], ranging from triple modular replication [24, 25], to bitstream "scrubbing" (i.e., reloading the original bitstream [26–28]), to having hard error correction coding built in to on-chip BRAMs [29, 30].

The other major concern for many embedded computing systems is low power consumption. In order to provide the design flexibility and density of these high-performance FPGAs, power consumption can be on the order of 10 W. Furthermore, neither Altera nor Xilinx provide a low power "sleep" state for their devices. This is particularly important for many embedded consumer applications that rely on battery power. Although Altera and Xilinx FPGAs consume considerably less power than an Intel high-performance processor (e.g., an i7) or a general-purpose graphics processing unit (GPGPU), without the ability to revert to a low power state when the device is not in use, battery-powered devices will have unacceptably short battery lives.

5.2.1.3 *Other SRAM-Based FPGAs* In 2008, SiliconBlue Technologies released a new low-power FPGA architecture aimed at the low-power embedded applications market (acquired by Lattice Semiconductor in 2011 [31]). Although they use SRAM-based configuration, their device architecture is much less complex and much smaller. Whereas a high-performance Virtex 6 provides more than 500,000 six-input LUTs for designers to use, the larger SiliconBlue devices are closer to 16,000 4-LUTs. Second, they provide significantly less on-chip memory on their high-performance series devices and support only a subset of the I/O standards relative to Altera and Xilinx's high-performance devices. While these choices in their device architecture reduce the power consumption of applications on their devices, the key difference is their inclusion of a low power mode for their devices where the clock is effectively disabled (0 kHz) and the device state remains locked until it starts clocking again.

In 2010, Tabula released their "Spacetime" architecture [32]. Like the larger FPGA vendors, their fabric is rich in heterogeneous resources. However, unlike the larger vendors, which provide a large physical fabric for all of their device resources, they generate multiple contexts for each physical resource and cycle through them to virtually extend the resources of the fabric. This is accomplished by providing a high core operating frequency (i.e., greater than 1 GHz) and then changing the resource contexts on every clock edge. Based on the number of contexts (i.e., 2/4/8), the effective operating frequency of the design is core frequency divided by the number of contexts. In some ways, this is akin to DPR; however, while each physical resource has multiple contexts, these are *statically* generated and remain unchanged for the duration of an application's execution. With respect to using these devices in embedded systems, their physical footprint is relatively smaller than a comparable Altera or Xilinx device; however, they operate at a considerably higher clock frequency. As such, there are currently no conclusive data as to which device technology is low power for a given application, or if there is a device that provides a consistently lower power solution.

Both Tabula and Lattice provide IP libraries that include soft processors. However, as both companies are relatively new, neither of them have had significant exposure within the research community exploring their capabilities and support for the traditional "processor plus hardware accelerators" embedded computing system architecture.

5.2.2 Flash-Based FPGAs

Originally, Actel sold only fuse-based technology FPGAs that were one-time programmable. However, after 2000, they introduced two families of reprogrammable, flash-based FPGAs—the ProASIC and IGLOO device families—and were eventually acquired by Microsemi in 2010. Although these devices are reprogrammable, they are nonvolatile, meaning that they retain their configuration even when powered down. This means that, unlike SRAM-based devices, they are active immediately on power up and provide better security for the bitstream as well as the firm-error immunity [33]. Furthermore, their devices consume less power than SRAM-based devices and, like SiliconBlue Technologies, they include a low power mode (called "Flash*Freeze"), which reduces power consumption to the order of microwatts to extend battery life. Their architecture is similar to the tradition of the previously mentioned FPGA vendors in that their devices provide CLBs (known as "Versatiles") and dual-port on-chip BRAMs. However, their CLBs have more limited functionality, with only three potential configurations: a 3-LUT, a D Flip-Flop with Clear or Set signal, or a D Flip-Flop with an Enable signal and a Clear or Set signal.

Like other FPGA vendors, Actel's IGLOO device allows users to instantiate a soft processor on the fabric. In fact, they support a 32-bit ARM Cortex-M1 processor. Relative to the NIOS and MicroBlaze soft processors (from Altera and Xilinx, respectively), the ARM core consumes a considerably larger percentage of the IGLOO's device resources at over 4000 tiles. As with the Tabula and SiliconBlue devices, the IGLOO architecture has not had significant exposure within the research community to explore its capabilities and support for the traditional "processor plus hardware accelerators" embedded computing system architecture. However, there are third-party software tool supports ranging from compilers, to debuggers, to real-time operating systems (RTOSs).

5.3 SOFTWARE SUPPORT

This section discusses the challenges of facilitating the description of FPGA designs through high-level synthesis as well as addresses the support for OSs that is now becoming a common feature for FPGA soft processors, reflecting their increased usage in embedded computing systems.

5.3.1 Synthesis and Design Tools

Currently, the most common design entry format for a circuit to be implemented on an FPGA is using a hardware description language such as Verilog or VHDL. To reduce the design time of complex circuits, designers will typically reuse IP modules from previous designs or those provided by their tool vendor. While this does improve productivity, the "gap" between the circuit size we can fabricate and the rate at which we can design these circuits is increasing [34]. As such, methods of describing circuits and, more importantly, SoCs at a higher level of abstraction and methods of efficiently transforming these descriptions into physical circuits (i.e., high-level synthesis) are important areas of research.

In the past, much of the research has focused on an illusive "C-to-gates" model, which would allow designers to describe their applications using an accessible software language; better yet, the ability to automatically convert any application written in C to a custom circuit. The general challenge in trying to generate a hardware circuit from software is that the clock's behavior in a software application is implicit to its description. Conversely, when an HDL is used to describe a hardware circuit, the clock's behavior is explicitly expressed. As such, HDLs such as HardwareC [35], HandelC [36], SpecC [37], and SystemC [38] support a restricted subset of the C language and include additional macros that allow users to explicitly describe the clock as needed to improve the efficiency (area and performance) of the design's mapping to the physical hardware.

More recently, Canis et al. [39] introduced the "LegUp" compiler, which translates a standard C program into a hybrid architecture containing a soft processor and custom hardware accelerators. As with previous works in this area, their objective is to increase the number of applications that can utilize FPGAs by making them accessible to software designers without drastically reducing the quality of the generated circuit relative to an HDL implementation of the system architecture. AutoESL AutoPilot [40] can be used to perform a similar, yet more comprehensive process for Xilinx FPGAs [7]. Along with being able to translate the synthesizable C/C++ code into synthesizable register-transfer level (RTL) while performing platform-based code transformations and synthesis optimizations, it also generates test bench wrappers, simulation scripts and estimates of FPGA resource utilization, timing, latency, and throughput of the synthesized design [40].

Both FPGA vendors and researchers have also been considering alternative high-level languages and frameworks for describing FPGA applications that are more accessible to software designers without significantly degrading the quality of the generated circuit, such as BlueSpec [41] and MATLAB [42]. One of the more promising high-level languages for describing the heterogeneous SoCs on FPGAs as well as for mapping applications to heterogeneous computing platforms that *include* FPGAs is OpenCL [43]. Both researchers and industry have noticed the potential for providing high-quality mappings of an application to hardware using OpenCL. Unlike C, OpenCL is an inherently parallel language and it allows parallelism to be declared by the programmer in the form of computational kernels. Also similar to HDLs, storage and movement of data is explicit, allowing programmers to manage their own memor(ies) [44]. Finally, it is aimed at heterogeneous compute platforms, facilitating the designers' ability to express problems in terms of processor software and custom computing hardware accelerators [44]. Given the nonprofit nature of the enterprise and the wide interest from industry, this framework has a great deal of promise for improving design productivity on FPGAs.

5.3.2   OSs Support

The greatest variety of OS support for FPGAs is provided by Altera and Xilinx; both of these companies have had soft processors for their devices for more than a decade. Originally, both the NIOS and MicroBlaze processor did not include memory management units (MMUs); as such, there was no support for virtual memory. Therefore, the original OSs supported by these soft processors were typically microkernels, such as Micriμm's μC [45], possibly with real-time support.

However, the MicroBlaze has had the option of including an MMU and has supported the operation of a full Linux kernel provided by a third-party vendor since 2007 [46]. In fact, the MicroBlaze ISA was added to the 2.4.37 version of the Linux kernel. More recently, MMU support has been included as an option for the NIOS processor. As such, they now also support a full Linux kernel as of 2011. Given that the embedded systems market is currently dominated by the Linux OS, this may facilitate the transition from a more traditional embedded systems computing platform to a complete SoC solution implemented on an FPGA.

5.4   FINAL SUMMARY OF CHALLENGES AND OPPORTUNITIES FOR EMBEDDED COMPUTING DESIGN ON FPGAs

FPGAs provide a unique set of opportunities for embedded computing systems. As they provide a configurable fabric large enough to support a processor plus custom hardware accelerators, they offer the opportunity to provide single-chip computing solutions without the costs of custom fabrication. They also provide more I/O pins for data than traditional microcontrollers, meaning that off-chip peripherals can be accessed independently and in parallel, as opposed to worrying about the scalability of adding numerous devices to a single shared bus architecture.

The latest generation of FPGA devices from both Altera (Stratix V) and Xilinx (Virtex 7) include embedded ARM processor cores in their fabric and support for DPR. These platforms provide the possibility of the traditional "processor + hardware accelerators" with the market's preferred embedded systems processor able to directly access a hardware fabric that can be configured with custom accelerators for an application. However, from a research perspective, the possibility of realizing a reconfigurable computer with OS support on a commercial device, potentially with multiple heterogeneous processors, is exciting.

While FPGAs are lower power options compared with high-end processors and GPGPUs, only a few of the devices provide the essential low-power mode needed for battery-operated applications. Another challenge is creating the initial system-level description. Although high-level synthesis is an active area of research, there is no clear winning "C-to-gates" solution, nor is there proof that this is the correct high-level approach to follow. Instead, commercial FPGA vendors, such as Altera, turned their focus to OpenCL and the opportunities suggested by its framework for describing application execution on heterogeneous platforms.

REFERENCES

[1] H. Nikolov, M. Thompson, T. Steanov, A. Pimentel, S. Polstra, R. Bose, C. Zissulescu, E. Deprettere, "Daedalus: toward composable multimedia MP-SoC design," in Proceeding of the 45th Annual Design Automation Conference (DAC'08), Anaheim, CA, June 2008.

[2] M. Horauer, F. Rothensteiner, M. Zauner, E. Armengaud, A. Steininger, H. Friedl, R. Pallierer, "An FPGA based SoC design for testing embedded automotive communication systems employing the FlexRay Protocol," in Proceeding of the Austrochip Conference, Villach, Austria, October 2004.

[3] V. Betz, J. Rose, A. Marquardt, Architecture and CAD for Deep-Submicron FPGAs, Kluwer Academic Publishers, Boston, 1999.

[4] O. Lehtoranta, E. Salminen, A. Kulmala, M. Hannikainen, T.D. Hamalainen, "A parallel MPEG-4 encoder for FPGA based multiprocessor SoC," in International Conference on Field Programmable Logic and Applications, Tampere, Finland, August 2005.

[5] K. Goossens, M. Bennebroek, J.Y. Hur, M.A. Wahlah, "Hardwired networks on chip in FPGAs to unify functional and configuration interconnect," in Proceedings of the Second ACM/IEEE International Symposium on Networks-on-Chip, Newcastle, UK, April, 2008.

[6] F. Lertora, M. Borgatti, "Handling different computational granularity by a reconfigurable IC featuring embedded FPGAs and a network-on-chip," in Annual IEEE Symposium on Field-Programmable Custom Computing Machines, Napa, CA, April 2005, pp. 45–54.

[7] Altera Corporation, Home Page. Available: http://www.altera.com.

[8] Xilinx Inc., Home Page. Available: http://www.xilinx.com.

[9] Introduction to the Altera Nios II Soft Processor. Available: ftp://ftp.altera.com/up/pub/Tutorials/DE2/Computer_Organization/tut_nios2_introduction.pdf.

[10] MicroBlaze Processor Reference Guide. Available: http://www.xilinx.com/support/documentation/sw_manuals/mb_ref_guide.pdf.

[11] P. Lysaght, B. Blodget, J. Mason, J. Young, B. Bridgford, "Enhanced architectures, design methodologies and CAD tools for dynamic reconfiguration of Xilinx

FPGAs," Invited Paper at the International Conference on Field Programmable Logic and Applications, Madrid, Spain, August 2006, pp. 1–6.

[12] K. Compton, S. Hauck, "Reconfigurable computing: a survey of systems and software," ACM Computing Surveys, 34(2), June 2002, pp. 171–210.

[13] S. Cadambi, J. Weener, S.C. Goldstein, H. Schmit, D.E. Thomas, "Managing pipeline-reconfigurable FPGAs," in ACM/SIGDA International Symposium on FPGAs, 1998, pp. 55–64.

[14] D. Deshpande, A.K. Somani, A. Tyagi, "Configuration caching vs. data caching for striped FPGAs," in ACM/SIGDA International Symposium on FPGAs, 1999, pp. 206–214.

[15] S.C. Goldstein, H. Schmit, M. Budiu, S. Cadambi, M. Moe, R. Taylor, "PipeRench: a reconfigurable architecture and compiler," IEEE Computer, 33(4), 2000, pp. 70–77.

[16] W. Luk, N. Shirazi, S.R. Guo, P.Y.K. Cheung, "Pipeline morphing and virtual pipelines," in W. Luk, P.Y.K. Cheung, M. Glesner, eds., Lecture Notes in Computer Science 1304—Field-Programmable Logic and Applications, Springer-Verlag, Berlin, Germany, 1997, pp. 111–120.

[17] J. Hauser, J. Wawrzynek, "Garp: a MIPS processor with a reconfigurable coprocessor," in Proceedings of the IEEE Symposium on Field-Programmable Custom Computing Machines, April 1997, pp. 24–33.

[18] J.E. Carrillo, P. Chow, "The effect of reconfigurable units in superscalar processors," in Proceedings of the Ninth ACM International Symposium on Field Programmable Gate Arrays, Monterey, CA, February 2001, pp. 141–150.

[19] Z.A. Ye, A. Moshovos, S. Hauck, P. Banerjee, "CHIMAERA: a high-performance architecture with a tightly coupled reconfigurable functional unit," in Proceedings of the 27th International Symposium on Computer Architecture, June 2000, pp. 225–235.

[20] W. Fu, K. Compton, "Scheduling intervals for reconfigurable computing," in FCCM, 2008, pp. 87–96.

[21] K. Compton, Z. Li, J. Cooley, S. Knol, S. Hauck, "Configuration relocation and defragmentation for run-time reconfigurable computing," Transactions on VLSI Systems, vol. 10, 2002, pp. 209–220.

[22] K. Rupnow, W. Fu, K. Compton, "Block, drop or roll(back): alternative preemption methods for RH multi-tasking," in IEEE Symposium on Field-Programmable Custom Computing Machines (FCCM), April 2009, pp. 63–70.

[23] J. Lach, W. Mangione-Smith, M. Potkonjak, "Low overhead fault-tolerant PFGA systems," IEEE Transactions on VLSI Systems, 6(2), February 1998, pp 212–221.

[24] F.L. Kastensmidt, L. Carro, R. Reis, Fault-tolerance Techniques for SRAM-based FPGAs, Springer Series: Frontiers in Electronic Testing, Vol. 32, June 2006.

[25] B. Pratt, M. Caffrey, J.F. Carroll, P. Graham, K. Morgan, M. Wirthlin, "Fine-grain SEU mitigation for FPGAs using partial TMR," IEEE Transactions on Nuclear Science, 55(4), August 2008, pp. 2274–2280.

[26] C. Bolchini, D. Quarta, M. Santambrogio, "SEU mitigation for SRAM-based FPGAs through dynamic partial reconfiguration," in Proceedings of the 17th ACM Great Lakes Symposium on VLSI, March 2007, pp. 55–60.

[27] C. Bolchini, A. Miele, M.D. Santambrogio, "TMR and partial dynamic reconfiguration to mitigate SEU faults in FPGAs," in Proceedings of the 22nd IEEE

International Symposium on Defect and Fault-Tolerance in VLSI Systems, September 2007, pp. 87–95.

[28] C. Carmichael, C.W. Tseng, "Correcting single-event upsets in Virtex-4 FPGA configuration memory," Xilinx Application Note, xapp1088, October 2009.

[29] Stratix V Device Handbook, Vol. 1: Device Interfaces and Integration, Chapter 2, "Mememory blocks in Stratix V devices." Available: http://www.altera.com/literature/lit-stratix-v.jsp.

[30] Virtex-6 FPGA Memory Resources: User Guide, Version 1.6, April 2011. Available: http://www.xilinx.com/support/documentation/user_guides/ug363.pdf.

[31] K. Morris, "Silicon symbiosis: Lattice acquires SiliconBlue," Electronic Engineering Journal, December 13, 2011. Available: http://www.eejournal.com/archives/articles/20111213-lattice/.

[32] K. Morris, "Tabula gets real: launches ABAX 3D FPGA family," Electronic Engineering Journal, March 23, 2010. Available: http://www.eejournal.com/archives/articles/20100323-tabula/.

[33] DataSheet: "IGLOO low power flash FPGAs with Flash*Freeze technology," Revision 22, September 2012. Microsemi Corporation.

[34] A. Ludwin, V. Betz, K. Padalia, "High-quality, deterministic parallel placement for FPGAs on commodity hardware," ACM/Sigda International Symposium on Field Programmable Gate Arrays, 2008, pp. 14–23.

[35] D.C. Ku, Giovanni De Micheli, "HardwareC: a language for hardware design." Technical Report CSTL-TR-90-419, Computer Systems Lab, Stanford University, California, August 1990. Version 2.0.

[36] Celoxica, http://www.celoxica.com. Handel-C Language Reference Manual, 2003. RM-1003-4.0.

[37] Daniel D. Gajski, Jianwen Zhu, Rainer Dömer, Andreas Gerstlauer, Shuqing Zhao, SpecC: Specification Language and Methodology. Kluwer, Boston, 2000.

[38] Thorsten Grötker, Stan Liao, Grant Martin, Stuart Swan, System Design with SystemC. Kluwer, Boston, 2002.

[39] A. Canis, J. Choi, M. Aldham, V. Zhang, A. Kammoona, J.H. Anderson, S. Brown, T. Czajkowski, "LegUp: high-level synthesis for FPGA-based processor/accelerator systems," ACM/SIGDA International Symposium on Field Programmable Gate Arrays, Monterey, CA, February 2011, pp. 33–36.

[40] J. Cong, B. Liu, S. Neuendorffer, J. Noguera, K. Vissers, Z. Zhang, "High-level synthesis for FPGAs: from prototyping to deployment," in IEEE Transactons on Computer-Aided Design of Integrated Circuits and Systems, 30(4), April 2011, pp. 473–491.

[41] Bluespec web site. Available: http://www.bluespec.com.

[42] MathWorks web site. Available: http://www.mathworks.com.

[43] Khronos Group web site. Available: http://www.khronos.org/opencl/.

[44] B. Gaster, L. Howes, D.R. Kaeli, P. Mistry, D. Shaa, "Heterogeneous Computing with OpenCL" Morgan Kaufmann, Waltham, MA, 2012.

[45] Micrium web site. Available: http://www.micrium.com.

[46] T.R. Halfhill, "MicroBlaze v7 gets an MMU." Microprocessor Report, November 13, 2007.

6 FPGA-Based Emulation Support for Design Space Exploration

PAOLO MELONI, SIMONE SECCHI, and LUIGI RAFFO

6.1 INTRODUCTION

Simulation has always been at the heart of design space exploration (DSE). Every decision that is taken during the exploration of different design alternatives has to be supported by some sort of simulation tool. The accuracy provided by this evaluation tool has a significant impact on the overall quality of the exploration process, reducing the uncertainty of the design choices. At the same time, DSE processes often require numerous architecture evaluations to be performed in order to converge at least to a suboptimal maximization of some performance function. Therefore, the time required for each architecture evaluation step has to be minimized to be able to repeat a high number of evaluation steps as required by the exploration algorithm. These two contrasting needs define a trade-off between simulation accuracy and simulation speed that constrains the choice of which simulation infrastructure to use to support the DSE.

Today, the majority of architectural simulation is still performed at maximum accuracy (i.e., cycle level) in software. Among the most famous simulators, many are still sequential, like SimpleScalar [1], Simics [2], Gem5 [3], or MPArm [4]. In order to respond to the increasing system complexity, both in terms of the number of processing units (multicore design) and component heterogeneity (accelerators, application-specific architectures), software simulation is evolving toward the exploitation of parallelism [5, 6],. Moreover, to deal with issues that are inherent to parallel software simulation (e.g., synchronization of the simulation units), new techniques are being developed. Some parallel software simulators give up on pure cycle-level accuracy and look at complex approaches, such as event-based simulation, statistical sampling, dynamic accuracy switching and simulation state rollback, to model complex, multicore

Embedded Systems: Hardware, Design, and Implementation, First Edition.
Edited by Krzysztof Iniewski.
© 2013 John Wiley & Sons, Inc. Published 2013 by John Wiley & Sons, Inc.

architectures that run full software stacks [7, 8]. Moreover, custom solutions have been created for particular classes of processor architectures, and specifically optimized to enable rapid DSE of such systems. For instance, in Ascia et al. [9] and Di Nuovo et al. [10], a software-based simulation and exploration framework targeting the optimization of a parametric very long instruction word (VLIW) microprocessor is presented.

Despite these innovations in the field, there is general consensus on the fact that classic software approaches are not able to provide cycle accuracy for complex multicore hardware–software (HW/SW) designs in reasonable simulation times anymore. A promising alternative approach aims at preserving cycle accuracy, resorting to the extremely high speeds that can be achieved by implementing the target architecture (or the most critical parts of it) on some kind of configurable hardware. This class of evaluation systems is often referred to as *hardware-based prototyping/emulation approaches*, as they rely on hardware execution of the target design or of a modified representation of it. Field-programmable gate array (FPGA) devices naturally provide this configurable hardware substrate, potentially enabling hundreds of megahertz of pure emulation speed already in the earliest stages of the design flow. In addition, their integration capacity, scalability, the relatively low cost, and the decreasing power consumption suggest that FPGAs are going to be the reference platform for hardware-based emulation for the next future [11].

This chapter will review the state of the art in FPGA-based hardware emulation and focus on the support that reconfigurable devices can offer to DSE problems (Section 6.2). Moreover, in Section 6.3, the chapter presents a framework, developed at the University of Cagliari, within the MADNESS European research project [12], intended to help the designers of on-chip multicore processing systems by adopting FPGAs to provide fast and accurate on-hardware emulation. Section 6.4 describes in detail how the framework has been enriched to be efficiently exploitable within DSE, reducing the overhead in terms of emulation time related with the FPGA implementation of the design points under evaluation. Finally, in Section 6.5, some use cases are presented.

6.2   STATE OF THE ART

FPGA devices offer a variety of different opportunities for hardware-based emulation. Among the many possible exploitations of such technology, two are the main trends that are currently followed in the adoption of FPGA devices for hardware-based emulation. The first approach descends from the area of application-specific integrated circuit (ASIC) design prototyping. In the past, configurable devices were generally intended (and still are) as a platform for HW/SW prototyping in the early stages of an ASIC design flow. The main reason for this was the low cost required to port and debug a design into a configurable device, as opposed to an early ASIC tapeout. Therefore, the first approach to FPGA-based emulation aims at implementing the design under test entirely on an FPGA chip, running all the software stack on it, and gathering all the necessary statistics and performance-related information that can be useful for the designer. The main limiting factor of such an approach is the amount of resources that are available on the modern devices, which has to be enough to contain the entire design. However, the expected growth in FPGA integration capability motivates this choice for the next years to come. Moreover, multichip and multiboard FPGA prototypes can be developed without particular effort in designing the off-chip communication.

The second approach to FPGA-based emulation comes from the area of software simulation, and includes all those techniques that use configurable hardware to accelerate software simulation. The principle behind this approach leverages the fact that hardware emulation is much faster than pure software simulation, and this increased speed comes along with full cycle accuracy with essentially no trade-offs involved. Therefore, for the most critical parts of the design, in terms of accuracy, the simulator can resort to fast and accurate hardware execution. This area is often referred to as hardware-accelerated cosimulation. The most complicated aspect of this technique regards the synchronization between the software and hardware worlds, involving timing adjustment and data communication.

6.2.1 FPGA-Only Emulation Techniques

The most important contribution to the field of large hardware FPGA-only platforms for emulation of complex systems is brought by the Research Accelerator for Multiple Processor (RAMP) project [13]. The RAMP project is a National Science Foundation (NSF)-funded open-source effort of a group of faculties, institutions, and companies (UC Berkeley, UT Austin, MIT, Stanford, University of Washington, Carnegie Mellon University, Intel, IBM, Xilinx, Sun, Microsoft) to create an FPGA-based emulation and computing platform that will enable rapid innovation in parallel software and multicore architecture. The reason that pushed the researchers to conceive the RAMP project are the need to cope with the high complexity and heterogeneity of modern systems-on-chip, the advances in FPGA integration capabilities, their high flexibility and relatively low cost, and the push toward modularity and intellectual property (IP) components reuse.

Several research activities have been condensed within the scope of this large project, including the first FPGA prototype of a transactional memory chip multiprocessor [14], a multiboard, thousand-core, high-performance system running scientific benchmarks on top of a message-passing communication library [15], and a Scalable Processor Architecture (SPARC)-based multithreaded prototype which implemented FPGA-based emulation by introducing the separation between functional and timing simulation [16]. Within the RAMP project, moreover, runtime configuration has been investigated to the extent of some cache-related parameters.

All these platforms shared the same underlying prototyping board, called Berkeley Emulation Engine (BEE), which has also been developed within the RAMP project. The BEE prototyping board contains five Xilinx Virtex-II Pro 70 FPGAs, the latest devices available at design time, each containing two PowerPC 405 hard cores and connected to four independent 72-bit double data rate (DDR2) banks, capable of a peak throughput of 3.4 Gbps.

Another interesting exploitation of pure FPGA-based hardware emulation has been proposed by Atienza and others ([17, 18]). They developed an automatic framework for full-system FPGA-based emulation and then used it to extract in real time execution traces and pass them to a thermal model that feedbacks control inputs to a thermal manager unit, implemented in hardware on the FPGA, which applies appropriate thermal management policies to reduce the thermal dissipation into the chip. Moreover, the paper estimates clearly the speed-up obtained by the FPGA-based emulation framework itself over standard software simulation as three orders of magnitude, without accounting for the FPGA synthesis and implementation phase.

Although this approach slightly differs from emulation since the thermal modeling and feedback phase is not part of any emulation process, but instead aims at regulating temperature in the actual FPGA die, it still provides some interesting insights on how the FPGA-based emulation can be used to obtain (real-time) activity traces. Those can then be used to estimate different metrics, not only purely functional (execution time, interconnection congestion, cache hit/miss rate, etc.) but also related to physical entities (power and energy consumption, thermal distribution within the die, etc.).

6.2.2 FPGA-Based Cosimulation Techniques

As already introduced in Section 6.2, cosimulation approaches use the FPGA substrate to accelerate a bigger software simulation infrastructure that runs on a host machine and communicates to the emulating hardware. Often, the FPGA device hosts a cycle accurate part of the simulation, which is particularly relevant to the system under test, while the software running on the host machine performs functional simulation and takes care of managing the simulation data structures into and out of the FPGA device. However, the idea of splitting a simulation into a cycle-accurate timed portion and a functional system model does not necessarily imply the use of software–hardware cosimulation infrastructures. In this chapter we are going to focus on solutions that employ both software and hardware modules to build a cosimulation system.

The FPGA-accelerated simulation technologies (FAST) emulation system [19] is an example of partitioned simulators. It defines a speculative functional model component that simulates the instruction set architecture (ISA) and a timing model component that predicts performance. The speculative functional model normally runs on a host machine on top of a standard software simulator, possibly parallelized, while the timing model is implemented in FPGA hardware for achieving high speed.

The functional model simulates the computer at the functional level, including the ISA and peripherals, and executes applications, the operating system (OS), and the Basic Input/Output System (BIOS) code. The timing model simulates only the microarchitectural structures that affect the desired metrics. For example, to predict performance, structures such as pipeline registers, arbiters, and associativity need to be modeled. On the contrary, because data values are often not required to predict performance, data path components such as arithmetic logic units (ALUs), data register values, and cache values are generally not included in the timing model. The functional model sequentially executes the program, generating a functional path instruction trace and pipes that stream to the timing model. Each instruction entry in the trace includes everything needed by the timing model that the functional model can conveniently provide, such as a fixed-length opcode, instruction size, source, destination and condition code architectural register names, instruction and data virtual addresses, and data written to special registers, such as software-filled translation lookaside buffer (TLB) entries.

ProtoFlex ([20, 21]) is a full-system FPGA-based simulation architecture. The two key concepts that characterize the ProtoFlex approach are hybrid software–FPGA simulation, realized with a mechanism called *transplanting*, and time-multiplexed interleaved simulation. The first instantiation of the ProtoFlex architecture incorporating the aforementioned concepts is the BlueSPARC simulator, which models a 16-core UltraSPARC III SMP server, hosted on the same BEE2 prototyping platform developed within the RAMP project. Hybrid simulation within BlueSPARC couples FPGA-based emulation with software simulation through the adoption of the VirtuTech Simics simulator [2]. Transplanting works by defining a *state* for the simulation common to the FPGA and software host platforms. This state gets "transplanted" to and from the FPGA when an acceleration is needed, and possibly through specific hardware placed on the FPGA.

The second major feature of the ProtoFlex simulation architecture is the virtualization of the simulation of multiple processor contexts onto a single fast engine through time multiplexing. Virtualization decouples the scale of the simulated system from the required scale of the FPGA platform and the hardware development effort. The scale of the FPGA platform is in fact only a function of the desired throughput (i.e., achieved by scaling up the number of engines). Time multiplexing is implemented in ProtoFlex by essentially borrowing the concept of simultaneous multithreading (SMT) from common multithreaded architectures, and by letting each pipeline simulate instructions from different target cores.

Giano [22] is an extensible and modular simulation framework developed at Microsoft Research for the full-system simulation of arbitrary computer systems, with special emphasis on the HW/SW codevelopment of system software and real-time embedded applications. It allows the simultaneous execution of binary code on a simulated microprocessor and of Verilog code on a simulated FPGA within a single target system capable of interacting in real time with the outside world. The Giano simulator has been used in different design cases. Its accuracy cannot be defined as cycle level, since it merely depends on the accuracy level of the modules description and partitioning between the FPGA and the host microprocessor.

6.2.3   FPGA-Based Emulation for DSE Purposes: A Limiting Factor

From the analysis of the state of the art we can conclude that FPGA-based emulation, both in the forms of pure hardware and cosimulation solutions, can certainly be beneficial for DSE, where even a small number of design variables result in a multidimensional design space that, in order to be explored, requires a high number of emulation steps.

Nevertheless, the achievable advantages are mitigated by the overhead introduced by the FPGA physical synthesis and implementation flow, which impacts on the number of feasible emulation runs that can be performed within an exploration loop. As a matter of fact, every time a hardware parameter is changed, the emulation platform needs to be resynthesized.

Several approaches are being developed, aimed at the reduction of the number of necessary synthesis/implementations, by looking at FPGA reconfiguration and programmability capabilities. For example, in Krasteva et al. [23], FPGA partial reconfiguration capabilities are used to build a framework for network-on-chip (NoC)-based emulation.

6.3   A TOOL FOR ENERGY-AWARE FPGA-BASED EMULATION: THE MADNESS PROJECT EXPERIENCE

Within the MADNESS project [12], an FPGA-based framework for the emulation of complex and large multicore processor architectures for embedded systems aimed at facilitating DSE has been developed and tested. The main purpose of the framework is to serve as an underlying architectural evaluation instrument during a DSE process, able to assess the performance of highly heterogeneous architecture configurations (the design points under evaluation). In order to target state-of-the-art embedded architectures and to expose to the designer all the needed degrees of freedom, design points can be composed by highly configurable processing elements and interconnection structures, such as NoC architectures and application-specific instruction-set processors (ASIPs).

Figure 6.1 provides a schematic view of the prototyping framework. The framework allows the easy instantiation of the desired system configuration and automatically generates the hardware description files for the FPGA synthesis and implementation, thus enabling rapid and accurate prototyping/ exploration of different complete architectures.

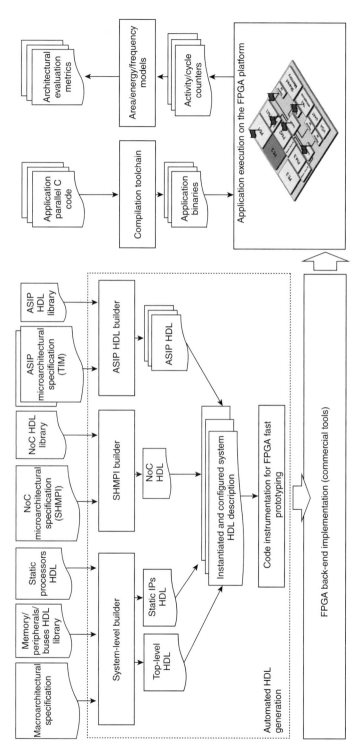

FIGURE 6.1. A view of the MADNESS architectural evaluation framework.

145

The hardware description language (HDL) generation phase involves three different HDL generators:

- A *system-level generator* that is in charge of constructing the top-level HDL, instantiating the static components (those components that are included in a library and do not require a dedicated generator to be customized) and the black boxes representing the ASIPs and the NoC architecture.
- An *NoC HDL generator* (called SHMPI builder [24]) that receives in input a high-level specification of the NoC structure and creates the related HDL, properly configuring and connecting the modules inside a dedicated HDL library. The Xpipes architecture ([25, 26]) has been adopted as the template for interconnection network infrastructures.
- An *ASIP HDL generator* [27], provided by Intel as an industrial partner in the MADNESS project. The tool enables the user to specify the needed processor configuration using a proprietary specification format, and to automatically obtain the HDL description of the processor architecture. The same specification is meant to be passed in input to the compilation toolchain to enable its retargeting on the basis of the instruction set features.

The HDL files created by the toolset are then linked together and instrumented, as will be explained more in detail in the following sections, to allow performance and switching activity extraction and to reduce the overhead related with the traversing of the back-end implementation step ([28, 29]).

The obtained HDL description is then passed to commercial tools for the FPGA implementation flow; the toolchain handles the configuration and the synthesis of the hardware modules. The software code is compiled by means of an integrated compilation toolchain that includes the retargeting compiler for the ASIPs. The application binaries can be uploaded to the FPGA platform where the architectural design point under configuration is implemented. In this way, it is possible to obtain, through FPGA prototyping, an accurate profiling of the target application on the architecture under evaluation. Moreover, cycle-accurate information on the switching activity can be collected and passed as input to analytic power and area models in order to investigate on a prospective ASIC implementation of the system.

6.3.1 Models for Prospective ASIC Implementation

Considering classic HW/SW embedded systems design flows, very often several iterations have to be performed before being able to meet the design closure and the assumptions made during the system-level design phase are verified after the actual back-end implementation. The effort required by this kind of process is increasing with the technology scaling. For example, capacitive effects related to long wires are very difficult to be pre-estimated only with the support of high-level models. Thus, introducing "technology awareness" already at system level would be crucial to increase the productivity and allow the reduction of the iterations needed to achieve a physically feasible and efficient system configuration. To this aim, the adoption of some kind of modeling is unavoidable to obtain an early estimation of technology-related features and to take technology-aware, system-level decisions.

The use of analytic models often requires the collection of maximally accurate (cycle-level) information on the considered system, thus leading to time-hungry cycle-accurate simulations. Within the proposed framework, the use of analytic models is coupled to FPGA fast emulation in order to obtain early power and area figures related to a prospective ASIC implementation, without the need to perform long software simulations. The FPGA-based emulation provides metrics to the analytic models for the estimation of the physical figures of interest. As an example, detailed information on the models related to the NoC components included in the framework can be found in Meloni et al. [30], referring to the power consumption and area occupation of the Xpipes NoC building blocks.

6.3.2   Performance Extraction

The framework provides the possibility to connect dedicated low-overhead memory-mapped event counters to specific logic for the evaluation of different performance metrics. The designer is allowed to insert specific performance counters at the relevant signals inside the system architecture; for example, at the processing core interface, at the switch output channel interface (in case a NoC structure is instantiated), at memory ports, and at the enable signals of the functional units inside the processor architectures. The declaration of the event counters to be instantiated in the system and their memory mapping to specific processors can be easily included in the specification files that are passed as input to the whole framework; dedicated scripts take care of the insertion and the connection of the necessary hardware modules and wires.

6.4   ENABLING FPGA-BASED DSE: RUNTIME-RECONFIGURABLE EMULATORS

It has already been mentioned in Section 6.2.3 that the benefits of FPGA-based hardware emulation are limited by the overhead introduced by the synthesis and implementation flows. In this section, a technique based on runtime reconfiguration of the emulation platform to reduce the number of FPGA implementation processes to be run is presented. Such technique envisions the synthesis, for a set of candidate architectural configurations, of a prototyping platform capable of reconfiguring itself via software to emulate multiple design space points under evaluation. The same kind of approach has been applied at the interconnection level and at the processor level to enable fast NoC topology selection and fast evaluation of multiple ASIP configurations.

6.4.1 Enabling Fast NoC Topology Selection

Given a set of interconnection topologies to be emulated, the approach envisions the possibility of identifying and reconfiguring what we call a *worst-case topology* (WCT). The basic idea behind this approach is to implement on the FPGA a topology that is overprovided with the hardware resources necessary to emulate all the topologies included in the predefined set of candidates; then, at runtime, each specific topology will be mapped on top of the implemented hardware of the WC one, exploiting dedicated software-based configuration mechanisms. If such an approach is feasible, then several emulation steps could be performed after a single FPGA synthesis and implementation run, resulting in a speedup of the whole topology selection process.

Specific HW/SW mechanisms supporting runtime reconfiguration need to be implemented. For instance, a processing core, connected to a specific switch in the WCT, might be connected to a different one in the topology under emulation. To avoid a new synthesis run, there must be the possibility of mapping the topology under test on top of the WCT, configuring the connections to avoid accounting for latencies introduced by the switching elements that are not included in the topology to be emulated. Thus, static direct zero-latency connection of two specific ports of a generic switch must be made configurable at runtime, resulting in the emulation of a combinational path bypassing a certain switch.

6.4.1.1 WCT Definition Algorithm The WCT is defined iteratively. As a first step, a switch is assigned to every core (processing element, memory or peripheral) to be included in the architecture. Thus, P being the number of cores to interconnect, P switches (referred to as "core switches" hereafter) are instantiated inside the topology. The collection of the switches to be included in the topology is then updated according to the analysis of every topology under evaluation. For each candidate configuration, the number of switches, S_{NC}, that equals to the number of switches that are not connected to any core or that are connected to more than one core ("noncore switches" hereafter), is calculated. After the analysis of the whole set of topologies to emulate, the collection includes a number of switches S_{WC}, that is, the sum of the number of cores to interconnect and the highest number of switches S_{NC} among all the N topologies to emulate:

$$S_{WC} = P + \max\{S_{NC}(i)\} \quad \text{for } i = 1, \dots, N. \tag{6.1}$$

The S_{WC} switches are initially connected in an all-to-all fashion, therefore the size, expressed in terms of the number of ports, is equal to S_{WC} for those switches that are not directly connected to a core and is equal to $S_{WC} + 1$

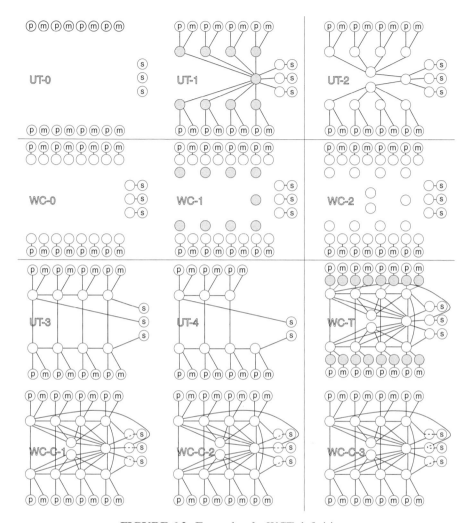

FIGURE 6.2. Example of a WCT definition.

for those that are directly connected to a core (e.g., upper-left sketch in Fig. 6.2).

The topologies, whether they are one of those to be emulated or the WC one, can be characterized by a connectivity matrix CM and two auxiliary vectors, SW_C and P_C, defined as follows:

$$CM_{ij} = \begin{cases} OP_{ij} & \text{if link between switch } i \text{ and switch } j \text{ exists;} \\ -1 & \text{if link between switch } i \text{ and switch } j \text{ does not exist,} \end{cases} \tag{6.2}$$

where OP_{ij} is the output port of switch i connected to switch j. We assume that the switch hardware is built in such a way that also OP_{ij} identifies the input port of switch i connected to switch j (i.e., the port is full duplex and has only one identifier). The size of the CM matrix obviously equals the number of switches present in the related interconnection topology.

The SW_C vector has size P, and its ith element indicates the number of the switch connected to the ith core of the topology. The P_C vector has size P, and its ith element indicates the number of the output port of the switch $SW\_C(i)$ that is connected to the ith core of the topology.

The matrices defined above must be identified for each topology under emulation, including the WC one, and can be used as inputs for the mapping algorithm. This process is in charge of generating the reconfiguration patterns, bypassing the switches, where needed, to ensure the correct emulation of the candidate topologies.

The algorithm scans all the cores included in the system for each of the candidate topologies to be emulated. For the ith core in the system, it first checks if the switch connected to the core is the same in the WCT and in the emulated topology by looking up in the SW_C_k and SW_C_{WC} vectors. If so, then the mapping of the considered core is straightforward and no bypass must be established. On the contrary, if the core is connected to different switches in the two topologies, then a bypass is needed from the switch connected to the core in the WCT to the switch connected to the core in the emulated topology, namely in switch $SW\_C_{WC}(i)$ from input port $P\_C_{WC}(i)$ to output port $CM_{WC}(SW\_C_k(i), SW\_C_{WC}(i))$.

Moreover, for every topology, the algorithm annotates, performing a comparison between CM_{WC} and CM_k, in which ports and links inside the WCT are actually used in at least one candidate configuration. The unneeded port and links can thus be removed from the WC template and the corresponding switch resized. Remaining switches that just directly connect one core with another switch (i.e., switches with two full-duplex ports) can be removed and replaced with direct connections (e.g., the switch connected to the node $N6$ in the lower-left sketch in Fig. 6.2).

Example 6.1: Definition of the overprovided topology able to emulate three different topologies under test (UT-1, UT-2, UT-3). The three topologies under emulation interconnect 19 cores (eight processors, eight memories, and three shared devices).

> *Step 0*: The first step of the algorithm is the analysis of the system population (represented in UT-0). The first version of the overprovided topology is obtained by assigning a switch to every core in the system (WC-0).
>
> *Step 1*: UT-1 is parsed; nine switches (highlighted) that are connected to more than one core are identified. WC-0 includes only "core switches," thus the overprovided topology is updated (WC-1) with the insertion of the highlighted nine switches.

Step 2: UT-2 is parsed; nine switches connected to more than one core and two switches that are not connected to cores are identified. The number of "noncore" switches in the WC-2 is thus increased to 11.

Step 3: UT-3 is parsed; eight "noncore" switches are identified. WC-2 already includes 11 "noncore switches," thus the WC-2 topology does not need to be updated. UT-4 shows that topologies featuring a different number of cores can prospectively be evaluated on the same prototyper.

Step 4: The switches inside the topologies are connected in an all-to-all fashion. The unused connections, never referenced by any topology under test, are then removed, as represented in WC-T.

Step 5: The unneeded "core switches" with two input ports and two output ports, that would be permanently set in bypass mode independently on the candidate topology under consideration, are removed and replaced with direct links. WC-C-1, WC-C-2, and WC-C-3 represent the eventual overprovided topology to be synthesized, highlighting, respectively, the bypass paths that are needed to emulate UT-1, UT-2, and UT-3.

6.4.1.2   *The Extended Topology Builder*   The previously mentioned SHMPI builder, in charge of creating the HDL description of the NoC architecture, has been modified to include the runtime reconfiguration capabilities. Figure 6.3 shows a block scheme of the modified framework.

To allow the fast prototyping of multiple candidate interconnect configurations inside the system, we allowed the tool to receive in input multiple

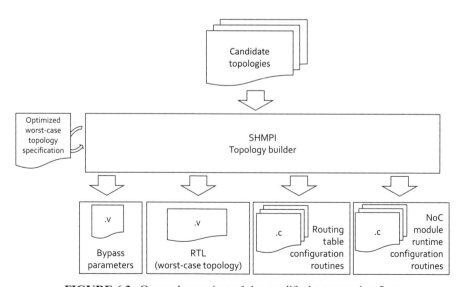

FIGURE 6.3. General overview of the modified prototyping flow.

topology specification files. After the definition and the analysis of the WCT, the tool is able to apply the mapping algorithm, as described in Section 6.4.2, to identify how to map the topologies under emulation on the WC template and to produce the information needed to program it accordingly. The identified reconfiguration patterns are produced in output as:

- parameters included by the HDL modules, defining which ports inside the switches must be equipped with the circuitry supporting the bypass
- C functions to be included for compilation that, when executed by the processors in the system, trigger the runtime reconfiguration

6.4.1.3 Hardware Support for Runtime Reconfiguration The HDL modules describing the switch hardware architecture have been enriched with the following reconfiguration capabilities:

- The reconfigurable switches can be set in bypass mode: a static direct connection between an input port and an output port builds a zero-latency combinational path among them that bypasses all the switching logic for that specific input/output (I/O) pair.
- The routing tables inside the network interfaces (NIs) have been made programmable, allowing the defining of different routing paths during emulation of different topologies.

Every switch in the WCT is programmable by a processor, which is automatically connected to the WCT through a direct point-to-point link. The processors in the system can thus enable the switch bypassing mechanism via software. In order to provide this capability, we added, for each switch port, several memory-mapped registers, namely two per output port and one per input port. The memory-mapped register at the input (e.g., *bypass_input_0* in Fig. 6.4) has to be written when the user wants to bypass the corresponding input buffer. A combinational path is thus established between the corresponding input port of the switch and the output circuitry. To complete the bypass path, at the considered output port, the processor connected to the considered switch has to write the following:

- one "input selection" register (e.g., *bypass_to_0* in Fig. 6.4), which specifies the ID of the input port that has to be directly bypassed with the considered output
- one "bypass enable" register (e.g., *bypass_0* in Fig. 6.4), which enables the bypass of the output buffer and forces the selection input of the corresponding crossbar branch to be statically driven by the content of the "input selection" register

The routing tables, originally implemented as lookup table read-only memories (LUT ROMs), have been made accessible for writing as memory-mapped

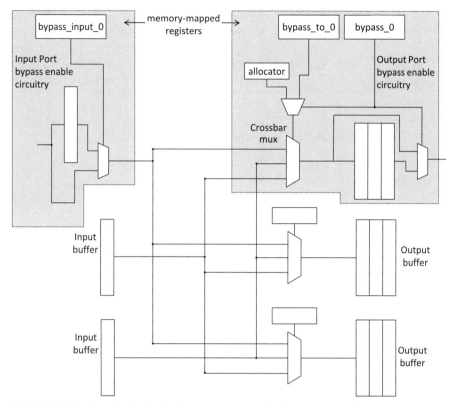

FIGURE 6.4. Sketch of the hardware resources dedicated to the implementation of the switch reconfiguration mechanism.

devices. A random access memory (RAM) memory has been developed, whose content is updated at the beginning of a new emulation with the routing information related to the topology under test, according to the software programming routines produced by the SHMPI topology builder (see Section 6.4.2). At runtime, the RAMs are accessed in read mode by the NIs, every time a network transaction required by the application is initialized, to obtain the routing information to be placed in the header flit.

6.4.1.4 Software Support for Runtime Reconfiguration As depicted in Figure 6.3, the runtime reconfiguration mechanism is handled with specific software modules. The SHMPI topology builder generates the software routines for the following:

- the configuration of the routing tables of all the cores included in the system: each topology to be emulated on top of the WCT needs a set of routing tables

- the configuration of the memory-mapped registers that implement, as described in Section 6.4.2, the bypass mechanisms in hardware

The routines generated at this step are then linked by the executables running on each processor and called right before the start of the actual application code.

6.4.2   Enabling Fast ASIP Configuration Selection

The reconfiguration-based technique for prototyping multiple design points on the same platform has been applied to the aim of speeding up the selection of the optimal ASIP configuration for a given application kernel. Within the MADNESS project, as mentioned, Intel's industrial design flow for customizable processors has been taken as reference. Such flow enables the construction of VLIW ASIPs, based on a flexible *Processor Architecture Template* (see Fig. 6.5). According to the template, every ASIP consists of a composition of substructures called *processor slices*, which are complete vertical datapaths, composed of elementary functional elements called *Template Building Blocks* such as the following:

- *Register Files*: These hold intermediary data in between processing operations, configurable in terms of width, depth, and number of read and write ports.
- *Issue Slots*: These are the basic unit of operation within a processor; every issue slot includes a set of function units (FUs) that implement the actually executable operations. Every issue slot receives an operation code from the instruction decoding and, accordingly, accesses the register files and activates the FUs.

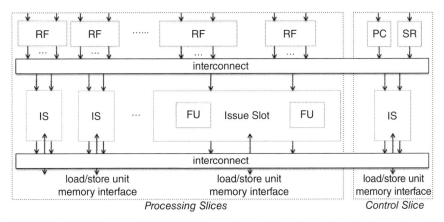

FIGURE 6.5. Reference VLIW ASIP template. RF, register files; IS, issue slot; PC, program counter; SR, status register.

- *Logical Memories*: The container for hardware implementing memory functionality.
- *Interconnect*: This is automatically instantiated and configured, implementing the required connectivity within the processor.

The approach enables a DSE covering the main degrees of freedom exposed by the processor architecture template (the only assertion we impose is that a *control* slice handling the program counter and status register update must be instantiated in the processor together with an arbitrary number of *processing* slices). Design points are processor configurations that instantiate an arbitrary number of processing slices and different parameterizations of the building blocks included in them. The design space under consideration is thus determined by the following degrees of freedom:

- $N_{IS}(c)$: the number of slices in configuration c;
- $FU\_set(x, c)$: the set of FUs in issue slot x, in configuration c;
- $RF\_size(x, c)$: the size (depth) of the register file associated with issue slot x, in configuration c;
- $n\_mem(c)$: the number of memories in configuration c.

The design flow identifies a *worst-case* configuration (WCC), overdimensioned to emulate all the configurations included in a predefined set of candidates. After its implementation on FPGA, each specific configuration is mapped onto the implemented WCC, activating/deactivating, at runtime, hardware sub-blocks when needed, exploiting dedicated software-based configuration mechanisms. The reference industrial flow was extended to provide the needed support for runtime configuration.

6.4.2.1 The Reference Design Flow   Figure 6.6 plots the baseline reference toolchain and indicates the simplest mechanism to perform the exploration of a given design space employing it. Every configuration to be evaluated during the DSE process is described using the proprietary description format. Every configuration description is passed to the ASIP HDL generator, which in turn analyzes it and provides as output the very high-level design language (VHDL) of the whole architecture. This HDL code is then used as input for the FPGA implementation phase that can be performed with commercial tools.

On the software side, in order to perform the evaluation of the architecture with the desired application, the target source code is compiled by means of an automatically retargetable compiler. The compiler extracts information about the underlying ASIP architecture from the same configuration specification provided in input by the DSE engine and retargets itself according to the considered ASIP configuration specification. After compilation, the program can be executed on the ASIP actually implemented on the FPGA.

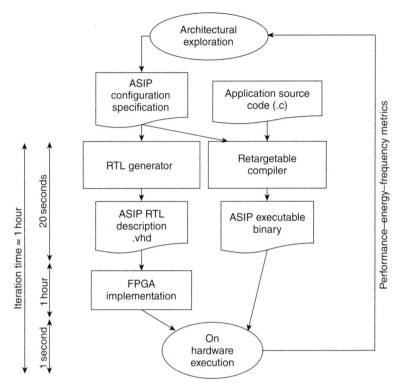

FIGURE 6.6. Baseline prototyping flow (evaluation time for N candidate architectures can be measured to be approximately N hours for a typical design case).

The left side of Figure 6.6 highlights the time necessary for traversing the entire baseline flow for each processor configuration. On a workstation equipped with a Core2 Q6850 processor running at 2 GHz, 8 GB of double data rate type three (DDR3) RAM memory and running Ubuntu 10 Linux OS with the Xilinx ISE FPGA synthesis/implementation toolchain for a Xilinx Virtex5 LX330 device, we experienced roughly 20 seconds for the platform VHDL generation and code compilation. The largest part of the time was consumed by the FPGA synthesis/implementation toolchain, which took about an hour to complete. This time has been measured with the FPGA device operating far from its resource capacity limit. Therefore, the basic scenario for DSE, employing an entire run for each of the N candidate configurations would require approximately N hours to complete.

6.4.2.2 The Extended Design Flow To allow fast prototyping of multiple candidate interconnect configurations inside the system, the baseline flow has been extended with a utility that analyzes the whole set of configurations

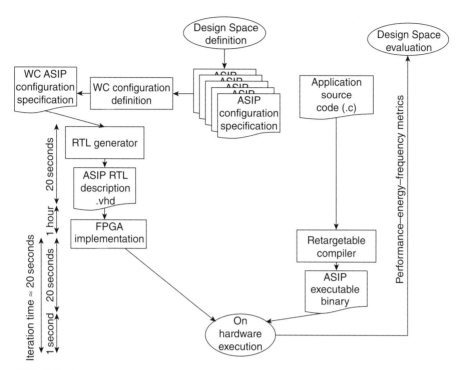

FIGURE 6.7. Prototyping flow extended with runtime reconfiguration capabilities (evaluation time for N candidate architectures is approximately 1 hour and $N \times 20$ seconds).

under prototyping (CUP), synthesizes the WCC, and creates the configurable hardware and the software functions needed to map each candidate configuration on top of the overdimensioned hardware.

Figure 6.7 shows the extended flow. By comparison with Figure 6.6, it can be noticed that multiple ASIP configuration descriptions can be passed as input to the flow. Based on such input sets, the extended flow identifies the WCC and creates its configuration description. As a consequence of this modification of the flow, Figure 6.7 shows how the time necessary to perform the evaluation of N different candidate ASIP configurations is now reduced to roughly $N \times 20$ *seconds* $+ 1$ *hour*. Reported times are meaningful examples of the duration of every evaluation step, as can be measured in real design cases. The precise numbers are obviously dependent on the application, on the hardware architectures, and on the system used for the implementation flow.

6.4.2.3 The WCC Synthesis Algorithm Algorithm 6.1 is the algorithm used to identify the WCC for the considered input set of candidate ASIP

ALGORITHM 6.1. WCC identification.

Input: A set of K candidate ASIP configurations
Output: Worst Case Configuration (WCC)
$N_{IS}(WCC) = 1$;
$RF\_size(1, WCC) = 0$;
$FU\_set(1, WCC) = \{\}$;
for *each candidate configuration c with c = 1, ..., K* **do**
    $N_{is}(WCC) = \max\{N_{is}(i)\}$;   with $i = 1, ..., c$
    for *each issue slot x of the configuration c* **do**
        $RF\_size(x, WCC) = \max\{RF\_size(x, i)\}$;   with $i = 1, ..., c$
        $FU\_set(x, WCC) = FU\_set(x, c) \cup FU\_set(x, i)$;   with $i = 1, ..., c$
    end
end

configurations. In the extended flow, all the design points under test must be provided to the flow at the beginning of the iterative process.

The WCC is defined iteratively while analyzing all the candidate configurations. At every iteration, it is updated according to the design point currently under analysis. At iteration N (i.e., parsing the N – th candidate configuration under test c):

- The number of issue slots inside c is identified and compared with previous iterations. A maximum search is performed, and then, if needed, the WCC is modified to instantiate $N_{IS}(WCC)$ issue slots. For every issue slot of every candidate configuration c, there must be one and only one corresponding issue slot in the WCC.

- For every issue slot x inside c, the size of the associated register file is identified and compared with previous iterations. A maximum search is performed, and then, if needed, the register file related to the issue slot x inside the WCC is resized to have $RF\_size(x, WCC)$ locations. Since there is one and only one issue slot in the WCC that corresponds to the issue slot x of c, the related register file in WCC can be identified without any ambiguity.

- For every issue slot x inside c, the set of FUs is identified and compared with previous iterations. The issue slot x inside the WCC is modified, if needed, to instantiate a set of FUs being the minimum superset of FUs used in previous configurations.

6.4.2.4 Hardware Support for Runtime Reconfiguration The software runtime reconfiguration capability is supported by two hardware modules, automatically generated and instantiated in the overdimensioned WCC architecture basing on the set of different input configurations that are passed to the exploration engine.

The first module is the *instruction adapter*, a programmable decoder that interprets and delivers every single chunk of the VLIW instruction to the relevant hardware element. For each candidate architecture in input, knowing the complete set of architecture parameters, the instruction bits can be split in subranges that identify specific control directives to the datapath. Examples of such bit ranges are operation codes (that activate specific FUs and specific operations inside the issue slots), index values (used to address the locations to be accessed in the register files), and configuration patterns (used to control the connectivity matrices that regulate the propagation of the computing data through the datapath). The width and the position of the boundaries between the bit ranges are not fixed but instead depend on the architectural configuration that must execute the instruction.

The configurable instruction adapter is in charge of translating the instructions produced by the compiler, which retargets itself for each candidate ASIP configuration, into an instruction executable on the WCC. All the sequences in the instruction related to a given slice of the configuration under evaluation are adapted in size and dispatched to the corresponding slice of the WCC. The value of each control directive is modified to ensure the instruction will provide the correct functionality on the overdimensioned prototype, despite the presence of additional hardware. All slices that do not exist in the configuration under test are disabled using dedicated opcodes.

Figure 6.8 shows an example of how the instruction adapter works. In the example, an instruction produced by the compiler for a configuration under test c requires the activation of the FU in charge of performing shift operations (*shu*) in *IS*1. Inside the candidate configuration c alone, the instruction decoder would statically split the VLIW instructions as it is stored in the program memory into different opcodes and pass each of them to the proper issue slot. Inside the issue slot *IS*1, only the *shu* FU would then be activated, basing on the opcode value.

In the extended flow, where the number of issue slots is potentially different (more issue slots are usually instantiated in the WCC) from the one of each candidate configuration, the instruction adaptation is necessary to execute the same instruction binary on the WCC. The adapter is adequately programmed via software through a memory-mapped register write in order to obtain information on the configuration identifier. According to this value, it then decodes the different instruction fields, generates a new (longer) instruction word and dispatches the new opcodes according to the mapping strategy. In the example of Figure 6.8, *IS*1 is mapped onto *ISm* in the WCC. The opcode originally targeted for *IS*1 is thus dispatched to *ISm*. Its value is translated to activate the *shu*, taking into account the architectural composition of *ISm* in terms of its worst-case FUs. Since the opcode values may differ from each candidate configuration to the WCC, the opcode width is adapted to the WCC

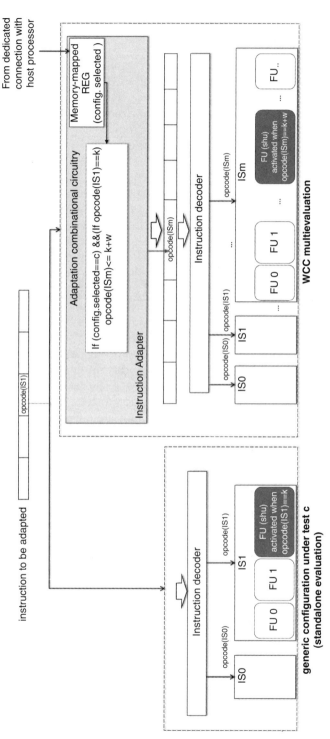

FIGURE 6.8. Example of instruction adapting. An instruction produced by the compiler for a configuration (config.) under test *c* requires the activation of the FU in charge of performing shift operations (shu) in IS1. In the extended flow, the same instruction can be executed on the WCC. The adapter, adequately programmed, decodes the instruction fields and dispatches them according to the mapping strategy. For example, IS1 is mapped onto ISm in the WCC. The opcode originally targeting IS1 is thus dispatched to ISm. Its value is translated to activate the shu, taking into account the composition of ISm. The opcode width is adapted to the WCC architecture. Similar dispatching/translation is applied by the adapter to the other instruction fields.

architecture. Similar dispatching/translation is applied by the adapter to the other instruction fields.

The second hardware module, the *memory router*, is introduced in order to support different connections between the pool of data memories instantiated in the WCC and the issue slots.

6.4.2.5 Software Support for Runtime Reconfiguration   Software support for reconfiguration is realized by simply writing a memory-mapped register, which stores a unique configuration identifier and acts as an architecture selector, directly accessible by a function call at C application level. The automatic flow provides, in the form of a simple application programming interface (API), the function that accesses this register. The value stored in the register, as already described in Section 6.4.2, controls the instruction adapter and the memory router to select one among the candidate configurations under emulation. The generated routines are suitable to be compiled and linked by the application executable file running on a host processor controlling the ASIP.

6.5   USE CASES

In this section we present a use case of the previously described runtime reconfiguration techniques. We plot the results obtained while performing the topology selection process over a set of 16 different system configurations. The considered system included eight processors, eight private memories, one shared memory, one hardware-based synchronization device (namely a test-and-set semaphore used to support mutual exclusion in case of concurrent accesses to shared memory), and one output peripheral (used to send in output the results related to performance extraction and switching activity estimation). The different topologies used to interconnect the cores inside the system were obtained from the application of a simple exploration algorithm that iteratively clusters the network switch to trade-off latency (the number of hops in the network increases with clustering) versus frequency (smaller switches have a smaller critical period). The application executed on the system is the well-known *radixsort* included in the Splash2 benchmark suite [31]. The adopted hardware FPGA-based platform features a Xilinx Virtex5 XC5VLX330 device, counting more than 2 million equivalent gates.

In Figure 6.9, we show the Pareto front resulting from the evaluation. The figures showing absolute time, power, and energy values are obtained by back-annotating the FPGA emulation results with the models described in Meloni et al. [30].

The register-transfer level (RTL) libraries synthesized for FPGA (for evaluation) are the same libraries that would be used for ASIC (for real production). Provided that the bypassing mechanism is adequately applied and given

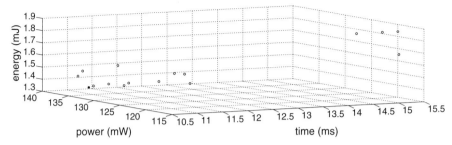

FIGURE 6.9. Use cases results. The Pareto graph represents the evaluation of 16 different topologies under emulation. Execution time is expressed accounting for the architectural maximum operating frequency, which is obtained via analytic modeling. Modeled power and energy consumption figures are also shown for the different topologies.

that the presence of unused ports inside the WCT do not affect the switch functionality, we can state that the prototyping does not insert any error in the estimation of "functional-related" (execution time, latency, switching activity) performances. Cycle/signal level accuracy is guaranteed by definition without the need of a test comparison (emulated vs. prospective implementation). The accuracy of the models described in Meloni et al. [30] is assessed in the paper to be lower than 10%, with respect to postlayout analysis of real ASIC implementations.

From the performed analysis, a designer could estimate the highlighted design point to be the optimal configuration for the target application from the point of view of energy consumption and actual execution time. To identify communication bottleneck or congestion/power hot spots inside the topologies, node-level detailed performances can be obtained, referring to each single port of the switches included in the topology under test.

All the presented data are obtained after traversing only one synthesis/implementation flow. The time needed to run synthesis and implementation with commercial tools increases with the size of the system and can be estimated in a matter of hours. Exploiting the runtime configuration capability enables one to try different interconnection topologies and configurations by performing only one actual FPGA synthesis and implementation flow plus several software compilations and FPGA programming, consuming approximately a matter of few minutes each, allowing a time saving that increases with the number of candidate topologies under prototyping (TUP).

The same kind of use case can be presented for ASIP configuration selection, more in detail reporting the architecture selection process over a set of 30 different ASIP configurations. The explored design points were identified by considering different permutations of the following processor architectural parameter values:

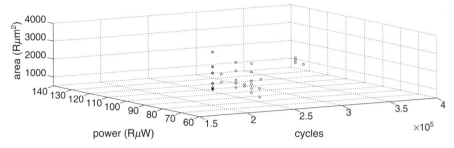

FIGURE 6.10. Use case results. The Pareto front represents the evaluation. Execution cycles, modeled power consumption (expressed in relative microwatts, RμW), and area occupation (expressed in relative square micrometers, Rμm^2) are shown.

- $N_{IS}(c)$: 2 or 3 or 4 or 5;
- $FU\_set(x, c)$: from 3 to 10 FUs per issue slot;
- $RF\_size(x, c)$: 8 or 16 or 32 entries, each 32 bits wide; and
- $n\_mem(c)$: 2 or 3 or 4 or 5.

A filtering kernel was compiled for every candidate configuration and the resulting binaries were executed on the WCC prototype and adequately reconfigured.

In Figure 6.10, we show the results of the evaluation obtained with respect to total execution time, area, and power dissipation. The energy results have been calculated assuming an identical clock frequency for all the considered ASIP configurations.

To avoid the disclosure of sensitive industrial information, we refer to "comparative" numbers for energy and area (e.g., Rμm^2 indicates relative square micrometers). Multiconstraint optimization can be effectively performed. For example, imposing a constraint on maximum execution time (e.g., 200K cycles), the user could identify a subset of candidates (in gray in the Pareto graph) satisfying the constraint. Then, among these, one could choose the best configuration with respect to power or area (highlighted in black).

Performance and power profiling at the functional unit level can also be obtained, referring to each single functional unit included in the configurations under test. As an example, we show in Figure 6.11 a plot reporting power consumption of each FU in a particular configuration during the execution of the already mentioned filtering kernel binaries.

The cycle-accurate correctness of the emulation of a candidate configuration with the WCC one is guaranteed by construction of the WCC architecture. In fact, every instruction that traverses the ASIP datapath, both in the candidate configuration and in the WCC, undergoes the same exact logic path. The

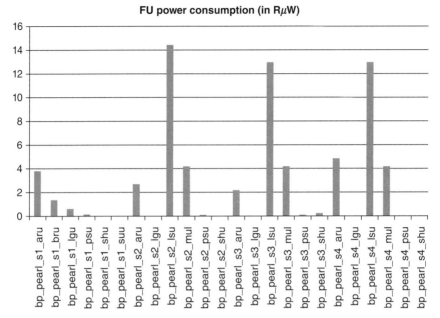

FIGURE 6.11. Power consumption for each FU in a particular configuration, composed of four issue slots, with register files of 16 entries, reported in relative microwatts (RμW).

WCC architecture does not insert any new pipeline stage in the instruction path with respect to the ASIP. What can change is only the operating frequency of the WCC and thus the resulting execution real time, due to the more complicated combinatorial logic (e.g.,: instruction adapter), but the emulated cycles per instruction (CPI) will be exactly the same (as a count of clock cycles).

6.5.1 Hardware Overhead Due to Runtime Configurability

As a consequence of the provision of runtime reconfiguration capabilities, it is easy to expect a degradation of the quality of results with respect to the hardware implementation of a single sample topology on the FPGA device. In particular, aspects related to the implementation quality that can be affected and potentially preclude the usability of the proposed approach are:

- the area occupation of the WCT, which determines whether or not the prototyping platform fits on one given target programmable device,
- its working frequency, that, if impacted—for example, by switches with a high number of ports or by long combinational paths traversing bypassed switches or the instruction adapter—can potentially increase the emulation time and reduce the benefits of on-hardware emulation.

The mentioned aspects have been properly taken into account while implementing the toolchain.

To assess the overhead introduced by the support for runtime programming of the network in terms of hardware resources, the programmable logic resources required for the implementation of the most hardware-hungry topology under test (when implemented for a standalone evaluation without support for runtime reconfiguration) and those required to implement the corresponding WC template are compared. Two NoC DSE runs involving, respectively, 4 and 16 TUP and an ASIP architecture selection process prototyping 30 design points have been considered.

As can be noticed from Table 6.1, the introduced device utilization overhead is limited in all cases and is controllable when the number of candidate design points increases.

We also report a comparison related with working frequency. Table 6.2 shows how the critical path is almost insensitive to the introduction of the support for rapid prototyping. When selecting the NoC topology, this is mainly due to the fact that the critical path is always bounded inside the switches, thanks to the WCT definition algorithm that limits the bypassed switches at the boundaries of the topology. When selecting the ASIP configuration, after the insertion of the instruction adapter and of the memory router, the critical path still resides inside more complex FUs (e.g., multiply and accumulate). A minimal increase (less than 0.1%) is due to unpredictable behaviors of the synthesis algorithm. The results show how the overhead is affordable when prototyping quite complex systems with state-of-the art commercial devices.

TABLE 6.1. Experimental Results Related with the Hardware Overhead Introduced by the Support for Fast Prototyping (FPGA Resource Utilization)

	Occupied Slices	Slice Registers	Slice LUTS
NoC exploration: largest TUP (4 topologies DSE)	17,327 (33%)	33,885 (16%)	44,673 (21%)
NoC exploration: WC (4 topologies DSE)	20,627 (39%)	41,313 (19%)	58,862 (28%)
NoC exploration: largest TUP (16 topologies DSE)	17,397 (33%)	34,487 (16%)	44,926 (21%)
NoC exploration: WC (16 topologies DSE)	21,815 (42%)	44,943 (21%)	64,696 (31%)
ASIP exploration: largest CUP	19,859 (37%)	6,923 (13%)	16,387 (31%)
ASIP exploration: WC (30 configurations DSE)	21,278 (40%)	6,931 (13%)	17,951 (34%)

TABLE 6.2. Experimental Results Related with the Hardware Overhead Introduced by the Support for Fast Prototyping (Critical Path)

	Critical path
Slowest TUP (4 topologies DSE)	10,902
WC (4 topologies DSE)	10,902
Slowest TUP (16 topologies DSE)	10,976 ns
WC (16 topologies DSE)	11,307 ns
Slowest CUP (30 configurations DSE)	9.809 ns
WC (30 configurations DSE)	9.817 ns

REFERENCES

[1] T. Austin, E. Larson, D. Ernst, "SimpleScalar: an infrastructure for computer system modeling," Computer, 35, February 2002, pp. 59–67. http://portal.acm.org/citation.cfm?id=619072.621910

[2] P. Magnusson, M. Christensson, J. Eskilson, D. Forsgren, G. Hallberg, J. Hogberg, F. Larsson, A. Moestedt, B. Werner, "Simics: a full system simulation platform," Computer, 35(2), February 2002, pp. 50–58.

[3] N. Binkert, B. Beckmann, G. Black, S.K. Reinhardt, A. Saidi, A. Basu, J. Hestness, D.R. Hower, T. Krishna, S. Sardashti, R. Sen, K. Sewell, M. Shoaib, N. Vaish, M.D. Hill, D.A. Wood, "The gem5 simulator," SIGARCH Comput. Arch. News, 39, August 2011, pp. 1–7. http://doi.acm.org/10.1145/2024716.2024718

[4] L. Benini, D. Bertozzi, A. Bogliolo, F. Menichelli, M. Olivieri, "MPARM: exploring the multi-processor SOC design space with SystemC," J. VLSI Signal Process. Syst., 41, September 2005, pp. 169–182. http://portal.acm.org/citation.cfm?id=1072958.1072962

[5] J.E. Miller, H. Kasture, G. Kurian, C. Gruenwald, N. Beckmann, C. Celio, J. Eastep, A. Agarwal, "Graphite: A distributed parallel simulator for multicores," 2010 IEEE 16th International Symposium on High Performance Computer Architecture (HPCA), Jan. 9–14, 2010, pp. 1–12.

[6] G. Zheng, G. Kakulapati, L. Kale, "BigSim: a parallel simulator for performance prediction of extremely large parallel machines," in Proceedings of the 18th International Parallel and Distributed Processing Symposium, 2004, April 2004, p. 78.

[7] E. Argollo, "COTSon: infrastructure for full system simulation," SIGOPS Oper. Syst. Rev., 43(1), 2009, pp. 52–61.

[8] A. Falcon, P. Faraboschi, D. Ortega, "Combining simulation and virtualization through dynamic sampling," in Proceedings of the IEEE International Symposium on Performance Analysis of Systems and Software, 2007, pp. 72–83.

[9] G. Ascia, V. Catania, M. Palesi, D. Patti, "A system-level framework for evaluating area/performance/power trade-offs of VLIW-based embedded systems," in Proceedings of the Asian and South Pacific Design Automation Conference, 2005, January 2005, Vol. 2, pp. 940–943.

[10] A.G. Di Nuovo, M. Palesi, D. Patti, G. Ascia, V. Catania, "Fuzzy decision making in embedded system design," in Proceedings of the 4th International Conference

on Hardware/Software Codesign and System Synthesis, CODES+ISSS '06, ACM, New York, NY, USA, 2006, pp. 223–228. http://doi.acm.org/10.1145/1176254. 1176309

[11] K. Underwood, K. Hemmert, "Closing the gap: CPU and FPGA trends in sustainable floating-point BLAS performance," in 12th Annual IEEE Symposium on Field-Programmable Custom Computing Machines—FCCM 2004, 2004, pp. 219–228.

[12] L. Raffo, MADNESS—Methods for Predictable Design of Heterogeneous Embedded Systems with Adaptivity and Reliability Support (2010). Available: http://www.madnessproject.org/

[13] D. Patterson, "RAMP: research accelerator for multiple processors—a community vision for a shared experimental parallel HW/SW platform," in Proceedings of the IEEE International Symposium on Performance Analysis of Systems and Software, 2006.

[14] S. Wee, J. Casper, N. Njoroge, Y. Tesylar, D. Ge, C. Kozyrakis, K. Olukotun, "A practical FPGA-based framework for novel CMP research," in FPGA '07: Proceedings of the 2007 ACM/SIGDA 15th International Symposium on Field Programmable Gate Arrays, 2007, ACM, New York, NY, USA, pp. 116–125.

[15] A. Krasnov, A. Schultz, J. Wawrzynek, G. Gibeling, P. yves Droz, "RAMP Blue: a message-passing manycore system in FPGAs," in 2007 International Conference on Field Programmable Logic and Applications, FPL 2007, 2007, pp. 27–29.

[16] Z. Tan, A. Waterman, R. Avizienis, Y. Lee, H. Cook, D. Patterson, K. Asanović, "RAMP Gold: an FPGA-based architecture simulator for multiprocessors," in Proceedings of the 47th Design Automation Conference, DAC '10, ACM, New York, NY, USA, 2010, pp. 463–468. http://doi.acm.org/10.1145/1837274. 1837390

[17] D. Atienza, P.G. Del Valle, G. Paci, F. Poletti, L. Benini, G. De Micheli, J.M. Mendias, "A fast HW/SW FPGA-based thermal emulation framework for multiprocessor system-on-chip," in Proceedings of the 43rd Annual Design Automation Conference, DAC '06, ACM, New York, NY, USA, 2006, pp. 618–623. http://doi.acm.org/10.1145/1146909.1147068

[18] P. Del Valle, D. Atienza, I. Magan, J. Flores, E. Perez, J. Mendias, L. Benini, G. De Micheli, "Architectural exploration of MPSoC designs based on an FPGA emulation framework," in Proceedings of XXI Conference on Design of Circuits and Integrated Systems (DCIS), 2006, pp. 12–18.

[19] D. Chiou, D. Sunwoo, J. Kim, N.A. Patil, W. Reinhart, D.E. Johnson, J. Keefe, H. Angepat, "FPGA-accelerated simulation technologies (FAST): fast, full-system, cycle-accurate simulators," in MICRO 40: Proceedings of the 40th Annual IEEE/ACM International Symposium on Microarchitecture, IEEE Computer Society, Washington, DC, USA, 2007, pp. 249–261.

[20] E.S. Chung, E. Nurvitadhi, J.C. Hoe, B. Falsafi, K. Mai, PROToFLEX: FPGA-accelerated hybrid functional simulator, in Proceedings of the International Parallel and Distributed Processing Symposium, March 2007, pp. 1–6.

[21] E.S. Chung, M.K. Papamichael, E. Nurvitadhi, J.C. Hoe, K. Mai, B. Falsafi, "ProtoFlex: towards scalable, full-system multiprocessor simulations using FPGAs," ACM Trans. Reconfigurable Technol. Syst., 2, June 2009, pp. 15:1–15:32. http://doi.acm.org/10.1145/1534916.1534925

[22] A. Forin, B. Neekzad, N.L. Lynch, "Giano: the two-headed system simulator," Tech. Rep. MSR-TR-2006-130, Microsoft Research (2006).

[23] Y.E. Krasteva, F. Criado, E. de la Torre, T. Riesgo, "A fast emulation-based NoC prototyping framework," in RECONFIG '08: Proceedings of the 2008 International Conference on Reconfigurable Computing and FPGAs, IEEE Computer Society, Washington, DC, USA, 2008, pp. 211–216.

[24] P. Meloni, S. Secchi, L. Raffo, "An FPGA-based framework for technology-aware prototyping of multicore embedded architectures," Embedded Syst. Lett. IEEE, 2(1), March 2010, pp. 5–9.

[25] F. Angiolini, P. Meloni, S. Carta, L. Benini, L. Raffo, "Contrasting a NoC and a traditional interconnect fabric with layout awareness," in Proceedings of the DATE '06 Conference, Munich, Germany, 2006.

[26] D. Bertozzi, L. Benini, "X-pipes: a network-on-chip architecture for gigascale systems-on-chip," IEEE Circ and Syst. Mgz., 4(2), 2004, pp. 18–31.

[27] SiliconHive, HiveLogic Configurable Parallel Processing Platform (2010). Available: http://www.siliconhive.com/flex/site/Page.aspx?PageID=17604

[28] P. Meloni, S. Secchi, L. Raffo, "Enabling fast network-on-chip topology selection: an FPGA-based runtime reconfigurable prototyper," in VLSI System on Chip Conference (VLSI-SoC), 2010 18th IEEE/IFIP, September 2010, pp. 43–48.

[29] P. Meloni, S. Pomata, G. Tuveri, S. Secchi, L. Raffo, M. Lindwer, "Enabling fast ASIP design space exploration: an FPGA-based runtime reconfigurable prototyper," VLSI Des., 2012, 2012.

[30] P. Meloni, I. Loi, F. Angiolini, S. Carta, M. Barbaro, L. Raffo, L. Benini, "Area and power modeling for networks-on-chip with layout awareness," VLSI-Design Journal, Hindawi Publications (ID 50285) (2007).

[31] J. Pal Singh, S.C. Woo, M. Ohara, E. Torrie, A. Gupta, "The SPLASH-2 Programs: characterization and methodological considerations," in Proceedings of the International Symposium on Computer Architecture, 1995.

7 FPGA Coprocessing Solution for Real-Time Protein Identification Using Tandem Mass Spectrometry

DANIEL COCA, ISTVÁN BOGDÁN, and
ROBERT J. BEYNON

7.1 INTRODUCTION

Since the first applications of mass spectrometry (MS) to the identification of organic compounds were reported in the 1950s [1], MS has emerged as the most sensitive and powerful tool for protein identification and quantification [2]. Advances in high-throughput, liquid chromatography–tandem mass spectrometry provide the technological means to measure protein expression on a proteome-wide scale. This should provide unprecedented insight into protein function [3], a more comprehensive characterization of gene expression. and will accelerate the discovery of disease-associated protein biomarkers [4].

While the majority of proteome-wide studies aim to characterize the protein expression at steady state, our ability to quantitatively characterize the dynamics of protein expression on a global scale is key to increasing our understanding of complex biological processes [5, 6]. Obtaining protein expression time series on this scale requires the generation and processing of huge data sets of tandem mass spectrometry (MS/MS), given that the analysis of a substantial proteome in a single experiment can generate hundreds of gigabytes of raw mass spectrometric data. Moreover, modern mass spectrometers have typical acquisition rates of 200 spectra per second, while the analysis of single spectra can take tens of seconds on a high-end computer workstation. Despite rapid increases in computer power, data analysis is still a major bottleneck in proteomics workflow.

In practice, initial processing of large amounts of mass spectrometric data generated by time-lapse proteomics experiments is best performed "near

Embedded Systems: Hardware, Design, and Implementation, First Edition.
Edited by Krzysztof Iniewski.
© 2013 John Wiley & Sons, Inc. Published 2013 by John Wiley & Sons, Inc.

instrument," where the end user has the option of adjusting the processing strategy according to results obtained in real time. An emerging strategy for improving the sensitivity and efficiency of protein identification involves the online optimization of data acquisition [7]. The closed-loop strategy can dramatically improve the quality of the data sets generated by MS/MS experiments, leading to a significant increase of the number of identified proteins per number of performed MS/MS analyses. This approach requires the identification of peptides and whole proteins to be carried out in real time as data become available. In addition, there is a clear opportunity for using proteome profiling information for real-time optimization and control of bioprocesses [8] and for clinical applications [9].

The computational demands posed by real-time proteomic analysis are difficult to meet by conventional computing systems currently available in proteomics laboratories.

High-performance computer clusters and grids, which are remotely accessible, are shared resources that are used to support a wide range of high-performance computing (HPC) tasks. They do not provide suitable solutions for real-time applications due to their relatively high internode communication latency and low bandwidth. Even for offline processing tasks, achieving significant computational speedup on cluster platforms requires significant computational resources. A grid implementation of a Basic Local Alignment Search Tool (BLAST) search algorithm, for example, achieved a 60-fold speed increase using 600 central processing units (CPUs) [10]. More recently, the parallel implementation on 100 processors of the protein identification software pFind [11] delivered a speedup of over 80-fold. Given the large costs associated with housing and powering large computer clusters (typical power consumption is now 10–20 per rack), which normally require dedicated and expensive connectivity and cooling infrastructure, there is clearly a need to improve efficiency and to reduce environmental and economic impact.

While the use of multicore processors has led to increased node performance and improved power efficiency dissipation, compared with an equivalent number of single-core processors, HPC clusters based only on general-purpose processing cores cannot meet the performance requirements of computationally intensive applications and, at the same time, achieve energy-efficient computing. Field-programmable gate array (FPGA) and graphics processing unit (GPU) coprocessors are seen as the only viable alternative to enhance the performance and the energy efficiency of supercomputer cluster nodes across a broad range of applications [12, 13].

In bioinformatics, FPGAs have been successfully used to accelerate DNA sequencing algorithms [14–18], gene sequence analysis [19], string matching [20, 21], sequence alignment [22], and to evaluate single-nucleotide polymorphism (SNP) predictions [23]. FPGAs have also been used to accelerate sequence database searches with MS/MS-derived query peptides [24], processing of mass spectrometric data [25], and to accelerate peptide mass fingerprinting [26]. For a more in-depth introduction to the FPGA-based computing field, the reader is referred to two comprehensive texts by Gokhale and Graham [27] and by Hauck and Dehon [28].

This work advocates the use of reconfigurable FPGA devices as coprocessing units to accelerate the protein identification algorithms implemented by the popular open-source software X!Tandem [29, 30]. The FPGA implementation, described in Bogdan et al. [31], has been integrated tightly with the X!Tandem software to enable existing users to exploit the benefits of FPGA technology without the need to change their established proteomic workflow.

X!Tandem allows the matching of MS/MS spectra against theoretical spectra generated from a protein sequence database by *in silico* protein digestion and peptide fragmentation. A number of parallel implementations of X!Tandem, which enable it to be run on multinode computer clusters, have been reported in recent years [32–34]. These implementations achieve speedups ranging from 28 to 36, using between 40 and 100 processors.

In contrast to the parallel implementations described above, which exploit coarse-grained parallelism between the compute nodes in a cluster, the FPGA acceleration solution presented here exploits massive fine-grained parallelism at the logic level.

The X!Tandem source code has been modified so that the entire database search is performed in FPGA. The search results are then passed back to X!Tandem to generate the final output file. Compared with the standard software X!Tandem implementation running on a dual processor system, the hardware-accelerated solution achieves a speedup of more than 100 when implemented on an FPGA system equipped with three Xilinx Virtex-II XC2V8000 devices.

The FPGA coprocessing solution could provide a proteomics laboratory with the same computing power as that available from an HPC cluster, allowing near-instrument data processing capabilities required by real-time, closed-loop proteomics studies, which cannot rely on remotely accessible, distributed computing resources.

7.2   PROTEIN IDENTIFICATION BY SEQUENCE DATABASE SEARCHING USING MS/MS DATA

Tandem mass spectrometers referred to in this chapter consist of two mass analyzers separated by an ion collision cell. This arrangement allows peptide ions of a given mass-to-charge ratio (m/z), measured by the first mass analyzer, to be selected, fragmented, and analyzed using the second mass spectrometer, yielding additional structural information for that particular ion [35], which is important, as the mass of the peptide alone is not enough to uniquely determine the identity of a peptide when these originate from a complex protein mixture.

Protein identification using MS/MS involves a number of processing stages, including protein isolation from a biological sample, enzymatic digestion of the resulting proteins, chromatographic separation, multistage MS analysis, and protein identification using either database searching or *de novo* sequencing [36].

Following enzymatic digestion, which is used to produce a subset of predictable protein fragments (peptides), liquid chromatography is used to separate individual peptides in the resulting peptide mixture. Subsequently, electrospray ionization is used to produce, directly from the analyte solution, intact gas-phase charged peptide molecules. These are injected in the first mass analyzer, which measures their m/z. Individual peptides, selected based on their m/z, are subsequently subjected to secondary fragmentation by collision with inert gas-generating peptide fragments that are analyzed using the second mass spectrometer.

The MS/MS product ion spectrum resulting from this analysis comprises of peaks that correspond to fragment ions from the selected precursor.

Proteolytic enzymes or proteases will break down a protein by cleaving the peptide at specific points, for example, after the occurrence of a specific amino acid residue. The end of a peptide which terminates with a free amine ($-NH_2$) group is known as the N-terminus, while the other end, which terminates with a free carboxyl group ($-COOH$), is known as the C-terminus. Peptide sequences are written starting from the N- to the C-terminus. The individual amino acid sites are labeled on the N-terminus as P1, P2, and so on and on the C-terminus as P1′, P2′, and so on [37–39]. In Figure 7.1, the last four cleavage sites using trypsin of the *ubiquinol–cytochrome–c reductase* (EC 1.10.2.2) *cytochrome c1 precursor—Paracoccus denitrificans* (C29413)—are shown.

During collision-induced dissociation, the charged peptides may fragment in three different ways to produce pairs of N- and C-terminus ions, (a_k, z_{n-k}), (b_k, y_{n-k}), and (c_k, x_{n-k}), where $k = 1,\ldots, n - 1$ indicates the number of amino acids in the fragment and n is the total number of amino acids in the peptide. Peptides produced by trypsin digestion ionized using the electrospray ionization technique, which are subjected to collision-induced dissociation, primarily yield b- and y- ion fragments.

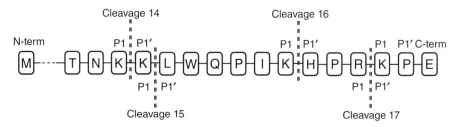

FIGURE 7.1. Cleavage example using trypsin.

The most common approach to determine the identity of the peptide based on the product ion spectrum is to compare the peptide fragment fingerprint with theoretical peptide fragmentation patterns derived from a database of known proteins [36].

This process involves the following computational steps:

1. the computation of theoretical peptide masses following *in silico* "digestion" of proteins in the database according to the cleavage rules associated with the specific proteolytic enzyme used in the experiment,
2. identifying the theoretical peptides that match the experimental precursor ion masses,
3. *in silico* computation of the theoretical product ion spectrum for every matching peptide, and
4. matching the theoretical product ion masses with the experimental product ion spectra, taking into account measurement errors. A statistical model is used to score peptide matches.

The resulting scores are ranked and further statistical analysis is performed to determine which of the matches indicate significant correlations. The remaining matches are considered to be random.

Although different search engines such as X!Tandem [17], Mascot [40], Sequest [41], Phenyx [42], and OMSSA [43] employ different scoring algorithms, all scoring schemes are based on the number of fragment mass matches, while some, including X!Tandem, also take into account the experimental peak intensities.

The FPGA implementation of the database search engine implements the X!Tandem scoring algorithm [44]. Like other protein identification programs, X!Tandem compares the acquired MS/MS spectra to a model spectrum of *b*- and *y*-ions obtained by *in silico* fragmentation of peptides from a protein database. A basic score, which represents the inner product between the experimental and model spectra,

$$S_1 = \sum_{i=1}^{n} I_i P_i,$$

is computed first. I_i denotes the intensity of an observed fragment ion in the MS/MS spectrum (normalized to a maximum of 100), and $P_i = 1$ for ions present in the model spectra and 0 otherwise.

Assuming a hypergeometric distribution, the initial score is augmented to account for the number of assigned *b*- or *y*-ions, n_b and n_y, as follows:

$$S_2 = S_1 n_b! n_y!$$

The peptide in the database with the highest score is assumed correct.

To indicate the significance of the peptide match, X!Tandem generates a histogram of all the scores corresponding to potential peptide matches and uses it to evaluate the probability that the highest score is a random match. The FPGA search engine only generates the S_2 score. The significance of the match is computed by the X!Tandem software.

7.3   RECONFIGURABLE COMPUTING PLATFORM

The digital search processors described in this chapter were implemented on a commercial off-the-shelf (COTS) multi-FPGA coprocessing system, consisting of a BenNuey motherboard that can host up to three BenDATA DIME-II modules from Nallatech Ltd. (http://www.nallatech.com). The BenNUEY board is a full-length PCI DIME-II motherboard equipped with a Xilinx Virtex-II XC2V8000 FPGA device and 4 MB onboard random access memory (RAM). This FPGA can be used to implement algorithms for smoothing raw mass spectra and for peak harvesting [25, 26]. The communication between the personal computer (PC) server and the FPGA system, via a standard Peripheral Component Interconnect (PCI) interface (32 bits, 33 MHz), is handled by a second, smaller FPGA (Xilinx Spartan-II) on the motherboard.

The motherboard has three DIME-II expansion slots, which allow users to configure additional system resources to meet processing, memory, and input/output (I/O) requirements.

The BenDATA DIME-II module has one user FPGA device (Virtex-II XC2V8000) and 1 GB of double data rate synchronous dynamic random access memory (DDR SDRAM), organized in four banks with a 64-bit wide data bus each. The total data bus width is 256 bits.

Each module is connected with the motherboard FPGA and with the other two modules via a 64 bit, 66 MHz local bus. This architecture enables the implementation of parallel searches at the FPGA level as well as across modules. The block diagram of the FPGA system is shown in Figure 7.2.

An important factor that has to be considered when choosing an FPGA platform is the communication overhead associated with data transfer between the PC and device, which should represent only a fraction of the actual execution time. The communication costs incurred by transferring data between hardware and software can be determined in advance given the architecture and performance specifications of the FPGA coprocessing system.

In this context, the FPGA computing platform adopted in this work is well suited for parallel searching a protein sequence database, as the BenDATA modules used to implement the database search engine provide large amounts of onboard memory, which allows an entire protein sequence database (of approximate size of 1.4 billion amino acid letters coded on 6 bits or 1.7 billion amino acid letters coded on 5 bits) to be stored in local memory, resulting in very low communication overhead. Moreover, the architecture of the module

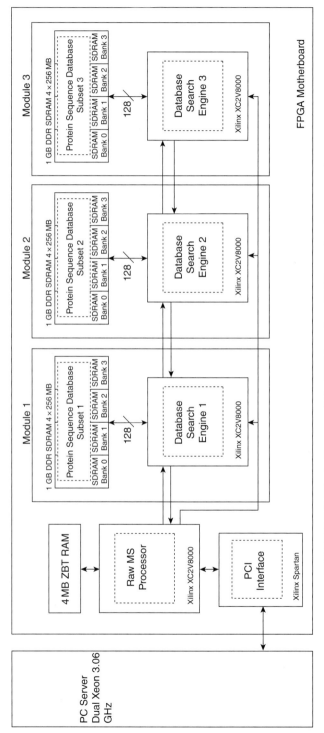

FIGURE 7.2. Block diagram of the FPGA system. ZBT, zero bus turnaround.

allows data transfers between memory and FPGA using a 256 bit-wide configuration at 100 MHz, allowing multiple protein sequences to be streamed out of the memory and processed in parallel by an array of search processors programmed on the FPGA fabric. The FPGA hardware implementation of the peptide matching algorithms implemented by the X!Tandem software is presented in the following sections.

7.4   FPGA IMPLEMENTATION OF THE MS/MS SEARCH ENGINE

The database search-for-matches algorithms implemented by all MS-based protein identification software solutions have relatively low data dependency, and so they are ideal candidates for parallel implementation in hardware. The design of the present implementation of the database search engine has been optimized to take into account the board architecture, the FPGA, and memory resources available, as well as to exploit pipelining and parallelism to produce an efficient implementation.

The MS/MS search engine identifies peptide spectra by searching a library of known protein sequence database. Each protein sequence has to be processed to compute peptide masses and product ion fragmentation models.

The search strategy adopted here involves dividing the database into a number of subsets P that are analyzed in parallel using an array of search processors. The entire protein library is searched using a single MS/MS spectrum (precursor mass and product ion m/z values) at a time. The advantage of this approach is that the entire computational flow can be implemented as a deep pipeline, which, once filled, produces a matching score every clock cycle. The approach has potential for online processing, when the identity of a peptide is required in real time, but is not efficient when large data files need to be processed offline as it requires one database pass for every individual spectrum.

7.4.1   Protein Database Encoding

The protein sequence database was encoded as described in Bogdan et al. [26] using 28 distinct 5-bit codes: 20 codes were used to represent the constituent amino acids, six codes were used to represent standard symbols adopted in the FASTA format, and two codes were used to mark the end of a protein sequence and the end of the database. By encoding the database using only 5-bit "characters," the database size was reduced by about 40%.

The encoded database occupies approximately 680 MB of the total 1 GB DDR SDRAM memory installed on the module.

The encoded database is loaded in the four memory banks of the FPGA coprocessing module organized into $4 \times 12 = 48$ columns of protein sequences that can be streamed out of the memory simultaneously for parallel processing. Each column contains between 45,500 and 107,150 complete protein sequences which are read out of the memory, starting from the N-terminal to the C-terminal.

Storing the database in the local module memory reduces the data communication overhead between the FPGA board and the host PC.

7.4.2   Overview of the Database Search Engine

A block diagram of the FPGA search engine is illustrated in Figure 7.3. The search engine, which in this study was implemented on a Xilinx XC2V8000 FPGA, has 12 search processors that work in parallel. Each search processor can compare up to 32 experimental product ion masses (the peptide mass fingerprint) with theoretical product ion pair masses generated by processing, on the fly, the protein database. In this particular case, there are 48 5-bit amino acid sequences that are streamed out in parallel from the memory and distributed to the 12 search processors so that each processor analyzes, in turn, four data streams.

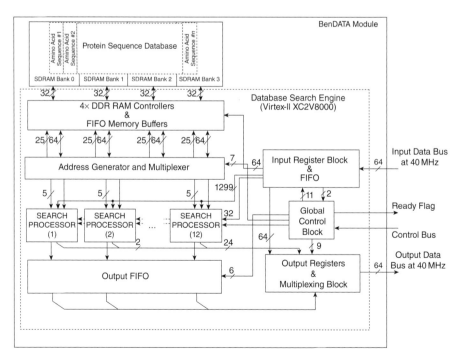

FIGURE 7.3. Block diagram of the MS/MS search engine.

The number of ion masses used to perform the comparison can be increased if needed, but this also increases the complexity of the design. For example, by doubling the number of ions that are compared in parallel from 32 to 64, the FPGA used in this study will only accommodate six search processors.

Four DDR SRAM controllers (one for each memory bank) provide memory refresh signals as well as address, data, and control signals for memory read and write operations. The DDR SRAM devices operate from a differential input, a 100 MHz master clock, which is independent from the clock used to drive the rest of the design. Each DDR controller transfers one 2×32 bit data word per clock cycle. The back-end user interface of each DDR controller consists of two 511 deep first in, first outs (FIFOs) for read and write operations. The address generator and multiplexing block coordinates read–write operations to/from the DDR SRAM controllers (via the FIFO memory buffers) and multiplexes the 48 data streams toward the 12 parallel search processors.

The DDR controllers' FIFOs are synchronized with the memory interface using the 100 MHz clock driving the search processors. Data are read in parallel, in blocks of 511 addresses, from the four banks of DDR SDRAM memory into the four output FIFOs. Read operations from the four output FIFOs to the search processors are synchronous. Data can be independently written to each of the four memory banks.

The design includes all necessary control and FIFO structures that implement 64-bit wide data transfer between the FPGA device on the motherboard and those on the plug-in modules at a rate of 320 MB/s. All arithmetic operations on the m/z values were performed using 32-bit unsigned fixed-point binary number representation of mass and abundance values, with 12 bits after the radix point.

7.4.3 Search Processor Architecture

Each protein sequence is processed sequentially by a search processor implemented in the module's FPGA. The search processor performs two basic operations: (1) *in silico* digestion of proteins in the database and computation of the theoretical masses of peptides and their fragmentation ions and (2) computation of the basic score.

Consequently, each search processor has two major functional blocks: an *in silico* protein digestion unit and a scoring module. A block diagram of the search processor is shown in Figure 7.4.

Each search processor reads one 5-bit coded protein sequence every clock cycle from one of the allocated data streams and passes it to the digestion unit.

The digestion unit is responsible for calculating the peptide masses according to the digestion rule stored in one of the input registers. The masses corresponding to the amino acid codes received from the memory, which are stored into a lookup table (LUT) as 32-bit fixed-point numbers, are added up sequentially, generating the theoretical ion fragment masses corresponding to

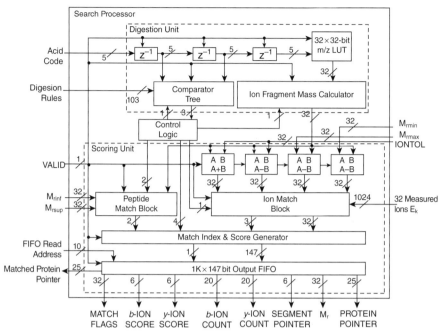

FIGURE 7.4. Block diagram of a peptide fragment search processor. M_{rinf}, minimum precursor mass; M_{rsup}, maximum precursor mass; M_r, precursor matching error.

the b-type (or alternatively a- or c- type) series, until a cleavage site, protein record delimiter, or the end of database code is received, which indicates that the end of a peptide has been reached. The complementary y-type (or alternatively z- or x-type) ion mass series is computed in parallel by subtracting the same amino acid masses from the mass of the precursor ion that is being matched.

The peptide matching unit compares the theoretical mass of each peptide in the database with the mass of the precursor ion. In parallel, the theoretical fragment ion masses corresponding to each peptide are compared with the experimental ion fragmentation spectra (up to 32 or 64 ions) to generate a basic matching score. The user can specify the matching tolerance of the precursor ions (IONTOL) and that of the ion fragments (M_{rmin} and M_{rmax}), which depend on instrument accuracy, which take into account the MS instrument precision and other sources of measurement errors.

When the 5-bit code which indicates the end of a protein sequence is found, a 25-bit protein record index, which uniquely identifies the matched protein, is generated and sent, together with the matching scores, to the output registers that store all search results. The data from these registers are transferred to the host computer over the 32-bit PCI bus, via the user FPGA on the main board. A global control block manages reading/writing data from/to the memory, loading data into the input registers, reading out the results from the output registers, and managing global and local reset signals between consecutive searches.

7.4.4 Performance

In order to validate the accuracy of the hardware implementation, reference C programs were developed for all the algorithms implemented in hardware to process the raw mass spectra and perform database searching and matching. The first validation step involved checking that both the hardware and software implementation generated identical results.

The MSDB database used in this study contains 3,239,079 protein entries with 1,079,594,700 amino acids. The encoded version occupies 67% of the available 1 GB RAM. The search engine design, including six processors, occupies 98% of logic slices available, 88% of the available block RAM, 33% of the digital clock managers, and 53% of external I/O pads of the Xilinx XC2V8000 user FPGA [45] available on the search module. The design has a clock frequency of 100 MHz which is limited by the clock frequency of the 1 GB onboard DDR SDRAM. The search processors perform one comparison of the experimental spectrum of 32 fragments each clock cycle. A complete database search is completed in less than 1 second.

Because of the parallel nature of the computations, the entire database can be divided into distinct subsets and loaded on separate FPGA modules, each with its own parallel database search engine. For example, the FPGA board used here accommodates three search modules which can perform a complete search in ~320 ms.

It is important to highlight the fact that the logic capacity and processing bandwidth of the latest FPGA devices has increased dramatically (13-fold gain in logic capacity, 15- to 22-fold gain in embedded storage capacity, 17- to 27-fold gain in embedded digital signaling processing [DSP] capability, four- to eightfold gain in data rate) compared with the device used in this implementation. By increasing the number of search processors, the current design of a search engine could easily be scaled up to take advantage of the dramatic increase in performance offered by the latest Virtex 7 devices (around 13-fold increase in number of processors per search engine and four- to eightfold increase in data rate depending on the type of RAM used).

7.5 SUMMARY

This chapter provided an overview of the MS-based protein identification algorithms and described the FPGA implementation of such algorithms on a coprocessing system equipped with Xilinx XC2V8000. The coprocessing solution is tightly integrated with X!Tandem, a popular open-source protein search software, for which several parallel implementations that run on computer clusters are already available.

The single-node FPGA system, which uses previous-generation FPGA devices, significantly outperforms the cluster implementations of the algorithms described in the literature.

Future implementations of the database search engine, running on Virtex-5 models of the BenNuey/BenData board and modules used in this work (http://www.nallatech.com), are expected to deliver significant performance improvements compared with the current implementation, thanks to changes in the fundamental architecture of the Virtex-5 chip components. With Virtex-5, Xilinx has replaced the four-input LUTs with six-input LUTs, has implemented a radically new, diagonally symmetric interconnect pattern, and enhanced DSP48E slices tuned to 550 MHz. This would imply doubling the number of processors per search engine and doubling the processing data rate.

The implementation described in this chapter is highly scalable, enabling increased performance through the addition of additional search modules. A motherboard equipped with three search modules delivers a match three times faster than a single-module motherboard. Each FPGA-accelerated system can be scaled up easier by adding up additional FPGA motherboards (two to four motherboards per system, for example). Furthermore, FPGA "clusters" can be set up for the ultimate performance. Given that a single FPGA system can deliver the performance of hundreds of conventional single-core microprocessors, an FPGA-enabled cluster could provide existing HPC clusters with a huge performance boost for a wide range of computing tasks.

ACKNOWLEDGMENTS

The authors gratefully acknowledge that this work was funded by the Biotechnology and Biological Sciences Research Council (BBSRC) (BB/F004745/1). The authors are also grateful for the support received from Xilinx Inc., who donated the devices and design tools used in this study.

REFERENCES

[1] J.H. Beynon, "The use of the mass spectrometer for the identification of organic compounds," Mikrochimica Acta, 44, 1956, pp. 437–453.

[2] L. Hugh, J.W. Arthur, "Computational methods for protein identification from mass spectrometry data," PLoS Computational Biology, 4, 2008, pp. 1553–7358.

[3] M. Gstaiger, R. Aebersold, "Applying mass spectrometry-based proteomics to genetics, genomics and network biology," Nature Reviews. Genetics, 10, 2009, pp. 617–627.

[4] D. Radulovic et al., "Informatics platform for global proteomic profiling and bio-marker discovery using liquid chromatography-tandem mass spectrometry," Molecular & Cellular Proteomics, 3, 2004, pp. 984–997.

[5] R.J. Beynon, "The dynamics of the proteome: strategies for measuring protein turnover on a proteome-wide scale," Briefings in Functional Genomics & Proteomics, 3, 2005, pp. 382–390.

[6] M. Mintz et al., "Time series proteome profiling to study endoplasmic reticulum stress response," Journal of Proteome Research, 7, 2008, pp. 2435–2444.

[7] T. Yokosuka et al., "'Information-Based-Acquisition' (IBA) technique with an ion-trap/time-of-flight mass spectrometer for high-throughput and reliable protein profiling," Rapid Communications in Mass Spectrometry, 20, 2006, pp. 2589–2595.

[8] E. Heinzle, "Present and potential applications of mass spectrometry for biopro-cess research and control," Journal of Biotechnology, 25, 1992, pp. 81–114.

[9] J.H. Atkins, J.S. Johansson, "Technologies to shape the future: proteomics applica-tions in anesthesiology and critical care medicine," Anesthesia and Analgesia, 102, 2006, pp. 1207–1216.

[10] T. Andrade et al., "Using Grid technology for computationally intensive applied bioinformatics analyses," In Silico Biology, 6, 2006, pp. 495–504.

[11] L. Wang et al., "An efficient parallelization of phosphorylated peptide and protein identification," Rapid Communications in Mass Spectrometry, 24, 2010, pp. 1791–1798.

[12] T. El-Ghazawi et al., "Is high-performance reconfigurable computing the next supercomputing paradigm?" in Proceedings of the 2006 ACM/IEEE Conference on Supercomputing, Tampa, FL, 2006. DOI:10.1109/SC.2006.38.

[13] R.G. Dimond et al., "Accelerating large-scale HPC applications using FPGAs," in IEEE Symposium on Computer Arithmetic, 2011, pp. 191–192.

[14] A.S. Guccione, E. Keller, "Gene matching using JBits," in Proceedings of the Reconfigurable Computing is Going Mainstream, 12th International Conference on Field-Programmable Logic and Applications, 2002, pp. 1168–1171.

[15] P. Guerdoux-Jamet, D. Lavenier, "SAMBA: hardware accelerator for biological sequence comparison," Computer Applications in the Biosciences, 13, 1997, pp. 609–615.

[16] R. Hughey, "Parallel hardware for sequence comparison and alignment," Com-puter Applications in the Biosciences, 12, 1996, pp. 473–479.

[17] D. Lavenier, "Speeding up genome computations with systolic accelerator," SIAM News, 31, 1998, pp. 1–8.

[18] H. Simmler et al., "Real-time primer design for DNA chips," Concurrency and Computation: Practice and Experience, 16, 2004, pp. 855–872.

[19] B. Fagin et al., "A special-purpose processor for gene sequence analysis," Com-puter Applications in the Biosciences, 9, 1993, pp. 221–226.

[20] Y.S. Dandass et al., "Accelerating string set matching in FPGA hardware for bio-informatics research," BMC Bioinformatics, 9, 2008, p. 197.

[21] A. Marongiu et al., "Designing hardware for protein sequence analysis," Bioinfor-matics (Oxford, England), 19, 2003, pp. 1739–1740.

[22] T. Oliver et al., "Using reconfigurable hardware to accelerate multiple sequence alignment with ClustalW," Bioinformatics (Oxford, England), 21, 2005, pp. 3431–3432.

[23] F. Panitz et al., "SNP mining porcine ESTs with MAVIANT, a novel tool for SNP evaluation and annotation," Bioinformatics (Oxford, England), 23, 2007, pp. 387–391.

[24] T.A. Anish et al., "Hardware-accelerated protein identification for mass spectrometry," Rapid Communications in Mass Spectrometry, 19, 2005, pp. 833–837.

[25] I.A. Bogdan et al., "Hardware acceleration of processing of mass spectrometric data for proteomics," Bioinformatics (Oxford, England), 23, 2007, pp. 724–731.

[26] I.A. Bogdan et al., "High-performance hardware implementation of a parallel database search engine for real-time peptide mass fingerprinting," Bioinformatics (Oxford, England), 24, 2008, pp. 1498–1502.

[27] M.B. Gokhale, P.S. Graham, Reconfigurable Computing: Accelerating Computation with Field-Programmable Gate Arrays, Springer, Dordrecht, 2005.

[28] S. Hauck, A. Dehon, Reconfigurable Computing: The Theory and Practice of FPGA-Based Computation, Elsevier, Burlington, MA, 2008.

[29] R. Craig, R.C. Beavis, "A method for reducing the time required to match protein sequences with tandem mass spectra," Rapid Communications in Mass Spectrometry, 17, 2003, pp. 2310–2316.

[30] R. Craig, R.C. Beavis, "TANDEM: matching proteins with tandem mass spectra," Bioinformatics (Oxford, England), 20, 2004, pp. 1466–1467.

[31] I.A. Bogdan et al., "FPGA implementation of database search engine for protein identification by peptide fragment fingerprinting," in Proceedings of the Seventh IASTED International Conference on Biomedical Engineering, Innsbruck, Austria, February 17–19, 2010.

[32] R.D. Bjornson et al., "X!!Tandem, an improved method for running X!Tandem in parallel on collections of commodity computers," Journal of Proteome Research, 7, 2008, pp. 293–299.

[33] D.T. Duncan et al., "Parallel tandem: a program for parallel processing of tandem mass spectra using PVM or MPI and X!Tandem," Journal of Proteome Research, 4, 2005, pp. 1842–1847.

[34] B. Pratt et al., "MR-Tandem: parallel X!Tandem using Hadoop MapReduce on Amazon Web Services," Bioinformatics (Oxford, England), 28, 2012, pp. 136–137.

[35] E. Hoffmann, "Tandem mass spectrometry: a primer," Journal of Mass Spectrometry, 31, 1996, pp. 129–137.

[36] B.M. Webb-Robertson, W.R. Cannon, "Current trends in computational inference from mass spectrometry spectrometry-based proteomics," Briefings in Bioinformatics, 8(5), 2007, pp. 304–317.

[37] N. Abramowitz et al., "On the size of the active site in proteases. II. Carboxypeptidase-A," Biochemical and Biophysical Research Communications, 29, 1967, pp. 862–867.

[38] I. Schechter, A. Berger, "On the size of the active site in proteases. I. Papain," Biochemical and Biophysical Research Communications, 27, 1967, pp. 157–162.

[39] I. Schechter, A. Berger, "On the size of the active site in proteases. III. Mapping the active site of papain; specific peptide inhibitors of papain," Biochemical and Biophysical Research Communications, 32, 1968, pp. 898–902.

[40] D. Perkins et al., "Probability-based protein identification by searching sequence databases using mass spectrometry data," Electrophoresis, 20, 1999, pp. 3551–3567.

[41] J. Eng et al., "An approach to correlate tandem mass spectral data of peptides with amino acid sequences in a protein database," Journal of the American Society for Mass Spectrometry, 5, 1994, pp. 976–989.

[42] J. Colinge et al., "OLAV: towards high-throughput tandem mass spectrometry data identification," Proteomics, 3, 2003, pp. 1454–1463.

[43] L. Geer et al., "Open mass spectrometry search algorithm," Journal of Proteome Research, 3, 2004, pp. 958–964.

[44] D. Fenyö, R.C. Beavis, "A method for assessing the statistical significance of mass spectrometry-based protein identification using general scoring schemes," Analytical Chemistry, 75, 2003, pp. 768–774.

[45] Xilinx, Virtex II Platform FPGAs: Complete Data Sheet. DS031, Xilinx Inc. (2007).

8 Real-Time Configurable Phase-Coherent Pipelines

ROBERT L. SHULER, JR., and DAVID K. RUTISHAUSER

8.1 INTRODUCTION AND PURPOSE

Field-programmable gate array (FPGA) development tools provide many computational functions, such as floating-point operations, as modules with the option of making them pipelines. Other user-implemented functions, such as filters and transforms, are easily and often implemented in FPGAs using a pipelined architecture. This allows high clock rates to be used for complex operations and high data throughputs to be obtained. This of course works by exploiting parallelism, in which each stage of the pipeline is simultaneously working on a different set of data. The data for a particular operation are said to flow through the pipe.

8.1.1 Efficiency of Pipelined Computation

Computations implemented in pipelined hardware can be very efficient. Not only are they fast, but inherently they have less overhead than sequential processors, and studies have shown that FPGA implementations of floating-point algorithms offer significant power advantages [1]. However, the effort to create the design, and especially the effort to modify and maintain it, can be much higher than alternative techniques. One method is to define a direct datapath, or systolic array—a connected graph of computational units and delay lines with everything timed so that the results of several pipeline units will appear at the correct clock cycle for input into another pipeline unit, until finally the desired results appear [2, 3]. It is hard to make changes to such a delicately timed arrangement, and particularly inefficient to reconfigure or modify the hardware and use it for something else. The operations and data-path are designed specifically for the desired operation. During runtime, the

Embedded Systems: Hardware, Design, and Implementation, First Edition.
Edited by Krzysztof Iniewski.
© 2013 John Wiley & Sons, Inc. Published 2013 by John Wiley & Sons, Inc.

entire systolic array has to be emptied of computations before it can be changed.

8.1.2   Direct Datapath (Systolic Array)

The direct datapath or systolic array approach typically produces the highest performance when compared to other options, but the design is not flexible and changes to the high-level algorithm require new iterations of a potentially time-consuming design effort. Often, particular computations occur only in a small fraction of the time, but the systolic array computational resources are wired for a specific computation. Hardware description language (HDL) tools will take advantage of if-then-else topology to recognize when inline fixed-point resources can be reused, but not module-based floating-point resources. In addition, real-time dynamic applications that interface with numerous sensor systems are characterized by sparse data arrivals from those systems. In this sparse data environment, fully pipelined designs are not used to their full capability.

8.1.3   Custom Soft Processors

Custom soft processors provide more flexibility to algorithm changes and better accommodate reuse of resources. But this approach has a more substantial initial design effort. Soft processors may have a (compile-time) configurable architecture or use custom instructions, each allowing the processor to be tailored to the application. These custom processors have been shown to perform competitively with general-purpose application-specific integrated circuit (ASIC) processors [4, 5].

8.1.4   Implementation Framework (e.g., C to VHDL)

Still more flexibility is afforded by a custom framework for implementation of applications, such as those tailored for floating-point computations. In Tripp et al. [6], the Trident compiler for floating-point algorithms written in C is described. The framework represents a substantial development effort with all the functionality of a traditional compiler: parsing, scheduling, and resource allocation. A custom synthesizer that produces very high-level design language (VHDL) code and custom floating-point libraries are also included. The phase-coherent pipeline approach does not require the complex components found in the Trident system, and is suitable for straightforward translation from C code to VHDL defining the pipeline design. The authors in Reference [7] present a VHDL autocoder to reduce the development time of floating-point pipelines. A computation is defined in a custom HDL-like pipeline description file, and the code for a single pipeline implementing the computation is produced. The approach requires a user to learn the author's custom HDL and does not attempt to share resources. A C++ to VHDL generation framework is presented in de Dinechin et al. [8], using an object-oriented approach to VHDL autocoding of arithmetic operations. All computations are subclasses of a class "operator," and a method of the class "operator" produces VHDL to implement a pipeline for the computation. Again a user must learn the author's syntax to define computations, and resource constraints are not addressed.

8.1.5   Multicore

To avoid the hardware system development overhead, a multicore approach is also available to application developers, when the performance of a single sequential processor does not meet performance requirements [9]. A multi-core approach trades the difficulty of a software parallelizing effort with the hardware development effort of a custom hardware solution. It must be determined on an application-specific basis which effort is greater, and a highly tailored parallel software solution may be no more flexible to algorithmic changes than a systolic array. In addition to the difficulty of algorithm development itself, current design environments for configurable hardware require substantial knowledge and expertise in hardware design that are not traditional skills of algorithm designers. These two characteristics can cause projects to prefer software and microprocessor-based solutions despite the performance potential of configurable hardware [10]

8.1.6   Pipeline Data-Feeding Considerations

Another difficulty with any pipelined solution is that if the pipelines cannot be continuously fed with data, their efficiency drops rapidly. This difficulty is particularly acute in a real-time system, because the acquisition of data may be at a much lower rate than what is natural for the clock frequency of a high-speed FPGA. There may be many computations that need to be performed on any one piece of data, resulting in a large set of pipeline units that are only fractionally utilized.

8.1.7   Purpose of Configurable Phase-Coherent Pipeline Approach

The purpose of this chapter is to present techniques for implementing pipelines that are similar to ordinary sequential programming techniques and, almost as easy, making the development and maintenance of pipelined algorithms more competitive with multicore approaches. These techniques will not embed the pipeline elements in a fixed systolic array. Instead, the pipeline element will be free to accept possibly unrelated data for an unrelated computation on the very next clock cycle, and a simple means will be given for tracking the flow of a computation and eliminating conflicts. This will enable a small number of computational units to implement a relatively complex "virtual" systolic array, or arrays, when the input does not arrive on each clock cycle.

Reconfigurable devices, such as FPGAs, can of course be used for either pipelined or multicore architectures, or a combination of the two. The techniques described here are not intended to completely eliminate the need for parallelism among multiple identical units or cores. But by increasing the work done by each module or core, and reducing the number of cores required, problems of memory and bus resource contention and algorithm subdivision can be greatly reduced.

8.2 HISTORY AND RELATED METHODS

The pipeline is a time-honored technique for dividing a computation into parts, so that new inputs may be accepted and outputs generated at much shorter time intervals than that required for the entire computation. Early supercomputers commonly used pipelines for floating-point computations, and medium- to high-performance modern microcomputers use pipelines for the process of fetching, decoding, executing, and storing the results of instructions. In computational uses, the most studied problem involves vector operations, in which the fetching of operands in a known sequence can be optimized to feed the pipe. For processor control uses, heuristic guesses as to the direction taken by branches allow the pipeline to be efficiently fed, without too much starting and stopping, as when the results of a computation affect the next input.

8.2.1 Issues in Tracking Data through Pipelines

In a pipeline implementation, the chief inconvenience is that the program logic, which feeds any particular pipeline, gets divorced from the program logic, which processes the outputs by some number of clock cycles, termed the latency of the pipe. Early on it was recognized that some method other than just counting cycles was desirable to keep track of the flow of data through a pipe. This not only makes debugging and maintenance easier, but in a case such as an instruction dispatcher, where operands are sent to either pipe or scalar units of different and possibly variable length, it makes the identification and management of results "possible."

Processors will often have a pipeline section and a scalar section. Scalar sections are optimized to produce results with lower latency, or with less total hardware, and may not accept unrelated data items on each clock cycle. Scalar sections often have variable latency, dependent on the operands presented. But scalar sections are managed within the *instruction pipeline*, and the techniques for managing them can be generally applied to reconfigurable pipelines.

8.2.2   Decentralized Tag-Based Control

Section 8.3.2 describes the use of a decentralized tag-based control scheme as a feature of the method described in this chapter. This approach bears some similarity to tagged token dataflow concepts that were an active topic of computer architecture research primarily in the 1980s. A handful of real systems were built, including the Manchester Dataflow Machine [11]. As described in Gurd et al. [11], dataflow computation was formally introduced in the late 1960s as an alternative to the Von Neumann computation model. Computations are ordered based on data dependencies, and control is decentralized by allowing operations to execute as soon as their inputs (tokens) are ready. Associating data tokens with tags is a means to handle re-enterant code, as would be used in multiple loop iterations, recursive flows, or multiple subroutine or function calls. The tag is a piece of information that associates a token with a particular context, or activation, of the operation that uses or produces it. In a dataflow machine, a matching operation is required to ensure that all the ready operands for a particular operation have the same context, or matching tag values before the operation can execute [12]. Dataflow computing has been reexamined more recently in the context of reconfigurable computing [13]. Tagged-token dataflow concepts specifically have also been used in a configurable hardware implementation for parallel execution of multiple loop iterations [14]. As will be described in detail in Section 8.3, the approach of this chapter uses tags to provide a context for a given data token to facilitate not only the reuse of a functional unit, but to allow a multistage pipeline to execute in parallel a number of unrelated contexts equal to the number of pipeline stages. Additionally, the local pipeline control in our method is simpler than a traditional dataflow computer, not requiring a matching unit, token queue, or overflow unit, for example.

8.2.3   Tags in Instruction Pipelines

The use of tags to manage an instruction pipeline with multiple computational units perhaps reached its peak in the IBM 360/91, which used Tomasulo's algorithm [15]. The tags are used to preserve dependencies in the instruction stream. Each operation is assigned a result tag. In the original Tomasulo implementation, there was a tag for each computational unit. An operation is placed in the queue for a computational unit, and the tags of the operations on which it depends are also placed in that queue. When those source operations are complete, as evidenced by a matching of completion tags to tags in the source register queue, the new operation commences. In turn, when it completes, its result tag is posted along with its result.

The algorithm has been somewhat generalized by Weiss and Smith [16], who observe that rarely are more than one or two computational units needed, except for long latency operations such as load (from memory), floating-point add, and floating-point multiply. Those are exactly the sort of operations we are interested in here. The present chapter is somewhat related to these techniques, but will expand by demonstrating for the reader (1) a method of simply coding tag-based constraints in VHDL that mimics ordinary sequential programming while actually executing in a highly parallel manner, and (2) a deterministic and simple method of allocating operands to computational units that increases reuse (utilization) and works well for real-time data having sparse data arrivals.

8.2.4   Similar Techniques in Nonpipelined Applications

The technique of tags may be familiar to firmware coders by other names. For example, with a Reed–Solomon decoder it is common to include "mark" bits, which allow such information as the state of frame sync to be passed along with the decoded data. This is necessary because the latency of data through the decoder is very large (minimum of one frame worth of data, in order to enable error correction) and often unknown to the programmer, or even variable. In order to resume the frame sync logic after the data are decoded, the state must be passed with the data. However, a Reed–Solomon decoder is not, strictly speaking, a pipeline unit. New data presented on each clock cycle are not unrelated operations, but part of a continuing unified data set. The Reed–Solomon decoder is dedicated to an operation, and cannot be reused from cycle to cycle for other operations the way a floating-point pipeline can be. So the tag control logic found in the decoder and filter designs is not reuse logic.

8.2.5   Development-Friendly Approach

The present chapter will address the issues of development time for floating-point arithmetic (or similar) algorithms in configurable hardware and ease of design modification with a framework that is simple in comparison to a software-to-hardware compilation system. The framework enables the definition of dynamically configured floating-point pipelines in HDL (VHDL will be used for examples) that follow the flow of a software implementation more closely than a systolic array, and are suitable for straightforward translation from an executable software implementation that can be used for verification of the design. The pipelines use a data tag control scheme that avoids the complexity of centralized pipeline control logic. The tag approach allows dynamic configuration of pipeline components on a cycle-by-cycle basis, and supports traditional fully pipelined datapath configurations, and several schemes for reuse of unallocated cycles in sparsely filled pipeline configurations. Reuse of unallocated cycles requires definite knowledge of where those cycles are. In one method of particular interest, resource constraints are addressed with a phase-coherent allocation approach that overloads pipeline stages with multiple operations, without the need for a scheduling algorithm or complicated control logic. This we call the *phase-coherent pipeline* approach, and it will be explained in the next section in detail.

8.3   IMPLEMENTATION FRAMEWORK

In this section, the key elements of the phase-coherent, dynamically configurable pipeline framework are described. An example design is used to illustrate how the concepts work together to provide the benefits of the approach.

8.3.1   Dynamically Configurable Pipeline

A dynamically configurable pipeline is a pipelined operation with an interconnect configuration that is selectable during circuit operation. This selectable input/output configuration between floating-point operations, temporary storage, and input/output devices is achieved with multiplexers on each port of the pipeline. As stated in Sun et al. [17], the resource overhead of using multiplexers to share computing resources is balanced by the reduction of resources achieved. An additional concern, particularly for reconfigurable computing, is development time [18]. Dynamically configurable pipelines provide a flexible datapath that is easier to modify than the fixed datapath of a direct implementation, reducing overall development and maintenance time.

So how does one implement a dynamically configurable pipeline? There are two methods. For an example, let us take the simple calculation A + B + C + D = sum. Define A, B, C, D, and sum as signals suitable for floating point, and assume that FADD is a floating-point adder module having two inputs and an output. Further define intermediate results, I1 and I2. We might then implement a direct datapath as follows in Figure 8.1:

The above code can be represented graphically as in Figure 8.2. (Here we show a latency of 3 for each adder, but this is arbitrary.)

```
        subtype fpvar is std_logic_vector(31 downto 0); -- floating point type
        signal A,B,C,D,sum,I1,I2 : fpvar;
begin
        fa1: FADD (A, B, I1);
        fa2: FADD (C, D, I2);
        fa3: FADD (I1, I2, sum);
```

FIGURE 8.1. Direct datapath VHDL code for A + B + C + D = sum.

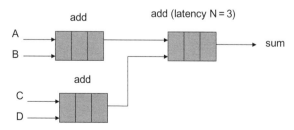

FIGURE 8.2. Direct datapath implementation of A + B + C + D = sum.

```
          subtype fpvar is std_logic_vector(31 downto 0); -- floating point type
          signal A,B,C,D,sum,I1,I2 : fpvar;
          type fpvars is array (integer range <>) of fpvar; -- array of floats
          signal FINA, FINB, FOUT: fpargs(1 to 3)
begin
          fa1: FADD (FINA(1), FINB(1), FOUT(1));
          fa2: FADD (FINA(2), FINB(2), FOUT(2));
          fa3: FADD (FINA(3), FINB(3), FOUT(3));
          FINA(1) <= A when DO_SUM_4 else … ;
          FINB(1) <= B when DO_SUM_4 else … ;
          FINA(2) <= C when DO_SUM_4 else … ;
          FINB(2) <= D when DO_SUM_4 else … ;
          FINA(3) <= FOUT(1) when DO_SUM_4 else … ;
          FINB(3) <= FOUT(2) when DO_SUM_4 else … ;
          Sum <= FOUT(3) when DO_SUM_4 else … ;
```

FIGURE 8.3. Flexible datapath implementation of A + B + C + D = sum.

There is no possibility that later we might perform some different computation with these units. To very simply make the configuration dynamic, suppose we have a mode variable DO_SUM_4 which is set to TRUE when the above computation is needed. Instead of hard-wiring the input signals A, B, C, and D to the adders, we wire the adders to their own dedicated signals FINA, FINB, and FOUT, and use a conditional multiplexer. We make the dedicated signals arrays for convenience in defining them. The implementation is then shown in Figure 8.3.

The code in Figure 8.3 produces exactly the same datapath configuration when DO_SUM_4 is TRUE. Otherwise, it produces some other configuration which for brevity is just indicated by the "..." symbol. If DO_SUM_4 was set to a constant value TRUE, then an optimizing synthesis system would even produce exactly the same hardware. But in the general case, multiplexers on the inputs would be inserted at the input to each FADD module.

This method, in a general case of a complex multiplexer, might have a timing problem. The time for the multiplex gets added to either the first or last stage of the FADD pipeline, depending on where registers are placed (usually added to the first stage as registers will be at the end of stages). This might require a nonoptimal reduction in clock rate to accommodate extra time at only one stage of the pipe. It is preferable, therefore, to add a register at the output of the multiplexer and dedicate an extra clock cycle to the multiplex operation. To perform this in VHDL, for example, we simply use a synchronous coding technique for the same logic, shown in Figure 8.4.

So far we have considered how to achieve complete reconfigurability of an ordinary datapath implementation. We don't mean reconfigurability by reloading the firmware! We mean instant reconfigurability based on programmed logic, just the way that a software program writes an IF statement and decides which of two computations to make.

8.3.1.1 *Catching up with Synthesis of In-Line Operations*   Note that modern synthesis tools already implement this kind of reconfigurability for

```
        subtype fpvar is std_logic_vector(31 downto 0); -- floating point type
        signal A,B,C,D,sum,I1,I2 : fpvar;
        type fpvars is array (integer range <>) of fpvar; -- array of floats
        signal FINA, FINB, FOUT: fpargs(1 to 3)
begin
        fa1: FADD (FINA(1), FINB(1), FOUT(1));
        fa2: FADD (FINA(2), FINB(2), FOUT(2));
        fa3: FADD (FINA(3), FINB(3), FOUT(3));

        process (CLOCK) begin
                if rising_edge (CLOCK) then
                        if DO_SUM_4 = TRUE then
                                FINA(1) <= A when DO_SUM_4 else … ;
                                FINB(1) <= B;
                                FINA(2) <= C;
                                FINB(2) <= D;
                                FINA(3) <= FOUT(1);
                                FINB(3) <= FOUT(2);
                                Sum <= FOUT(3);
                        else
                                . . .
                        end if;
                end if;
        end process;
```

FIGURE 8.4. Synchronous flexible datapath implementation of A + B + C + D = sum.

```
        if DO_SUM_4 then
                I1 <= A + B;
        else
                I1 <= C + D;
        end if;
```

FIGURE 8.5. Example of code that typically is synthesized with resource reuse.

built-in operations like + and −. For example, consider the code sample in Figure 8.5.

Unless the programmer sets special options otherwise, the synthesizer will use only one adder and a multiplexer for the above code, because the two computations are mutually exclusive based on the DO_SUM_4 control variable. What we have done so far is to "catch up" with modern synthesis of inline operators when using arithmetical modules.

8.3.1.2 Reconfiguration Example with Sparse Data Input

Suppose that the inputs A, B, C, and D only change occasionally. Perhaps they are data from a sensor or a bandwidth-limited input device. It is reasonable, then, to use only one FADD module for the entire computation. Some type of control logic is required to configure the adder. If we are deriving the control logic by "counting clock cycles" it could be very brittle, hard to change. In the following example, we show only how to configure the computation using one adder, not

```
if DO_AB_START then
        FINA(1) <= A;
        FINB(1) <= B;
elsif DO_AB_RESULT then
        I1 <= FOUT(1);
elsif DO_CD_START then
        FINA(1) <= C;
        FINB(1) <= D;
elsif DO_CD_RESULT then
        FINA(1) <= I1;
        FINB(1) <= FOUT(1);
elsif DO_FINAL_RESULT then
        Sum <= FOUT(1);
end if;
```

FIGURE 8.6. Single shared adder implementation of A + B + C + D = sum.

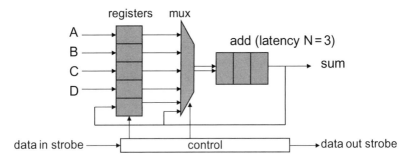

FIGURE 8.7. Single adder implementation of A + B + C + D = sum.

the control logic. Some control variables DO_xxx are assumed to be "somehow" set. (In the next section, we will solve the control problem.) FADD definitions and the process loop setup are not shown, for brevity. Figure 8.6 is the code for a single shared adder module.

The single adder datapath and multiplexer are represented graphically in Figure 8.7.

Notice that the adder is only partly utilized in this example, since each phase of the computation waits on the previous results, and the multiplexer (mux) adds a stage of latency. That means the adder is available for other computations much of the time *IF* a sufficiently flexible control strategy is available. A possible control strategy for just the computation above, in terms of counting clock cycles, is illustrated by the following:

- at data strobe, initiate add A + B, save C and D
- after one clock, initiate C + D
- after three clocks, save A + B in bottom register

- after four clocks, initiate final sum
- after seven clocks, signal data out strobe
- accept data every five cycles, highly dependent on particular computation

It would be very tedious to change such a strategy, or to reuse available cycles for other work. Two new concepts are required to make such advantages feasible.

8.3.2   Phase Tag Control

A dynamically configurable datapath allows the inputs of an operation to be consuming operands and results to be produced for different computations potentially every clock cycle. Further, operations such as floating-point arithmetic typically require latencies greater than 1 to function at high enough clock frequencies to meet the performance requirements of applications. A control scheme is required to route operands and results between pipelines and other resources with the correct timing. In contrast to using a scheduling algorithm and global control approach, a distributed phase tag control method is used.

8.3.2.1   Tags   In the phase tag control method, a tag word is associated with a set of inputs. The tag is assigned to a buffer that has the same latency as the operation consuming the inputs. The buffer is called a phase keeper. The output of the phase keeper is tested to determine when the outputs of the operation are valid for the inputs associated with the tag. One tag can be used to control the input/output configuration of many floating-point units, providing all units have the same latency. For coding convenience, and to handle occasional units with different latency, phase keepers were built in to all our floating-point units. If they are not used, the HDL tool flow removes them automatically. The functional units output two phase tag words. The ready tag (.r), usually with a latency of one clock cycle, is used to signal that a pipeline is ready to accept data. The completion tag (.p) indicates that an operation is finished. The tags are used to control the inputs and outputs of a pipeline. The content of the tag is used by an algorithm developer to indicate which step of the computation is associated with the operands or results. The phase tag control approach supports both dense and sparsely allocated pipelines. Examples of several design cases follow.

8.3.2.2   Data Entity Record Types   In order to keep coding at a high level, tags and operands are grouped into record structures, so that the entire record structure may be set with a single assignment statement. The definitions in Figure 8.8 are assumed to be kept in a package and used in all subsequent examples:

```
-- type definitions
subtype fpvar    is std_logic_vector(31 downto 0);
subtype fpphase  is std_logic_vector( 7 downto 0);
subtype fpfunc   is std_logic_vector( 1 downto 0);
type fparg is record
      a : fpvar;
      b : fpvar;
      p : fpphase;
      f : fpfunc;
end record;
type fpout is record
      o : fpvar;
      r : fpphase;
      p : fpphase;
end record;
type fpvars   is array (integer range <>) of fpvar;
type fpphases is array (integer range <>) of fpphase;
type fpargs   is array (integer range <>) of fparg;
type fpouts   is array (integer range <>) of fpout;

-- function code definitions for FADDP
constant ADD : std_logic_vector := "00";
constant SUB : std_logic_vector := "01";

-- phase tag constants
constant NOPH : fpphase := (others => '0');
constant ZPH  : fpphase := (others => 'Z');
constant RDYPH: fpphase := x"ff";
constant ERRPH: fpphase := x"fe";
```

FIGURE 8.8. Example package code for tag-based pipeline control.

The definitions in Figure 8.8 provide for 32-bit floating numbers, 8-bit tags, and a 2-bit operation code (used, for example, to determine whether the adder module does an add or subtract operation, as some modules can be configured to do both). Also included is a two-operand input record, with tag and function code, and an output record with two tags, .r and .p as explained above. The tags are completely arbitrary, as long as their use is unambiguous. They could be defined as mnemonics, but for brevity we will simply use two digit hexadecimal notation, such as x"10". One can think of the tags as being analogous to line numbers in BASIC.

8.3.2.3 *Tag Shells*   Instead of directly using the FADD module, a shell is written called FADDP that processes the tag information and maps the record parameters to the scalar parameters of FADD. Since we are not using a function code for add or subtract, it is assumed that FADD has an additional input of type fpfunc (not used in previous examples). In order to share the FADDP unit among several modules, there is also a provision for tristating its output. The complete definition of FADDP is shown in Figure 8.9.

```
entity FADDP is port (
            clk    : in   std_logic;
            arg    : inout   fparg;
            res    : out fpout
); end FADDP;
architecture behavior of FADDP is
      signal op : std_logic_vector(5 downto 0);
begin
      op <=   xcADD when arg.f = ADD
            else xcSUB when arg.f = SUB
            else xcLT  when arg.f = LT
            else xcLE  when arg.f = LE;
      fa: FADD port map(arg.a, arg.b, op, clk, res.o);
      fp: PKEEP port map(clk, arg.p, res.p);
      arg <= ("ZZZZZZZZZZZZZZZZZZZZZZZZZZZZZZZZ"
            ,"ZZZZZZZZZZZZZZZZZZZZZZZZZZZZZZZZ","ZZZZZZZZ","ZZ");
      res.r <= arg.p when arg.p /= NOPH else RDYPH;   -- always ready
end behavior;
```

FIGURE 8.9. VHDL shell code for a tag-controlled and shared pipelined adder.

8.3.2.4 Latency Choices   The definition in Figure 8.9 is written for a FADD latency of 2, not of 3 as we have previously been showing it. Floating-point module latency is quite arbitrary. It is a matter of how many sets of latches the user requests the core generator insert into a combinatorial logic flow. In general, latches are overhead, and the user will want to run tests to determine the minimal latency that will support the clock rate for the application at hand. A latency of 2 supports a 125 MHz clock rate on a Virtex 5 FPGA, which is also the same clock rate as gigabit Ethernet byte transfers, and was convenient for several applications. What we will see shortly is that it is fairly easy to change the latency when using tags—just rewrite the shells for the floating-point modules for the correct number of phase keeper stages—since the follow on computations are triggered by tags exiting the pipe, not by counting clock cycles. (Later, in the question of reuse, of course latency may change the interval in which a module can be reused.)

8.3.2.5 Example   The first phase tag example will be a recoding of the three-adder configurable datapath, using tags and an input strobe instead of the hypothetical DO_SUM_4 control variable. The problem with DO_SUM_4 is that it configures the entire datapath at once. If new inputs arrive which need to undergo some different type of processing, this global mode variable cannot begin the new configuration until all the old data have exited the pipe. But with tags, the configuration information travels along with the data. Instead we assume a signal DATA_STROBE_4 is set for one clock cycle when the inputs A, B, C, and D are available. The arrays of input and output records for the adders are AI and AO, respectively. The code is shown in Figure 8.10.

In Figure 8.10, three adders are connected such that the first unit (1) adds inputs A and B in parallel with the second unit (2) that adds inputs C and D.

```
        signal AI : fpargs(1 to 3) := (others=>NOARG);
        signal AO : fpouts(1 to 3);
begin
        ad1 : FADDP  port map (clk, AI(1), AO(1));
        ad2 : FADDP  port map (clk, AI(2), AO(2));
        ad3 : FADDP  port map (clk, AI(3), AO(3));
process (clk) begin if rising_edge (clk) then
        for i in 1 to 3 loop AI(i) <= NOARG; end loop;  -- create strobes
        if DATA_STROBE_4 then
                AI(1) <= (A, B, x"10", ADD);             -- initiate adds
                AI(2) <= (C, D, x"10", ADD);
        elsif AO(1).p = x"10" then                       -- 1st adds done
                AI(3) <= (AO(1).o, AO(2).o, x"10", ADD); -- final add
        elsif AO(3).p = x"10" then                       -- result ready
                sum <= AO(3).o;
        end if;
end if; end process;
```

FIGURE 8.10. Phase tag implementation of A + B + C + D = sum, without reuse.

The outputs of the first two adders are wired to the inputs of the third adder (3), typical of a direct implementation. Instead of an explicit state machine to control the output of the pipeline when the result of the computation is valid, the phase tag appearing at the output of the third adder, AO(3).p, is tested to determine when to store the result in registered signal sum.

8.3.2.6 Strobes Note that at the beginning of the process loop, all input tags are set to NOPH, which is the phase value of NOARG. In fact, all operands are set to zero. Functionally, this means all the tags will be strobes, valid for only one cycle. The tag will be NOPH *unless* it is set to something else later in the process loop, according to the semantic rules of VHDL. Aesthetically, this means that all operands will be zero when not in use. There is no functional value to unused operands being zero, but it makes it very easy to look at a simulation output and trace values.

8.3.2.7 Tag Values The arbitrary tag x"10" is used to identify this operation all the way through the datapath. Whenever it emerges at the output of a functional unit, the sequential logic can use it to decide how to dispose of the also emerging result.

8.3.2.8 Coding Overhead The number of lines of code required is, oddly enough, about the same as when directly coding the configurable three-adder design. We do not count the package overhead, since it occurs only once in a design. There is one extra IF statement per operation in the tag approach. But the operand record allows both input operands to be conveniently assigned in one line of code, giving one less line of code per operation, resulting in no net change in coding lines. In addition, there is much greater logical clarity in the tag implementation as to the exact condition under which the result appears.

8.3.2.9 Reusing Functional Units Adders 1 and 2 inputs are *only* used when DO_SUM_4 is true. At *any other time* these adders can be used for unrelated purposes. All that is required is to assign a tag other than x"10" for those other operations, and the logic above will ignore those tags.

Reusing the third adder is not so straightforward. True, it is available as long as the tag x"10" is not emerging from adders 1 and 2, but this is emerging from the adders, not from the user's control logic, so it is not (directly) determined by the user. In the next section, we will learn how to also easily reuse this third adder, without worrying about what is emerging from the first two adders.

8.3.2.10 Tag-Controlled Single-Adder Implementation The single-adder version of Figures 8.6 and 8.7, shown with the control algorithm unspecified, can easily be controlled by tags. Figure 8.11 shows the block diagram architecture.

Figure 8.12 shows the same code as Figure 8.6, with the hypothetical control variables replaced by tag references.

Three tag values are used: 1, 2, and 3. Notice how each control variable is mapped to a tag on one of the two tag outputs: input ready (.r) or output ready (.p). The operation is initiated by adding A and B with a tag of 1. As soon as this input has been accepted, "1" appears on the .r output. It actually appears on the same clock cycle where the data are presented, so that control logic

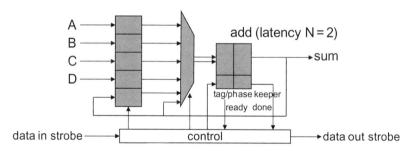

FIGURE 8.11. Single-adder A + B + C + D = sum with tag-based control.

```
if DO_SUM_4 then
      AI(1) <= (A, B, x"10", ADD);
elsif AO.p = x"10" then                    -- was DO_AB_RESULT
      I1 <= AO(1).o;
elsif AO.r = x"10" then                    -- was DO_CD_START
      AI(1) <= (C, D, x"20", ADD);
elsif AO(1).p = x"20" then                 -- was DO_CD_RESULT
      AI(1) <= (I1, AO(1).o, x"30", ADD);
elsif AO(1).p = x"30" then                 -- was DO_FINAL_RESULT
      Sum <= AO(1).p;
end if;
```

FIGURE 8.12. Single shared adder implementation with tag-based control.

clock	trigger	logic	adder pipe stage 1	adder pipe stage 2
1	DATA STROBE	A1 & B1 to adder, tag #1		
2	ready #1 tag	C1 & D1 to adder, tag #2	process A1+B1	
3			process C1+D1	process A1+B1
4	done #1 tag	save A1+B1		process C1+D1
5	done #2 tag	C+D, A+B to adder, tag #3		
6	DATA STROBE	A2 & B2 to adder, tag #1	process final sum	
7	ready #1 tag	C2 & D2 to adder, tag #2	process A2+B2	process final sum
8	done #3 tag	store or forward result	process C2+D2	process A2+B2

FIGURE 8.13. Timing for single-adder A + B + C + D = sum.

may query it to determine what to do on the next cycle. And that is to present inputs C and D with a tag of 2. Now two operations are in the pipe, A + B and C + D. When each comes out of the pipe, its tag appears on the .p output, and this triggers another part of the computation. A state variable I1 is used to store A + B until it is needed. When tag = 3 appears at the output, the result "sum" is stored. A timing diagram for this process appears in Figure 8.13.

8.3.2.11 Interference Notice that with a two-stage adder, giving a total latency number of L = 3, the earliest a new data set can be accepted is every fifth cycle. If it is not available on the fifth, then the next opportunity is every eighth cycle. To use any other starting time less than 9 (the eighth cycle after last start) will cause interference (contention) in this single-adder pipeline. If the right data input is available, either by coincidence or by buffering, then utilization of the adder is over 50% (a second point in shaded text is shown in the figure). But accomplishing such utilization is very tricky, leading to brittle and hard-to-maintain code. As one might guess from our use of the terms "interference" and "phase coherence," in the next section we will show how to get rid of many cases of interference and make the construction of pipes with reused elements much more flexible.

8.3.3 Phase-Coherent Resource Allocation

Phase-coherent pipeline allocation is a simple means to allow pipeline stage sharing that enables different computations to be allocated to the same functional units. The method requires that results associated with a particular computational sequence all emerge at a constant phase, that is, at a constant multiple of a minimum unit latency L. If a unit does not naturally have this latency, it must be padded with enough empty pipeline stages. The multiplex stage is included in the latency value. The data introduction interval (DII) should be equal to or greater than L. The pipe can be said to have L independent phases. For maximum reuse, successive data inputs are allocated to different phases, until all phases are used, and only then are conflicts possible.

Under these conditions, a simple algebraic relationship can be used to compute the period of time that units can be reused as follows.

8.3.3.1   *Determining the Reuse Interval*   Given a dynamically configurable pipelined functional unit with a latency of L, each pipeline stage, $S_p(n)$, can process a datum of an independent computation. The reuse interval, I_R, is defined as the number of clock cycles in which units can be reused freely. This interval is computed as shown in Equation (8.1):

$$I_R = DII \cdot L. \qquad (8.1)$$

8.3.3.2   *Buffering Burst Data*   If an application has a suitable average data arrival rate, but sometimes input data points come in bursts with an inconvenient DII, then a first in, first out (FIFO) buffer can be used to smooth out the arrival rate and allow a better reuse interval.

For maximum reuse, the interval can be applied separately to each functional unit. Figure 8.14 shows a timing diagram of phase-coherent pipeline allocation for the computation of the single-adder version of prior code examples (functional block diagram is the same as Figure 8.11, as only the control logic changes). In Figure 8.14, the DII = 3, L = 3, and I_R = 9. The d_n variables represent a set of input operands for the four add operations at DII = n. As shown, the phase offsets for allocation are implemented with a one- or two-cycle store of the incoming data value at the input of the unit, as shown in clock cycles 3, 6, and 7. This is required because the input stage of the unit is

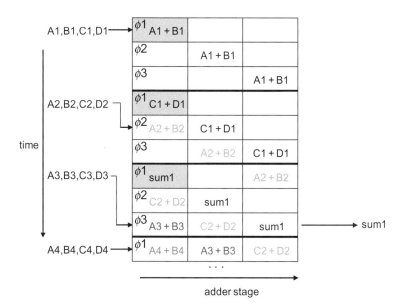

FIGURE 8.14. Timing diagram for phase-coherent single-adder A + B + C + D = sum.

busy processing prior input data at these cycles. The single-cycle latency ready tag is tested to determine when the input stage is available. Also shown in Figure 8.14 is that by cycle 8, when the third input data set is consumed, each pipeline stage of the unit is processing data. Intermediate results or temporary state variables—for example, the intermediate addition result A + B—do not benefit from the phasing scheme and must either be used within the DII or copied every DII clocks. Alternately they could be retained in no-op pipe units.

The VHDL that implements the example of Figure 8.14 is shown in Figure 8.15. As shown, the code implementing the phase-coherent allocation method is straightforward and suitable for generation by an autocoder. The phase tags control each stage of the computation as well as the cycle the adder unit is free to accept new data. The dynamically configurable inputs and outputs allow the same unit to process each computation stage within the reuse interval.

8.3.3.3   Allocation to Phases   What the timing diagram implies, and what we will show below, is that there is a simple automatic method for allocating incoming data to a "phase." As long as the design DII is not violated, data can arrive at any time. Once introduced at a particular phase, the computation proceeds without regard for interference. The total latency is increased from seven to nine clock cycles, but conflicts have disappeared, and if data are arriving at the DII, then 100% utilization is obtained after only eight clock cycles.

The main difference between the code of Figure 8.15 and previous examples is the addition of a recirculating phase-keeper loop using the module PKEEP, and a state variable "waiting" to indicate if an incoming data point is waiting to be allocated to a phase.

8.3.3.4   Harmonic Number   The PKEEP module is a simple no-operation pipe that delays a phase tag from input to output for L cycles, to exactly match the minimum latency of functional modules. The latency of functional modules must be then $N*L - 1$, where N is any integer $1, 2, \ldots$ and so on. For the adder, $N = 1$ and a three-cycle latency L is obtained. The PKEEP module matches this. For a divider implemented in one of our test applications, we used $N = 4$ so that the latency of the divider was 11, for a total latency of $L = 12$. "N" is thus the "harmonic number" of the module. If modules comply with this requirement, then they will always produce results in the same "phase" in which the related operation was initiated, and processing these results immediately will not interfere with any other ongoing computations. All that is required is to observe the DII. Note above that rules for state variables are more restrictive, as they must be consumed within L cycles, or else allocated to pipe modules for deferral to the time at which the information is needed in the usual way that state information is carried along in pipe computations.

8.3.3.5   Allocation Algorithm   The input data A, B, C, and D in this example are assumed to be persistent until overwritten by another data point, and thus equivalent to state variables. If this is not so, then for the allocation of data to pipe phases, strobed inputs must be copied to state variables.

```
        signal PI : fpphases(0 to 1) := (others=>NOPH);
        signal PO : fpphases(0 to 1);
        signal sum_ab : fpvar := ZERO;
        signal waiting : boolean := FALSE;
begin
        pk0 : PKEEP  port map (clk, PI(0), PO(0));
        ad0 : FADDP  port map (clk, AI(0), AO(0));
process (clk) begin
        if rising_edge(clk) then
                waiting <= FALSE;
                PI(0) <= PO(0); -- default to a re-circulating phase tag
                if data_strobe or waiting then
                        if PO(0) = RDYPH then   -- wait for an available pipe phase
                                PI(0) <= x"10";  -- 1st operation this re-use interval
                                AI(0) <= (A, B, x"10", ADD);-- start A+B
                        else
                                waiting <= TRUE; -- if pipe busy, just wait
                        end if;
                end if;
                if PO(0) = x"10" then -- when a+b completes, save and start next
                        PI(0) <= x"20";
                        AI(0) <= (C, D, x"20", ADD);
                        sum_ab <= AO(0).o;  -- state varble valid L clocks or DII/L stages
                end if;
                if PO(0) = x"20" then -- when c+d completes, initiate final sum
                        PI(0) <= x"30";
                        AI(0) <= (sum_ab, AO(0).o, x"30", ADD);
                end if;
                if PO(0) = x"30" then  -- when final sum completes ...
                        PI(0) <= RDYPH;  -- vacate this phase
                        sum3 <= AO(0).o; -- consume result
                end if;
        end if;
end process;
```

FIGURE 8.15. VHDL code for phase-coherent single-adder A + B + C + D = sum.

The input PI and output PO phase tags associated with the PKEEP module contain recirculating tags, which are all "ready" (RDYPH) initially. Every phase is available. When the input strobe is detected, the allocation proceeds by attempting to "grab" the current phase if it is available. If not, then the WAITING flag is set, and on the next clock cycle, the grab is attempted again, and so forth until an available phase is found. This simple allocation algorithm adds only four lines of code (not counting signal definitions) to the previous code examples. All of the module and type definitions can be kept in a package file so they do not clutter up the application.

8.3.3.6 *Considerations*   For phase-coherent allocation to be useful, the DII must be greater or equal to the minimum unit latency L. Otherwise, consider using tag-based control by itself. We found ourselves using simple tag-based control when processing batches of startup parameters, and phase-coherent allocation when processing the main data stream.

For state variables to be useful, DII must be at least 2L. Otherwise the valid persistence before the state variable is overwritten is effectively zero. In that case, delay pipes must be defined to carry state information.

Mode variables to identify and configure major operational configurations of a computation are useful. Without mode variables, the number of different tag values will proliferate greatly. With mode variables, one starts over on tag numbers in each configuration. Care, of course, must be taken to be sure that no data are in the pipe when switching modes. The general code structure will be to write "if MODE1 = TRUE then . . . end if," and include all the computation and tag checking and strobe formation for configuration MODE1 inside that IF statement.

8.3.3.7 External Interface Units External interface modules, such as memory or communications, are problematic because one cannot just instantiate as many copies of them as one would like. They are a fixed resource!

If multiple banks of memory are available, these can be used simultaneously to access different pieces of data—but obviously not the same data. Different types of data have to be allocated to the different banks to allow simultaneous memory access. For example, in a terrain map generation algorithm, we stored terrain height in one memory bank, and its weighted value for averaging with new points in another memory bank. Often the vendor-supplied memory interface will map all banks to a single address space, and this module has to be replaced to provide parallel access. However, in pipelined applications, parallel memory access can be critical to achieving performance goals.

Latency through the memory subsystem could be considerable, and in the case of READ operations, has to be padded to comply with phase coherence requirements just as any other functional unit. If the application might read just-written values from a previous data point, then conflicts between the reuse interval and the WRITE latency can arise, and one must manually analyze these, just as with any application.

Some memory systems can be clocked at higher rates than the application logic. If a 2× clocking is allowed, then in theory one can instantiate a second memory access module referring to the same data space, for independent use.

The authors found it useful to instantiate all board interface modules at the top level, and to provide some interface modules at the top level, which give a pipeline friendly access to the board interfaces, maintaining tags, and so forth. The signals associated with the client interfaces are then passed to the application module as arrays, much like the arrays associated with floating-point module inputs and outputs.

8.4 PROTOTYPE IMPLEMENTATION

The authors tested the approach described in the previous section in support of the National Aeronautics and Space Administration's (NASA) Automated Landing and Hazard Avoidance Technology (ALHAT) project [19]. ALHAT is developing a system for the automatic detection and avoidance of landing hazards to spacecraft. The system is required to process large amounts of terrain data from a Light Detection and Ranging (LIDAR) sensor, within strict power constraints. The processing is an example of a sparse data arrival, real-time, dynamic, three-dimensional computational application ubiquitous in aerospace systems. Details of the algorithms supporting ALHAT for landing hazard detection and avoidance are provided in Johnson et al. [20]. The general approach is to produce a regular grid of surface elevation data in the coordinate frame of the landing site from the LIDAR range samples. This elevation map is then analyzed for surface slope and roughness and compared to thresholds for these parameters to identify hazards. The processing stages for LIDAR scan data are coordinate conversion, regridding, and hazard detection. The first two stages are currently demonstrated in the prototype design. The computations implemented are summarized in this section.

8.4.1 Coordinate Conversion and Regridding

As described in Johnson et al. [20], the coordinate conversion stage converts each LIDAR range sample from scanner angle and range coordinates to Cartesian coordinates. The computation is shown in Equation (8.2), where pr_x, pr_y, and pr_z are components of the converted point; t_x, t_y, and t_z are the components of the sensor position vector; p_x, p_y, and p_z are the components of the range sample; and q_1, q_2, q_3, q_4 are components of a quaternion vector for the coordinate rotation:

$$pr_{x,y,z} = 2 \begin{bmatrix} (q_3q_3 + q_0q_0 - 0.5)p_x + \\ (q_0q_1 + q_3q_2)p_y + \\ (q_0q_2 + q_3q_1)p_z \end{bmatrix} + t_{x,y,z} \tag{8.2}$$

In the regridding stage of the computation, converted range samples are projected into a grid cell of the elevation map, and a bilinear interpolation scheme is used to update the elevation of each vertex of the cell containing the projected point. The elevation of the projected point weighted by the distance from the point to the vertex is added to the current weighted elevation for that vertex. Updates to the weighted elevations and the weights for each vertex of the grid cell containing a projected point are made using the computation shown in Equation (8.3). In Equation (8.3), r and c are the row and column numbers of the elevation map grid cell vertices, respectively.

$$\begin{aligned}
u &= r - \lfloor r \rfloor; \\
v &= c - \lfloor c \rfloor; \\
W(r, c) +&= (1-u)(1-v); \\
E(r, c) +&= (1-u)(1-v)pr_z; \\
W(r+1, c) +&= u(1-v); \\
E(r+1, c) +&= u(1-v)pr_z; \\
W(r, c+1) +&= (1-u)v; \\
E(r, c+1) +&= ((1-u)v)pr_z; \\
W(r+1, c+1) +&= uv; \\
E(r+1, c+1) +&= (uv)pr_z;
\end{aligned} \tag{8.3}$$

8.4.2 Experimental Setup

The prototype design is tested on a Xilinx® Virtex™-5 FX130T FPGA hosted on an Alpha Data ADM-XRC-5TZ mezzanine card. The Virtex-5 family has a radiation tolerant version, providing a path to space flight certification of the design. The ADM-XRC-5TZ board has 48 MB of static random access memory (SRAM) across six banks, used for storing the elevation and weight data for the elevation map. The prototype design uses two SRAM interfaces that bundle two SRAM banks each. The interfaces are designed to use one bank for even addresses and one for odd. This approach makes it possible to run the SRAM interface at twice the design rate to reduce memory latency. Currently all interfaces operate at the same clock frequency.

A Gigabit Ethernet interface is included for command and data input and output. The prototype is designed to run at the Ethernet clock speed of 125 MHz. To avoid buffering of the input data, the prototype is designed with a DII of 12 clock cycles. The 12 cycles is derived from each LIDAR sample consisting of three single-precision floating-point components of 4 bytes each.

With this prototype platform, the coordinate conversion and regridding computations were implemented within a few days using the phase-coherent pipeline approach. A full-rate elevation map computation is verified (input LIDAR samples are converted and regridded within 12 clock cycles) on this prototype. These results show that the LIDAR data can be processed in real time, or faster than real time.

8.4.3 Experimental Results

A comparison of resources used between various implementations of the example computation presented in Section 8.3 is shown in Table 8.1. The resource values are reported from the Xilinx synthesis tools. The static implementation is a direct wiring of the adders and datapath to realize the

TABLE 8.1. Resource Comparison between Different Implementations of Computation Sum = A + B + C + D

Resource	Static	Implementation: Three Instantiations of Sum = A + B + C + D	
		Dynamically Configurable/ Phase Tag Control	Phase Coherent
Slice registers	1198	753	342
LUTs	4768	1755	705
Slices	1653	694	309
LUT/flip-flop (FF) pairs	4706	2010	834
Minimum period	6.6 ns	6.6 ns	6.6 ns

computation as shown in Figure 8.1. The dynamically configurable/phase tag control implementations are designed as presented in Figures 8.10, 8.12, and 8.15. The phase-coherent implementations are designs applying the phase-coherent method to each case represented by Figures 8.10, 8.12, and 8.15. The floating-point units are implemented using the Xilinx CORE Generator™ tool. DSP slice resources are used in the multiplier units but not shown in Table 8.1. Comparing the lookup table (LUT) resources between the static and phase-coherent implementations shows the phase-coherent pipeline method yielding an 85% reduction in resources. This means a given FPGA can hold the equivalent of about seven times as many source lines of floating-point application equivalent code using the phase-coherent method as using traditional datapath methods. If a particular design does not approach resource limits, phase-coherent reuse reduces design size, resulting in faster place and route.

8.5   ASSESSMENT COMPARED WITH RELATED METHODS

The qualitative relation between flexibility (ease of making change) and complexity (special expertise and initial setup) for the methods discussed is shown in Figure 8.16.

The two methods with the most attractive combination of high flexibility in making changes and low complexity in initial design effort are the phase-coherent pipeline and commercial C to HDL translators that support pipelined modules (chiefly floating point).

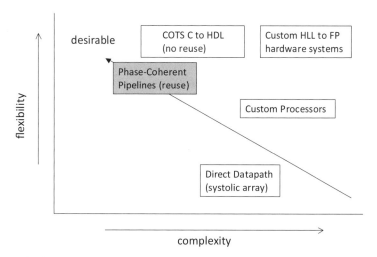

FIGURE 8.16. Comparison of methods discussed. Commercial off-the shelf (COTS) C to high-level design language (HDL); custom high-level language (HLL) to floating-point (FP) hardware systems; custom processors; direct datapath; and phase-coherent pipelines.

The resource-sharing approach for phase-coherent pipelines has some similarities to those developed for ASIC synthesis algorithms. Hwang et al. [21] define the DII as the period in clock cycles between new data arrivals at the input of a pipeline. A DII equal to 1 represents fully pipelined operations. As discussed in Section 8.3, a DII greater than 1 is required for phase-coherent resource sharing. Resource-sharing approaches are described in Wakabayashi et al. [22] and more recently for FPGAs in Mondal and Memik [23], where analysis of a data flow graph of the computation and heuristics must be used. The phase-coherent allocation approach is governed by simple algebraic relationships and does not require complex analysis or heuristics. Phase-coherent allocation does not perform dynamic scheduling and does not require any scoreboarding method [24] or hardware to check for structural or data hazards.

A hand-coded systolic array and MATLAB-based FPGA implementation of the coordinate conversion stage of processing LIDAR scan data for an automatic landing hazard detection system is compared in Shih et al. [25]. Fixed-point arithmetic is used. As previously discussed, a hand-coded systolic array is not easily adaptable to algorithm design changes. The MATLAB solution is more easily developed and adapted, but is most suitable to the processing of streaming data and would not be an effective approach for other computations, such as hazard detection.

The method described in this chapter achieves substantial improvements in the ease of both development and resource reuse for pipelined computations on configurable hardware. Using this HDL method, declarations and wiring are simplified, and operand/result assignments are easily mixed with other synchronous code. The HDL reads like and corresponds closely to a software-specified algorithm. This allowed rapid design of the prototype and should allow fast response to algorithm changes. The HDL is suitable for straightforward translation from an executable software definition that can be used for algorithm verification. This reduces the gap between the expertise required to design configurable implementations and that of typical algorithm designers.

The difficulty of resource reuse is reduced with a data tag control scheme and phase-coherent allocation method that replace the need for complex global scheduling, heuristics, or cycle-dependent logic. Sparse data arrival in real time is efficiently allocated to pipeline stages, reducing design size, and place and route times.

REFERENCES

[1] G. Govindu et al., "Area, and power performance analysis of a floating-point based application on FPGAs," in Proceedings of the Seventh Annual Workshop on High Performance Embedded Computing (HPEC 2003), http://citeseer.ist.psu.edu/viewdoc/summary?doi=10.1.1.157.9977.

[2] K. Sano et al., "Systolic architecture for computational fluid dynamics on FPGAs," in 15th Annual IEEE Symposium on Field-Programmable Custom Computing Machines, 2007. FCCM 2007, April 2007, pp. 107–116.

[3] S. Qasim et al., "A proposed FPGA-based parallel architecture for matrix multiplication," in IEEE Asia Pacific Conference on Circuits and Systems, 2008. APCCAS 2008, November 2008, pp. 1763–1766.

[4] D. Goodwin, D. Petkov, "Automatic generation of application specific processors," in Proceedings of the 2003 International Conference on Compilers, Architecture and Synthesis for Embedded Systems (CASES '03), ACM, New York, NY, USA, 2003, pp. 137–147.

[5] D. Rutishauser, M. Jones, "Automatic tailoring of configurable vector processors for scientific computations," in T.P. Plaks, ed., Proceedings of the 2011 International Conference on Engineering of Reconfigurable Systems and Algorithms (ERSA'11), CSREA Press, Las Vegas, NV, 2011.

[6] J. Tripp et al., "Trident: an FPGA compiler framework for floating-point algorithms," in International Conference on Field Programmable Logic and Applications, 2005. August 2005, pp. 317–322.

[7] G. Lienhart et al., "Rapid development of high performance floating-point pipelines for scientific simulation," in 20th International Parallel and Distributed Processing Symposium, 2006. IPDPS 2006, April 2006, p. 8.

[8] F. de Dinechin et al., "Generating high-performance custom floating-point pipelines," International Conference on Field Programmable Logic and Applications, 2009. FPL 2009, September 2009, pp. 59–64.

[9] Tilera Corporation, "Tilera has solved the multi-processor scalability problem." Available: http://www.tilera.com/ (accessed 12/22/11).

[10] C. Villalpando et al., "Investigation of the Tilera processor for real time hazard detection and avoidance on the Altair Lunar Lander," IEEE Aerospace Conference, March, 2010, pp. 1–9.

[11] J.R Gurd et al., "The Manchester Prototype Dataflow Computer", Communications of the ACM, 28(1), January 1985, pp. 34–52.

[12] Arvind, R.S. Nikhil, "Executing a program on the MIT tagged-token dataflow architecture," IEEE Transactions on Computers, 39(3), March 1990, pp. 300–318.

[13] J.S. Luiz, "Execution of algorithms using a Dynamic Dataflow Model for Reconfigurable Hardware—a purpose for Matching Data," in The 6th International Workshop on, System-on-Chip for Real-Time Applications, December 2006, pp. 115–119.

[14] H. Styles et al., "Pipelining designs with loop-carried dependencies," in Proceedings of the 2004 IEEE International Conference on Field-Programmable Technology, December 6–8, 2004, pp. 255–262.

[15] R. Tomasulo, "An efficient algorithm for exploiting multiple arithmetic units," IBM Journal of Research and Development, 11(1), January 1967, pp. 25–33.

[16] S. Weiss, J. Smith, "Instruction issue logic in pipelined supercomputers," IEEE Transactions on Computers, C-33(11), November 1984, pp. 1013–1022.

[17] W. Sun et al., "FPGA pipeline synthesis design exploration using module selection and resource sharing," IEEE Transactions on Computer-Aided Design of Integrated Circuits and Systems, 26(2), February 2007, pp. 254–265.

[18] J. Villarreal et al., "Designing modular hardware accelerators in C with ROCCC 2.0," in 18th IEEE Annual International Symposium on Field-Programmable Custom Computing Machines (FCCM), May 2010, pp. 127–134.

[19] C. Epp et al., "Autonomous Landing and Hazard Avoidance Technology (ALHAT)," in IEEE Aerospace Conference, March 2008, pp. 1–7.

[20] A. Johnson et al., "LIDAR-based hazard avoidance for safe landing on Mars." Available: http://trs-new.jpl.nasa.gov/dspace/ (accessed 6/3/2011).

[21] K.S. Hwang et al., "Scheduling and hardware sharing in pipelined data paths," in Proceedings of the IEEE International Conference on Computer-Aided Design. ICCAD-89, November 1989, pp. 24–27.

[22] S. Wakabayashi et al., "A synthesis algorithm for pipelined data paths with conditional module sharing," in Proceedings of the IEEE International Symposium on Circuits and Systems, vol. 2, May 1992, pp. 677–680.

[23] S. Mondal, S. Memik, "Resource sharing in pipelined CDFG synthesis," in Proceedings of the ASP-DAC 2005. Asia and South Pacific Design Automation Conference, 2005, vol. 2, January 2005, pp. 795–798.

[24] D.A. Patterson, J.L. Hennessy, Computer Architecture: A Quantitative Approach, 3rd ed., Morgan Kaufmann Publishers, San Francisco, CA, 2003.

[25] K. Shih et al., "Fast real-time LIDAR processing on FPGAs." Available: http://www.informatik.uni-trier.de/~ley/db/conf/ersa/ersa2008.html (accessed 6/3/2011).

9 Low Overhead Radiation Hardening Techniques for Embedded Architectures

SOHAN PUROHIT, SAI RAHUL CHALAMALASETTI, and
MARTIN MARGALA

9.1 INTRODUCTION

As device sizes continue to shrink into the nano regime, the reliability of circuits, devices, as well as the architectures that use these becomes a very critical issue. Perhaps one of the most critical impacts of technology scaling is the significant reduction in the device threshold values. In other words, the amount of critical charge required for changing the state of a device or a circuit node from 1–0 or 0–1 has dropped considerably. As a result, modern high-performance logic and memory circuits become increasingly susceptible to transient errors caused due to charged particles striking sensitive circuit nodes. These phenomena associated with faulty behavior of the circuit due to particle strikes are known as soft errors. Soft errors can be classified as single event errors (SEEs) and single event upset (SEU) [1]. The former consists of a permanent defect or error induced by a particle strike and is an issue more efficiently handled at the device level. SEUs, on the other hand, involve a temporary flipping of states of the circuit and can be rectified using circuit-level as well as top-level architectural techniques.

After being initially observed by Intel in their memory designs, this problem of SEUs was thought of as being restricted to space applications. Space-based electronics are constantly required to work in environments with high particle concentration, resulting in large doses of radiation being bombarded on the chips. Considering reliability as the key cost function in space electronics, techniques suggested to handle radiation-induced errors in such circuits

Embedded Systems: Hardware, Design, and Implementation, First Edition.
Edited by Krzysztof Iniewski.
© 2013 John Wiley & Sons, Inc. Published 2013 by John Wiley & Sons, Inc.

focused solely on reliability without much attention being paid to the performance, power, and area penalties incurred in the process. While SEUs were primarily a problem in space electronics over a decade ago, recently they have started manifesting themselves increasingly in common consumer electronics. As a result, most electronic circuits targeting areas such as consumer electronics, automobiles, avionics, and so on require to be protected against these errors. Furthermore, since some of these markets are extremely cost sensitive, it is imperative for the solutions to have low overhead in performance, power, and area. Despite this need, solutions targeting SEU mitigation have still relied on the use of time or space redundancy to incorporate robustness against SEUs.

In this chapter, we discuss the design of a radiation-hardened reconfigurable array. Reconfigurable computing is a rapidly evolving field with a wide range of applications in consumer as well as space sectors for data and media processing applications. Combining the developments in silicon technology scaling and hardware–software codesign paradigm, reconfigurable embedded computers have managed to put hundreds of processing cores and large amount of memory, and thus, hundreds of millions of transistors, on to a single chip. Naturally, it can be argued that these systems face similar reliability issues as faced by other very-large-scale integration (VLSI) systems in advanced technology nodes and hence are obviously susceptible to soft errors. However, a property unique to the reconfigurable computing domain is the strong emphasis on low design cost, design reusability, minimal time to market, and stringent design budgets. As a result, traditional techniques based purely on redundancy no longer remain economical solutions for hardening these architectures. At the same time, it is desired to achieve 100% SEU protection in them. As a result, these architectures need to be hardened using techniques that minimize the performance, power, and area penalties while still providing 100% SEU mitigation capability.

Recently published works show some interesting circuit techniques for building SEU-hardened circuits. Zang et al. [2] present a novel design scheme for radiation-hardened sequential element design. Choudhary et al. [3] use transient filter insertion to incorporate SEU tolerance. At the circuit level, Nagpal et al. [4] propose to use the Code Word State Preservation logic (CWSP) to mitigate errors induced due to charged particles of a specific profile. Zhou and Mohanram [5] propose a technique for transistor sizing to bolster the capacitance charge at critical nodes and make it difficult to upset the state of the circuit due to random radiation strikes. Similarly, Zang and Shanbag [6] present a logic design style to combat soft errors by placing critical transistors in isolated wells. It can be noted that all the abovementioned techniques use device-level redundancy to tackle the issue of soft errors. While these techniques tackle SEU mitigation as a circuit design problem, they can be considered local solutions for specific charge profiles in specific types of combinational circuits. As a result, while they provide 100% SEU tolerance for the charge profiles they are designed for, they do not provide a global, all-encompassing solution to the soft error problem. Moreover, due to their localized treatment of the SEU problem, the techniques may fail under random process–voltage–temperature variations. Traditional methods to tackle this issue include the use of radiation-hardened fabrication processes or designing systems using radiation-hardened field-programmable gate arrays (FPGAs).

In this chapter, we discuss a combination of circuit-level and logic-level techniques for implementing SEU hardening in reconfigurable architectures. From an architectural standpoint, the crux of the hardening mechanism proposed here is an instruction rollback technique. In addition to this, three circuit-level techniques for protection against SEUs in combinational circuits have also been proposed. All the techniques mentioned here were implemented in the reconfigurable cell (RC) of the recently proposed multimedia-oriented reconfigurable array (MORA) architecture [7]. We also discuss the interprocessor communication during such a rollback. The chapter provides a discussion on the impact of each of the circuit-level techniques on the total throughput/area/power consumption of the MORA processor. We also discuss the impact of a rollback on throughput of the MORA array when implementing a two-dimensional discrete cosine transform (2D DCT) algorithm. To put the proposed techniques in perspective, their impact on circuit performance was also compared with recently proposed SEU mitigation schemes.

This chapter is organized as follows. Section 9.2 presents our recently proposed exploratory work on low-cost circuit and logic reorganization techniques for SEU mitigation. The implementation of the instruction rollback scheme on the reconfigurable array and interprocessor communication during rollback are presented in Section 9.3. Section 9.4 presents a discussion on the impact of the various schemes and comparison results with recently proposed techniques. Finally, some concluding remarks follow.

9.2 RECENTLY PROPOSED SEU TOLERANCE TECHNIQUES

The most traditional techniques for soft-error mitigation, however, have still relied on architectural reorganization of reliability. Perhaps the most well-known method of SEU mitigation has been triple modular redundancy (TMR) [8]. TMR uses duplicated or even triplicated logic to account for soft errors. The three cores feed into a majority circuit called voter, and thus ensure correct result despite a particle strike in one of the cores. This technique is based on the assumption that a soft error is an isolated event, and that a charged particle strike cannot occur at more than two locations at the same instant. Although effective, this technique implies 100–200% overhead in area and power of the circuit being hardened, and hence is an expensive albeit reliable method of mitigating soft errors. Another popular technique targeted specially for memories involves the use of error-correcting codes (ECCs) to check and correct transient faults due to particle strikes [9]. More sophisticated techniques rely on checkpointing at regular intervals and use watchdog processors to rollback to the correct checkpoint in the event of a particle strike. Despite their reliability, traditional architectural solutions have always suffered from a large overhead in area, power, and throughput of the system.

To tackle some of these issues, we proposed several circuit-level techniques that aim to provide 100% protection against soft errors while at the same time incurring minimum penalties in area, performance, and power. These techniques have been described in the next few subsections.

9.2.1   Radiation Hardened Latch Design

The proposed design is based on the popular true single-phase clock (TSPC) and clocked complementary metal–oxide–semiconductor (C²MOS) flip-flop topologies. The design is 100% SEU tolerant and performs better than the original TSPC and C²MOS designs in terms of area, speed, and delay. The flip-flop shown in Figure 9.1 was built in IBM's 65 nm CMOS process. As shown in the figure, the input stage resembles that of a conventional simplified TSPC register. In the event of a particle strike, however, there is a possibility of one of the transistors incorrectly switching stages. To minimize the possibility of data being corrupted due to a particle strike, the flip-flop uses two identical copies of this stage. The outputs of this state are sent as inputs to the final clocked stage. This stage resembles the output stage of the C²MOS register. All the flip-flops discussed here, that is, TSPC, C²MOS, and the proposed

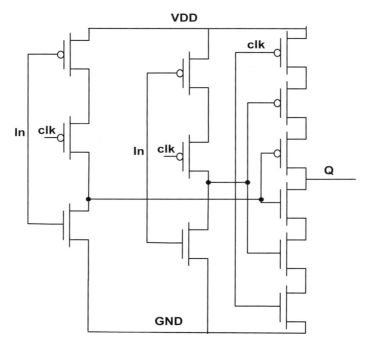

FIGURE 9.1. Radiation-hardened latch design.

TABLE 9.1. Comparison Results of Proposed Latch with Published Works

Flip-Flop	Number of Transistors	Flip-Flop Timing Metrics (ps)			Power (μW)	Area (μm^2)
		Set-Up	Hold	Clock (Clk)-Q		
Proposed	12	61	11	46.750	1.972	6.5621
TSPC	26	112	12	138.43	6.679	17.505
C²MOS	22	33	38	90.366	4.286	17.5132
Knudsen and Clark [18]	46	–	–	–	–	89.1
Haghi and Draper [19]	26	–	–	139 ps	–	59.07

–, data not reported.

design, are basically chains of inverters controlled by the clock signal. In the proposed design the output stage of the C²MOS flip-flop was modified to form a three-input, clocked control element. As a result, from the circuit operation point of view, in the absence of a soft error, the flip-flop works identical to the TSPC and C²MOS designs. In the event of a particle strike, though, the functionality of the C-element ensures that the incoming particle strike does not change the data being latched into the flip-flop and keeps the output unchanged. Simulation results presented in Table 9.1 show the proposed design to outperform several other popular flip-flops. A comparative study performed with prominent radiation-hardened flip-flops and the redundancy-based approach was found to outperform contemporary radiation-hardened flip-flop designs by a significant margin.

9.2.2 Radiation-Hardened Circuit Design Using Differential Cascode Voltage Swing Logic

Differential cascode voltage swing logic (DCVSL) [10] was originally proposed as a ratioed logic style to completely eliminate static currents and provide rail-to-rail output swing. The DCVSL topology requires that each input be available in both its original as well as complemented forms. The circuit consists of two cross-coupled p-type metal–oxide–semiconductor (PMOS) transistors and two corresponding pull-down networks (PDNs). Both the PDNs receive complementary input signals, thereby simultaneously producing the output and its complement. The two pull-up devices are connected in a feedback-style fashion and thus combine differential logic and positive feedback. This technique also allows sharing of transistors between the two PDNs, thereby reducing the area overhead in duplicating the PDNs. The SEU-tolerant design style presented in this section is based on the DCVSL design methodology, but modified to incorporate SEU inhibiting properties. Figure 9.2 shows a general-circuit design scheme using the proposed technique.

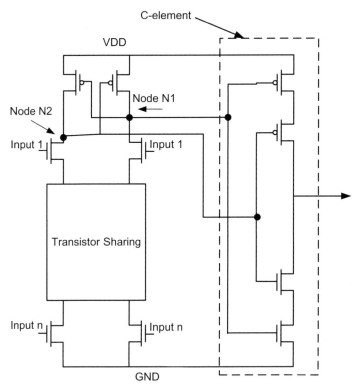

FIGURE 9.2. DCVSL-based SEU-tolerant circuit design style [11].

The proposed scheme [11] looks very similar to the original DCVSL topology. A major difference, however, is that this technique relies on the use of duplicating the PDNs as opposed to conventional DCVSL, which uses complementary PDNs. As a result, both the output nodes provide the same logic function OUT_BAR. As in the case of conventional static CMOS circuits, this OUT_BAR needs to be inverted in order to produce the final output. The final-stage inverter is replaced by a different inverter topology. The final stage shown in the figure is known as the C-element. This circuit works as an inverter only when both the output nodes are at identical logic levels. If the two inputs to the inverter are different or complementary, the circuit prevents the output from changing state. Figure 9.3 shows a two-input AND gate implementation using the proposed style. It should be noted that although intuitively for the basic AND, OR, and exclusive OR (XOR) gates this topology seems to just duplicate the PDNs, the technique allows for using shared transistors while implementing larger logic functions. Similar to DCVSL, this approach allows the designer to limit the overhead while implementing large logic functions. Another important advantage offered by this design style comes through the

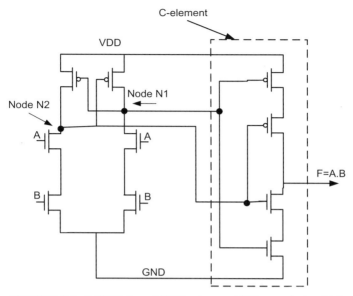

FIGURE 9.3. DCVSL-based SEU-tolerant AND gate design [11].

reduction of a full PMOS pull-up network. The two PMOS transistors shown in the figure act as keeper transistors for each of the individual PDNs. The elimination of a dual PMOS pull-up network provides significant advantages in terms of area and overall transistor density of the circuit.

The proposed design requires the designer to identify the devices connected to the ground rails and the output nodes of the circuit. The basic idea is to duplicate these devices instead of placing them in separate wells. As a result, the technique requires the realization of the required logic function through duplicating the transistors connecting to ground and the output node. The C-element used at the output doubles as an inverter as well as a soft-error mitigator. The inputs to the C-element are the signals from the two output nodes N1 and N2. The circuit works as an inverter only when both the inputs are identical. Considering the nature of particle-induced errors, we can assume that the error will almost always occur at only one of the pull-down paths. As a result, the values of N1 and N2 will always be different in the event of an SEU. In this scenario, the C-element maintains its previous output level, thereby preventing the error from propagating to the following stages. The two cross-coupled PMOS devices, apart from acting as loads for the individual PDNs, also double as keepers to restore the charge at the output node and ensure correct circuit operation.

To measure the effectiveness of the proposed technique, several arithmetic circuits were implemented in the IBM 90 nm process. These included the design of the basic logic gates, an 8-bit Manchester Carry Adder, and 8-bit multiplier. Table 9.2 shows the corner simulation results for the designed

TABLE 9.2. Corner Simulation Results for Circuits Designed using the Proposed SEU-Tolerant Design Scheme

Corner	Fast Slow		Slow Fast		Fast Fast		Slow Slow	
Circuit	Delay	Power	Delay	Power	Delay	Power	Delay	Power
AND	36 ps	3.59 μW	41 ps	3.57 μW	36 ps	3.74 μW	46 ps	3.57 μW
XOR	72 ps	20.01 μW	75 ps	19.7 μW	67 ps	20.58 μW	75 ps	19.73 μW
MCC	157 ps	0.273 mW	168 ps	0.28 mW	149 ps	0.32 mW	187 ps	0.26 mW
Multiplier	398 ps	0.35 mW	431 ps	0.33 mW	387 ps	0.371 mW	453 ps	0.33 mW

MCC, Manchester Carry Chain.

circuits. The next step in the evaluation involved testing for soft errors. The circuits were tested for SEU tolerance by injecting current pulses of varying magnitude and pulse widths ranging from a few picoseconds up to 500 ps (comparable to common clock pulses used in modern digital circuits). The circuits provide 100% SEU tolerance over all the pulse widths for currents up to1.2 mA, which validates the proposed design style. An interesting point to note is that most circuit-level hardening techniques and some at the architectural level work effectively only when custom-designed for specific current profiles. Nagpal et al. [4], for example, present two sets of simulation results for two different charge values and rise and decay times of the current pulse (error pulse). As a result, even though 100% SEU tolerance is claimed, it is valid over a limited range of specific expected current profiles. Similarly, the transient filter insertion proposed by Choudhary et al. [3] also uses an estimated error current profile as a guideline to dictate the design of the filter gates. Such circuits, therefore, have limited portability and need to be redesigned while migrating from one environment to another. An advantage with our proposed approach is that the circuit design process remains relatively similar to standard CMOS logic design while still maintaining SEU robustness over a high range of possible charges. Our simulation results showed that the proposed circuits begin to show errors when the injected current reaches approximately 1.3 mA for a pulse width of 400 ps or greater. Due to the relatively low delay overhead introduced by the technique and especially the significant area savings involved, it is possible to introduce a second level of hardening at the architectural level, and further increase the system robustness at a relatively lower cost.

9.2.3 SEU Detection and Correction Using Decoupled Ground Bus

Soft errors or SEUs occur due to a charged particle striking the source or drain node of a transistor. A charged-particle strike injects additional external energy into a device. If this energy exceeds the critical charge of the device, the device incorrectly switches state. From a circuit point of view, as shown in Figure 9.4, a soft error can be modeled as a current source connected in

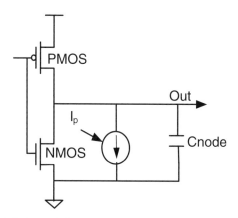

FIGURE 9.4. Modeling the particle strike on a transistor [12].

parallel across an n-type metal–oxide–semiconductor (NMOS) or PMOS transistor [12]. As shown in the figure, a particle strike at the drain node of the NMOS causes a momentary short-circuit path between VDD and ground (GND) and can be observed as a distinct spike in the current waveform of the circuit. The current in case of a particle strike is higher is magnitude as the particle strike generates more charge carriers due to the ionization process, thereby generating a higher current. Also, as digital circuits switch at very high speeds, normal short-circuit current pulses are of extremely low magnitude. Hence it is possible to detect an error signal by monitoring these current pulses in the circuit current waveform. In the proposed technique, a soft-error signal gets detected by monitoring the voltage on the ground bus of the circuit. For this purpose, a single NMOS transistor is connected between the ground terminal of the circuit and the global ground bus, as shown in Figure 9.5, thereby creating a virtual ground. The transient activity in the circuit is now reflected onto this decoupled ground bus, which in turn can be monitored to detect soft errors. Purohit et al. [13] use this concept to combine decoupled ground lines of several circuit blocks together to simultaneously detect soft errors in the circuit by monitoring the decoupled ground line through a voltage comparator. However, connecting a large number of circuits together to a common decoupled ground line adds a lot of noise to the observed voltage signature. Hence, careful partitioning of the circuit is essential to extract a pulse during a particle strike. This requires each circuit partition to be monitored through a separate comparator. Using this error pulse as a logic input, the logical function of the circuit being realized can be modified to account for an error pulse. However, before incorporating this error signal as a design variable, it is important to understand the impact of adding such a device on the performance, power, and area characteristics of the circuit being hardened. We simulated several standard logic gates, adders, and multipliers to evaluate the impact of adding an extra pull-down device (PDD) in the circuit. The results are presented in

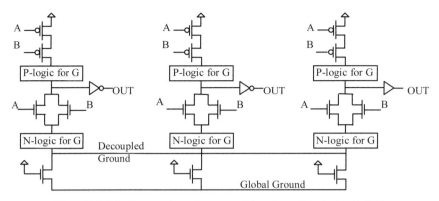

FIGURE 9.5. Decoupling the ground bus in a digital circuit [13].

Table 9.3. From the table, it is clear that while for smaller two-input AND, OR, and XOR, one ends up paying higher penalties in circuit performance and power, these penalties fade away as the gates are cascaded or used together to implement bigger structures. It should also be noted that in each circuit, the extra PDD has no significant impact on the area of the circuit. Table 9.4 shows an evaluation of the minimum current value, decay time required by popular design styles to produce an erroneous result, and the corresponding voltage fluctuation obtained on the decoupled virtual ground. The data from this preliminary analysis suggest that it is possible to reliably use the decoupled ground signal as a variable in the logic design step and realize a robust self-repairing digital circuit with SEU mitigation capabilities.

In our proposed design scheme [14], we use an approach similar to the design of finite state machines while realizing Boolean functions of circuits that need to be hardened. In our design scheme, we use two main inputs to control incorrect transitions due to soft errors. The first is the error signal obtained on the local decoupled ground denoted as E, while the second is the previous output state of the circuit represented as F_P. The main concept utilized here is that the circuit checks on the status of the ground bus. If there is an error, the circuit is designed to retain the value of its previous output until the transient error has passed, that is, the circuit remains in its previous state as long as the error signal is a logic 1. Using this approach, every logic circuit to be realized can now be represented as a state machine. Figure 9.6 shows the design of the two-input OR gate represented using the proposed design scheme. It can be observed that the SEU tolerance has been built into the logic function being realized. After reducing the truth table using Karnaugh maps, the OR function can be represented in our approach as:

$$OR_{\text{hardened}} = \bar{E} \cdot (A + B) + P \cdot E.$$

Repeating the similar steps for standard two-input AND, XOR, as well as full adder circuits yields the following expressions:

TABLE 9.3. Performance Penalty due to Insertion of Decoupling Transistor

Parameter Circuit	Delay			Power			Area		
	Original (ps)	Extra PDD (ps)	Penalty (%)	Original (μW)	Extra PDD (μW)	Penalty (%)	Original (μm²)	Extra PDD (μm²)	Penalty (%)
AND_2input	35	40	14.28	1.24	1.488	20	5.714	7.82	36.38
OR_2input	38	44	15.78	1.72	2.502	45	5.35	6.142	14.80
XOR_2input	68	74	8.82	4.09	5.551	35.72	16.66	22.81	36.91
Full adder	112.1	125	10.70	12.08	14.107	16.77	84.46	91.295	8.09
8-bit MCC	186	205	10.21	172	188	9.3	654.44	716.12	9.42
8-bit multiplier	309	328	6.14	697	886	27.11	2197.35	2410.5	9.70

TABLE 9.4. Minimum Voltage and Pulse Widths Required to Cause Node Upsets

Pulse Width (ps)	Peak Value I_o			
	0.1 mA	0.2 mA	0.23 mA	0.3 mA
100	XX	XX	XX	$\sqrt{}$, 0.6 V
200	XX	XX	XX	$\sqrt{}$, 0.5 V
300	XX	XX	$\sqrt{}$, 0.55 V	$\sqrt{}$, 0.8 V
400	XX	XX	$\sqrt{}$, 0.7 V	$\sqrt{}$, 0.8 V
500	XX	XX	$\sqrt{}$, 0.7 V	$\sqrt{}$, 0.8 V

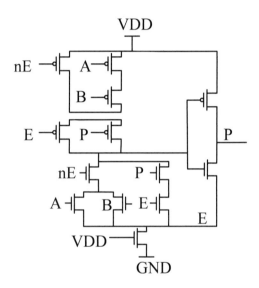

FIGURE 9.6. SEU-tolerant OR gate using self-correcting design methodology [14].

$$AND_{hardened} = \bar{E} \cdot (A \cdot B) + P \cdot E,$$

$$XOR_{hardened} = \bar{E} \cdot (A\bar{B} + \bar{A}B) + P \cdot E,$$

$$FullAdder\_Sum_{hardened} = \bar{E} \cdot (A \oplus B \oplus C) + P \cdot E,$$

$$FullAdder\_Carry_{hardened} = \bar{E} \cdot (AB + BC + AC) + P \cdot E.$$

Based on these expressions, the generalized expression for any logic function F therefore becomes,

$$F_{hardened} = \bar{E} \cdot (F) + P \cdot E.$$

Thus, from the above expressions, it is clear that the proposed technique uses the error signal from the decoupled ground line as a control signal to allow only correct transitions by the circuit being hardened. In the event of an error, the circuit retains its state until an indication through the ground line that the erroneous charge injection phenomenon has passed. This technique not only allows the circuit from making incorrect transitions but also prevents any additional dynamic power dissipation in the circuit, due to the particle strike.

9.3   RADIATION-HARDENED RECONFIGURABLE ARRAY WITH INSTRUCTION ROLLBACK

As seen from the discussion above, several techniques exist at the architectural and circuit levels to protect systems against soft errors. However, it is also clear that an optimum solution would exist only by combining circuit and architectural techniques to combat soft errors. To this end, we combined the SEU detection mechanism based on a decoupled ground bus along with register reorganization to implement the recently proposed MORA architecture with 100% SEU tolerance and minimum penalties in area, performance, and throughput. This section presents the implementation details of the MORA array using a combination of circuit and logic reorganization schemes.

9.3.1   Overview of the MORA Architecture

The top-level architecture of the MORA–coarse-grained reconfigurable architecture (CGRA) [7] is depicted in Figure 9.7. It can be seen that the workhorse of the proposed structure is a scalable 2D array of identical RCs organized in 4×4 quadrants and connected through a hierarchical reconfigurable interconnection network. The array uses identical RCs to support both binary arithmetic and control logic, thus reducing control complexity and enhancing data parallelism. The MORA reconfigurable array does not use a centralized random access memory (RAM) system. Instead of employing a centralized shared memory, data storage is partitioned among the RCs by providing each of them with their own local, internal data memory. In this way, by adopting the processing-in-memory (PIM) approach [15], a high memory access bandwidth is supplied to efficiently exploit the inherent massive parallelism of multimedia and digital signal processing tasks. This solution significantly reduces the data transfers through the interconnection network, with evident benefits in terms of energy consumption. These are mainly due to the lower energy consumption assured by the locality of data access (especially when repeated accesses to the same data are required), and the greater energy efficiency of block-based, as opposed to word-at-a-time, data transfers. Similarly, every RC is also provided with an independent but identical control unit, with independent instruction memory. This allows each RC to function as a tiny processor, working independently from the remaining processors in the array. Thus, once an application has been mapped onto the array, each RC receives its local data and local instruction set and performs its operations individually,

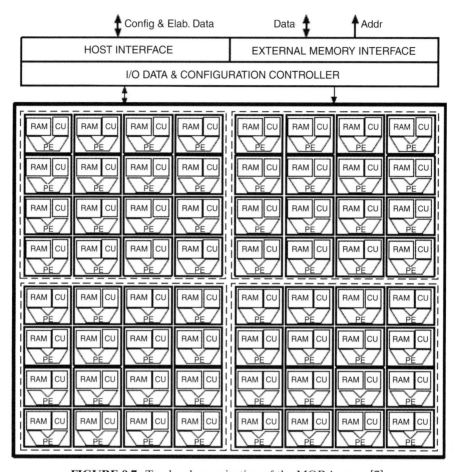

FIGURE 9.7. Top-level organization of the MORA array [7].

unless limited by serialization in the application algorithm, when the inherent data dependency is reached.

The interconnection network is organized into two levels. The level 1 interconnections assure a single clock-cycle nearest neighbor data exchange within each 4×4 quadrant. In contrast, latches included into the programmable switches pipeline the data transfer through the level 2 interconnections. In this way, all mapped algorithms can always operate at the maximum supported clock rate. It is worth noting that the proposed organization can efficiently support vector operations. In fact, the structure of the RCs and the pipelined level 2 interconnections allow computational tasks to be mapped with a programmable degree of pipelining. This means that complex computations can be easily broken into several steps, each performed within one maximum frequency clock cycle. The configuration of the entire array is driven by a centralized controller that also manages the external data exchange by directly accessing the memory space of the RCs through standard read and write operations. On the contrary, the internal data flow is controlled in a distributed manner through a handshaking mechanism which drives the interactions between the RCs. It should also be noted that, thanks to the general input/output (I/O) system interface, the proposed CGRA can be easily integrated with external and/or embedded microprocessors and systems. The designed I/O interface system includes a *Host Interface* and an *External Memory Interface*. The former is used to specify the device configuration and for I/O data transfer, whereas the latter is provided to supplement the on-chip memory (when needed). The RC shown in Figure 9.8 is the fundamental building block of the array, and is designed to execute all of the required arithmetic operations for data flow-oriented digital signal processing and multimedia applications. It also provides the user with all the necessary control and storage resources. The generic RC exploits the 8-bit granularity. The 8-bit granularity was chosen to maximize the achievable flexibility and to minimize the area requirement. Moreover, the chosen granularity perfectly matches the operands word lengths mostly involved in the target tasks. Higher word lengths, such as 16, 32, and 64 bits, can be easily supported by connecting more RCs together in single instruction, multiple data (SIMD) fashion. Furthermore, the arithmetic and logic unit inside the RC is based on an extremely general algorithm which can be easily extended to handle higher width operands [16]. Owing to this extendibility, and the fact that the control mechanism is independent of the width of operands supported by the computational logic, the RC and, in turn, the entire MORA array can be easily extended for supporting

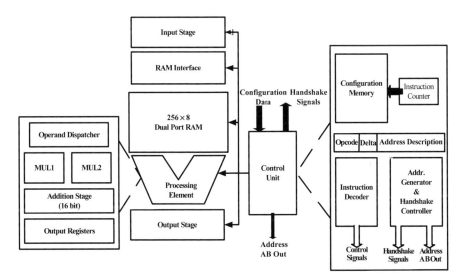

FIGURE 9.8. Organization of the MORA RC [7]. MUL, multiplication.

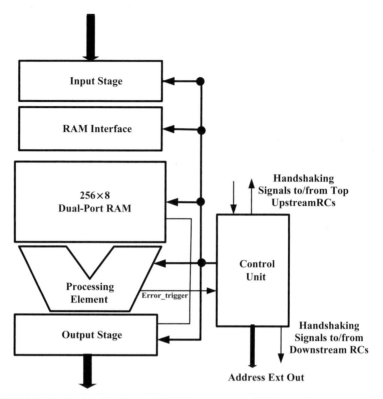

FIGURE 9.9. Radiation-hardened MORA processor with instruction rollback scheme [17].

application algorithms which require higher granularity. It is worth noting that this extendibility comes without any significant changes to the programming model, which allows the hardware as well as software design flows to be updated for future generation applications, with a significantly low turnaround time.

As depicted in Figure 9.9, each RC consists of three main building elements:

- 8-bit processing element (PE) for arithmetic and logic operations
- 256 × 8-bit dual-port static random access memory (SRAM) acting as an internal data buffer, and,
- control unit with a 16-word SRAM-based instruction memory for RC-wide synchronization and handshaking control during inter-processor communication.

Each RC has two possible operating states: *loading* and *executing*. When the RC is in the *loading* state, packets of data can be input through the input ports and stored into the internal dual-port SRAM. As soon as all the required operands are available, the RC evolves into the *executing* state. When the generic RC is in the *executing* state, it can operate in four different modes: a feed-forward, a feedback, a route-through, and a dual route-through. In the feed-forward mode, the PE processes packets of data coming from the internal memory and dispatches the results to one or more RCs using one or both output data ports. In the feedback mode, processing results are internally stored to be used by the same cell during future instruction executions. In the abovementioned operation modes, the PE can be dynamically configured to execute different fixed-point arithmetic operations on internal memory data or on constant data provided by the control unit, each requiring only a single clock cycle for execution. The possibility of dynamically reconfiguring each PE within the same application running allows a very high hardware utilization, which means reducing connectivity requirements. This is a very important feature especially for reconfigurable hardware, which typically dedicates the vast majority of the area occupied for the interconnection structures. It is worth noting that each *RC* can be also used as a route-through cell. This operation mode is particularly useful to achieve effective physical applications mapping.

The MORA RC was designed as a low-cost, resource-efficient architecture for media processing applications. The RC followed a hardware–software codesign approach to achieve the most efficient utilization of array resources to ensure the most cost-effective realization for media processing algorithms. Similarly, all our techniques for radiation hardening also aim to achieve complete SEU tolerance with minimum penalties in area, power, and performance. Thus, both the target architecture as well as the proposed SEU tolerance schemes follow a common underlying theme of achieving the most cost-efficient solution. The following subsection discusses the implementation of the SEU tolerance schemes on the MORA processor.

9.3.2   Single-Cycle Instruction Rollback

Figure 9.10 presents an overview of register organization in the proposed rollback scheme. As shown in the figure, the concept involves the use of duplicate registers in the pipeline stage to store copies of the data being processed. The scheme can be visualized as two sets of registers working with the same set of data but out of phase by one clock period. In the event of a soft error being detected at any time during program execution, this register organization allows the system enough time to roll back safely through a multiplexer and repeat the instruction, this time using data through the duplicated pipeline. In modern high-performance architectures which employ deep pipelines, this scheme can be inserted at extremely low hardware cost, allowing every stage of the pipelined system to issue a rollback, inform the neighboring stages of the rollback, and thus ensure that only the correct result is written back into the memory. This scheme treats SEUs as a problem of logic design rather than

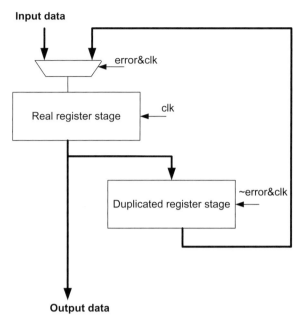

Input data

error&clk

Real register stage

clk

Duplicated register stage

~error&clk

Output data

FIGURE 9.10. Register organization in proposed rollback scheme [17].

a purely electronic one. As a result, the scheme is easily compatible with several circuit-level techniques for soft error detection. In this scheme, we achieved 100% SEU detection coverage using a hierarchical voltage monitoring scheme for error detection [13]. As reported in Purohit et al. [13], we were able to show 100% SEU tolerance at the cost of 10% delay, 9.7% area, and 21% extra power. The proposed technique can also be integrated with built-in current sensors (BICS)-based SEU detection. Table 9.5 shows the impact of applying the proposed scheme on various components of the MORA array.

Several techniques in the past have proposed radiation hardening through the use of duplicate registers. These involve techniques like CWSP, which employ a duplicate register in conjunction with a preprogrammed delay element. The hardening component in this technique lies outside the critical evaluation path and hence contributes only 0.5% extra delay. However, this technique relies on the use of delay elements programmed according to the precise charge profile, and hence the scheme does not provide a global solution. A rollback approach proposed in Shazli and Tahoori [9] employs ECCs and achieves rollback without using duplicate registers. However, the ECC results in multiple and variable number of clock cycles being needed to implement the error detection process, after which the entire processor is stalled and the instructions rolled back. In our proposed technique, we merely duplicate the registers and initiate a rollback immediately in the next clock cycle as soon as an error has been detected. This not only avoids excessive clock

TABLE 9.5. Performance Variation of Different Circuit Components using Hierarchical Voltage Monitoring and Decoupled Ground Bus

Parameter Circuit	Delay			Power			Area		
	Original (ps)	Extra PDD (ps)	Penalty (%)	Original (μW)	Extra PDD (μW)	Penalty (%)	Original (μm²)	Extra PDD (μm²)	Penalty (%)
AND	35	40	14.28	1.24	1.488	20	5.714	7.82	36.38
OR	38	44	15.78	1.72	2.502	45	5.35	6.142	14.80
XOR	68	74	8.82	4.09	5.551	35.72	16.66	22.81	36.91
Full adder	112.1	125	10.70	12.08	14.107	16.77	84.46	91.295	8.09
8-bit MCC	186	205	10.21	0.172	0.188	9.3	654.44	716	9.42
Multiplier	309	328	6.14	0.697	0.886	27.11	2197.35	2410.5	9.70

cycle overhead, but also reduces the penalties incurred in the event of false positives. Moreover, all the circuits in a particular stage of the pipeline, including the pipeline and duplicate registers, can be simultaneously monitored, allowing the system to easily detect and recover from multibit errors.

9.3.3 MORA RC with Rollback Mechanism

Figure 9.2 shows the top-level organization of the hardened MORA RC [17]. MORA consists of a 2D array of RCs arranged in an 8 × 8 configuration. Each RC is a small 8-bit processor with an 8-bit arithmetic and logic unit, finite-state machine (FSM)-based control unit, a small instruction memory, and a 256 × 8 dual-port data memory. In Purohit et al. [13], we described the hardening of the pipelined arithmetic unit using the proposed register organization in conjunction with hierarchical voltage monitoring system. For hardening each individual processor, however, it is necessary to harden each individual component as well as to control their interaction in the event of an error being detected in one of the components.

The control unit of the MORA RC consists of an instruction memory, address generators, and a state machine to serve as the brain of the processor. Depending on the status of program execution, the state machine issues signals to the instruction memory and the address generators for synchronizing instruction flow and data processing. The architecture has been described in the previous section. In order to implement radiation hardening, the state machine and, subsequently, the control unit have to be able to process information from error detectors all over the RC and decide between continued program execution or instruction rollback. In the event of a soft error, when rollback is initiated, the state machine in the standard MORA RC counts the number of times each instruction is executed, the addresses of the operands, the destination address for the results, instruction decoding, and external handshaking signals with other RCs in the array. For each instruction, this information is stored as the current state of the control unit FSM. To have a hardened system, we extended each RC to have a radiation-hardened control unit capable of monitoring itself, the RC it belongs to, as well as collaborating RCs for soft errors during program execution. In other words, to realize a hardened RC, we implemented a hardened control unit with soft-error-aware state transition control.

A standard state machine design consists of a register stage and combinational stage. Figure 9.11 shows the state machine hardened by duplicating the current state and output registers. As described earlier, the proposed scheme uses multiplexers and duplicate registers controlled by an error-aware enable signal. The state machine in the hardened RC uses the same concept by monitoring the RC for errors and using the information to control state transition and hence program execution. This means that in the event of an error being detected, the control unit issues a rollback throughout the RC and delays state transition until the entire sequence of instructions associated with a particular

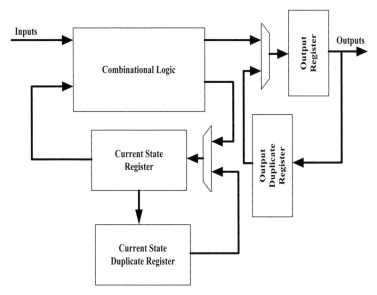

FIGURE 9.11. SEU-tolerant FSM with duplicated registers [17].

state has been correctly executed, written into memory or transferred to a collaborating RC.

It should be noted that the error detection scheme targeted here is the one proposed in Purohit et al. [13] using hierarchical voltage monitoring. All the circuits in the RC, including the PE, control unit, and memory are implemented using the same technique. As a result, the control unit is also able to monitor itself for radiation-induced errors, resulting in 100% reliable rollback control.

MORA is designed to work as a stream processing architecture. This means that from the point of view of communication between the processor cores, we can consider data transmission between processors working similar to a large pipelined system. The only difference is that data transfers occur at precisely controlled time instances instead of every clock cycle. Neighboring RCs in the MORA architecture operate in alternate Execute and Load (Idle) modes. During the execution mode, the RC processes data from its data memory based on the controls initiated by its control unit. During loading state, the RC remains idle. This allows the neighboring RC in execution mode to use the memory of the idle RC as a scratchpad memory, or to provide the idle RC with data required by it when it switches back to execution mode. The RCs achieve this through two sets of handshaking signals:

- *data_rdy*: This signal is used to signify data have been written into the downstream RCs memory bank, allowing it to start processing the new data (completed execution mode).

- *rc_rdy*: This signal is used to inform the upstream RC that downstream RC is ready to accept data (ready in loading mode).

This stream-based operation of the MORA processors allows the architecture to work in a globally asynchronous locally synchronous (GALS) fashion. This means that every RC can proceed independently with its individual program execution and only communicate with its neighbors when critical data have to be exchanged. From a soft-error point of view, it allows all the neighboring RCs to continue with their program execution even when one of them has detected a particle strike and initiated a rollback. The only bottleneck appears when a soft error is detected and rollback initiated at the exact instance of data transfer between the RCs. Intuitively one would expect incorrect data to be written into the idle RC. However, in reality, in the event of a soft error, the control unit of the transmitting RC only sends out a *data_rdy* signal when the whole data are written into downstream RC. This ensures that only correct, coherent data are sent out to the receiving RC while at the same time avoiding an extra memory access to rewrite the correct data. What makes this idea particularly interesting is that the proposed scheme achieves this at the expense of just a single additional idle cycle at the receiving RC. We can thus say that the stream-based data transfer between the RCs provides inherent compatibility with the proposed instruction rollback mechanism.

9.3.4 Impact of the Rollback Scheme on Throughput of the Architecture

Whenever we include extra circuitry and reorganize the architecture to accommodate SEU tolerance, there are bound to be penalties in terms of area, power, and performance. It is also important to analyze the scheme from the point of view of penalties in throughput. Each RC in the MORA array executes two types of instructions: processing based (arithmetic, logic, and shifting) and memory based (moving across memory, data movement within memory). Additionally, each instruction execution can be decomposed into three main parts: memory read, execute, and memory write. Consequently, any arithmetic, logic, or shifting instruction requires access to the PE and hence will need all the three phases of execution. In case of purely memory-based instructions, however, we only need the memory read and write operations. Being random errors, there is the possibility of an error occurring at either of these stages of instruction execution. It can be argued that irrespective of the stage of instruction execution where the error occurs, there is always an overhead of only one clock cycle to issue the rollback and recover from the error. Thus we can conclude that a soft error during a processing-type instruction translates into four clock cycles for complete correct execution while in case of memory-only operations, it would take three cycles.

With the overhead latency determined for the individual instructions, the total latency L for a program execution on the MORA array (accounting for instruction rollback penalties) can be modeled as follows:

$$L = \Sigma_i (T_L^i + O_R^i) \times N^i,$$

where:

- T_L^i is the latency of each individual instruction, where *i* is the index of the instruction,
- O_R^i is the rollback overhead of the individual instructions, and
- N^i is the number of times an error occurred for the same instruction in the system.

We evaluated the overhead in latency in the event of a varying number of detected errors. We programmed our architecture to perform 8×8 2D discrete wavelet transform (DWT) and 2D DCT. Figure 9.12 shows the plot of latency overhead with an increase in the number of soft errors within each sample of DWT and DCT. The MORA array executes one 8×8 block of DWT and DCT in 99 and 292 clock cycles, respectively. The worst-case latency overhead observed were around 60–80% for DWT and 17–25% for DCT depending upon the error location: processing logic or memory. This overhead is in case of multiple errors occurring consecutively, which by itself seems to be a rare occurrence given the random and isolated nature of soft errors. It should also be noted that due to the simultaneous detection of even multiple errors in each stage, the latency overhead depends on the instances of error occurrence and not on the number of bits in error in a single cycle.

It is also important to understand how the proposed hardening technique fares when compared with more popular hardening techniques such as TMR.

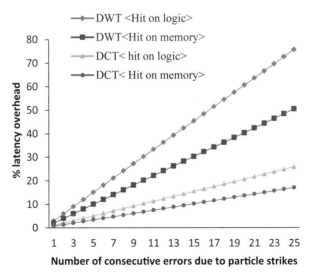

FIGURE 9.12. Latency overhead for different number of consecutive errors when executing 8×8 2D DCT and 2D DWT [17].

Assuming TMR operates with identical latency irrespective of whether an error occurs or not, we can observe that for execution of a single sample of DWT, irrespective of the error rate, the TMR-configured MORA array achieves one sample per 99 clock cycles, which is identical to the normal MORA latency. However, configuring MORA as a TMR system means using up to three times the number of RCs to execute the same operation. This means that while a DWT requires 10 RCs on a non-TMR MORA array, each sample will require 30 RCs if configured as TMR. Our proposed scheme, which uses only 10 RCs, similar to the ordinary MORA array, is thus able to process six DWT samples in parallel compared with only two lanes in parallel achievable using TMR. Thus, despite the overhead in latency imposed by our technique, we can demonstrate significant throughput advantages, as well as better area utilization than a TMR implementation.

9.3.5 Comparison of Proposed Schemes with Competing SEU Hardening Schemes

The previous sections presented circuit- and system-level techniques proposed to protect reconfigurable embedded architectures from soft errors caused due to charged particles striking critical nodes in the circuits. All the schemes discussed above were implemented by us to evaluate their resilience to soft errors as well as to analyze the impact of these techniques on the area, power, and performance of the circuits. As a result, all the circuit-level schemes were incorporated into the MORA reconfigurable array and simulated to evaluate their feasibility. Table 9.6 presents a summary and comparative analysis of the area, power, and performance overheads incurred when each of the proposed schemes were incorporated into the MORA processor. To put things in perspective, we have also included comparison results with other circuit and architectural hardening schemes recently proposed in literature. It can be observed from the table that all the schemes presented here are charge independent and offer 100% SEU mitigation capability. Moreover, since the circuit-level techniques and architectural reorganization are independent of each other, the instruction rollback mechanism can be implemented at the logic level, in conjunction with one of our proposed circuit-level techniques, to provide additional protection against SEUs. A unique characteristic of the schemes presented here is their independence to the charge profiles of the striking particles. As a result, while several previously published schemes offer charge-dependent protection, the techniques described in this chapter provide generalized solutions for SEU mitigation.

9.4 CONCLUSION

This chapter highlighted several circuit as well as architectural techniques for designing reconfigurable architectures with built-in radiation tolerance.

TABLE 9.6. Comparative Summary of SEU Mitigation Techniques

Technique	Overhead (%)			Nature of Technique
	Area	Delay	Power	
DCVSL based	−26	7	28	Charge independent
Logic reordering	4.73	29.43	13.8	Charge independent
Decoupled ground	9.7	10	21	Charge independent
Nagpal et al. [4]	42.33	0.54	Not reported	Charge dependent
Zhou and Mohanram [5]	38.3	3.8	27.1	Charge dependent
Anghel et al. [20]	17.60	28.65	Not reported	Charge dependent
TMR [8]	100	23	Not reported	Charge independent
Bulk BICS [21]	29	Not reported	100	Charge independent

Research in this area has revealed that circuit as well as architectural techniques have some trade-offs associated with them in terms of additional area, power, and performance overheads. Hence, designing an SEU-tolerant digital system involves achieving the optimum balance between reliability and the associated overheads. The techniques discussed in this chapter are intended to give readers insights into designing radiation-tolerant reconfigurable architectures with relatively low overheads than traditional techniques. These techniques are particularly applicable and feasible to the area of reconfigurable embedded computing by providing reasonable levels of security against soft errors at very low design costs and with a high degree of reusability. All the techniques discussed in this chapter focus on redesigning the circuits and logic instead of implementing brute-force redundancy through methods such as TMR. More importantly, it is the mix of circuit and architectural techniques which help to achieve reliable low-cost SEU protection.

An important point to be noted is that the circuit-level techniques discussed in this chapter are also fully compatible with traditional approaches such as TMR, ECC, watchdog processors, check-pointing algorithms, and so on. Especially considering the fact that several reconfigurable embedded systems consist of multiple cores, many of which remain idle, it is possible to use appropriate hardware–software codesign approaches to configure (or reconfigure) systems to incorporate architectural reliability measures. For instance, unused cores may be used to redundantly perform a task mapped on other sets of cores (TMR), or may be used as watchdog processors to monitor for error occurrences, and so on. Finally, it should be noted that while reliability is still the critical issue, the increased demand for reconfigurable and embedded processors has also placed equal importance on the cost of these reliability features, and in the future, only a combination of circuit-level and architecture-level schemes supported by a suitable software framework will help achieve multicore reconfigurable embedded processors with low cost and high reliability.

REFERENCES

[1] R. Baumann, "Radiation-induced soft errors in advanced semiconductor technologies," IEEE Transactions on Device and Materials Reliability, 5(3), 2005, pp. 305–316.

[2] M. Zang et al., "Sequential element design with built in soft error resilience," IEEE Transactions on Very Large Scale Integration Systems, 14(12), 2006, pp. 1368–1378.

[3] M. Choudhary et al., "Soft error reduction using circuit optimization and transient filter insertion," Journal of Electronic Testing: Theory and Applications, 25(2–3), 2009, pp. 197–207.

[4] C. Nagpal et al., "A delay efficient radiation-hard digital design approach using CWSP elements," in Proceedings of IEEE Design Automation and Test in Europe (DATE), 2008, pp. 354–359.

[5] Q. Zhou, K. Mohanram, "Gate sizing to radiation harden combinational logic," in Proceedings of Computer Aided Design of Integrated Circuits and Systems, 2006, pp. 155–166.

[6] M. Zang, N. Shanbag, "A CMOS design style for logic circuit hardening," in Proceedings of IEEE 43rd Annual Reliability Physics Symposium, 2005, pp. 223–229.

[7] S. Chalamalasetti et al., "MORA: architecture and programming model for a resource efficient reconfigurable processor," in Proceedings of 4th NASA/ESA Conference on Adaptive Hardware Systems, 2009, pp. 389–396.

[8] R. Oliviera et al., "A TMR scheme for SEU mitigation in scan flipflops," in Proceedings of 8th International Symposium on Quality Electronic Design, 2007, pp. 905–910.

[9] S. Shazli, M. Tahoori, "Transient error detection and recovery in processor pipelines," in Proceedings of 24th IEEE International Symposium on Defect and Fault Tolerance in VLSI Systems (DFT 2009), Chicago, IL, USA, October 7–9, 2009, pp. 304–312.

[10] L. Hellar, "Cascade voltage swing logic: a differential CMOS logic family," in Proceedings of IEEE International Solid State Circuits Conference, 1984, pp-16–17.

[11] S. Purohit et al., "An area efficient design methodology for SEU tolerant digital circuits," in Proceedings of the 2010 IEEE International Symposium on Circuits and Systems (ISCAS 2010), Paris, France, May 31st–June 2nd, 2010, pp. 981–984.

[12] E. Neito et al., "Using bulk built-in current sensors to detect soft errors," IEEE Micro, 26(5), 2006, pp. 10–18.

[13] S. Purohit et al., "Low overhead soft error detection and correction scheme for reconfigurable pipelined data paths," in Proceedings of 5th NASA/ESA Conference on Adaptive Hardware Systems (AHS 2010), Anaheim, CA, USA, June 15–18, 2010, pp. 59–65.

[14] S. Purohit et al., "Design of self correcting radiation hardened digital circuits using decoupled ground bus," in Proceedings of 20th ACM Great Lakes Symposium on VLSI, Rhode Island, USA, May 16–18, 2010, pp. 405–408.

[15] D.G. Elliott et al., "Computational RAM: implementing processors in memory," IEEE Design & Test of Computers, 16(1), 1999, pp. 32–41.

[16] S. Purohit et al., "Design space exploration for power-delay-area efficient coarse grain reconfigurable data path," in Proceedings of the IEEE International Conference on VLSI Design, New Delhi, India, 2009, pp. 45–50.

[17] S.R. Chalamalasetti et al., "Radiation hardened reconfigurable array with instruction roll-back," IEEE Embedded Systems Letters (Special Issue on Reliable Embedded Systems), 2(4), 2010, pp. 123–126.

[18] J.E. Knudsen, L.T. Clark, "An area and power efficient radiation hardened by design flip-flop," IEEE Transactions on Nuclear Science, 53(6), 2006, pp. 3392–3399.

[19] M. Haghi, J. Draper, "The 90 nm Double-DICE storage element to reduce Single-Event upsets," in MWSCAS '09. 52nd IEEE International Midwest Symposium on Circuits and Systems, 2009, 2009.

[20] L. Anghel et al., "Evaluation of a soft error tolerance technique based on time or space redundancy," in Proceedings of 13th Symposium on Integrated Circuits and Systems Design, 2000, pp. 236–242.

[21] Y. Tsiatouhas et al., "A hierarchal architecture for concurrent soft error detection based on current sensing," in Proceedings of the 8th IEEE International Online Testing Workshop (IOLTW), 2002, pp. 56–60.

10 Hybrid Partially Adaptive Fault-Tolerant Routing for 3D Networks-on-Chip

SUDEEP PASRICHA and YONG ZOU

10.1 INTRODUCTION

Advances in complementary metal–oxide–semiconductor (CMOS) fabrication technology in recent years have led to the emergence of chip multiprocessors (CMPs) as compact and powerful computing paradigms. However, to keep up with rising application complexity, it is widely acknowledged that solutions beyond aggressive CMOS scaling into deeper nanometer technologies will be required in the near future, to improve CMP computational capabilities [1]. Three-dimensional integrated circuit (3D-IC) technology with wafer-to-wafer bonding [2, 3] has been recently proposed as a promising candidate for future CMPs. In contrast to traditional two-dimensional integrated circuit (2D-IC)-based chips that have a single layer of active devices (processors, memories), wafer-to-wafer bonded 3D-ICs consist of multiple stacked active layers and vertical through-silicon vias (TSVs) to connect the devices across the layers. Multiple active layers in 3D-ICs can enable increased integration of cores within the same area footprint as traditional 2D-ICs. In addition, long global interconnects between cores can be replaced by shorter interlayer TSVs, improving performance and reducing on-chip power dissipation. Recent 3D-IC test chips from IBM [2], Tezzaron [3], and Intel [4] have confirmed the benefits of 3D-IC technology.

A major challenge facing the design of such highly integrated 3D-ICs in deep submicron technologies is the increased likelihood of failure due to permanent and intermittent faults caused by a variety of factors that are becoming more and more prevalent. Permanent faults occur due to

Embedded Systems: Hardware, Design, and Implementation, First Edition.
Edited by Krzysztof Iniewski.
© 2013 John Wiley & Sons, Inc. Published 2013 by John Wiley & Sons, Inc.

manufacturing defects, or after irreversible wearout damage due to electromigration in conductors, negative bias temperature instability (NBTI), dielectric breakdown, and so on [5, 6]. Intermittent faults, on the other hand, occur frequently and irregularly for several cycles, and then disappear for a period of time [7, 8]. These faults commonly arise due to process variations combined with variation in the operating conditions, such as voltage and temperature fluctuations.

On-chip interconnect architectures are particularly susceptible to faults that can corrupt transmitted data or altogether prevent it from reaching its destination. Reliability concerns in sub-65 nm nodes have in part contributed to the shift from traditional bus-based communication fabrics to network-on-chip (NoC) architectures that provide better scalability, predictability, and performance than buses [9, 10]. The inherent redundancy in NoCs due to multiple paths between packet sources and sinks can greatly improve communication error resiliency. It is expected that 3D-ICs will employ 3D NoCs, much like today's 2D-IC-based CMPs are employing 2D NoCs [4, 11, 12]. To ensure reliable data transfers in these communication fabrics, utilizing fault-tolerant (FT) design strategies is essential. In particular, FT routing schemes are critical to overcome permanent and intermittent faults at design time and runtime. In the presence of intermittent or permanent faults on NoC links and routers, routing schemes can ensure error-free data flit delivery by using an alternate route that is free of faults. As a result, FT routing schemes for 2D NoCs have been the focus of several research efforts over the last few years [13–25]. However, to the best of our knowledge, no prior work has yet explored FT routing schemes for 3D NoCs.

In this chapter, we explore fault-resilient 3D NoC routing and propose a new FT routing scheme (4NP-First) for 3D mesh NoCs that combines turn models together with opportunistic replication to increase successful packet arrival rate while optimizing energy consumption. Our extensive experimental results indicate that the proposed 4NP-First FT routing scheme has a low implementation overhead and provides better fault resiliency than existing dimension-order, turn-model, and stochastic random walk-based FT 2D NoC routing schemes that were extended to 3D NoCs.

10.2   RELATED WORK

NoC routing schemes can be broadly classified as either static (also called deterministic) or dynamic (also called adaptive). While static routing schemes [26] use fixed paths and offer no fault resiliency, dynamic (or adaptive) routing schemes [13–25] can alter the path between a source and its destination over time as traffic conditions and the fault distribution changes. The design of adaptive routing schemes is mainly concerned with increasing flit arrival rate, avoiding deadlock, and trying to use a minimal hop path from the source to the destination to decrease transmission energy. Unfortunately, these goals are often conflicting, requiring a complex trade-off analysis that is rarely addressed in existing literature.

In general, existing adaptive FT routing schemes (for 2D NoCs) can be broadly classified into three categories: (1) stochastic, (2) fully adaptive, and (3) partially adaptive. *Stochastic* routing algorithms provide fault tolerance through data redundancy by probabilistically replicating packets multiple times and sending them over different routes [13]. Examples of such schemes include the probabilistic gossip flooding scheme [14], directed flooding [15], and *N*-random walk [15]. The major challenges with these approaches are their high energy consumption, strong likelihood of deadlock and livelock, and poor performance even at low traffic congestion levels. *Fully adaptive* routing schemes make use of routing tables in every router or network interface (NI) to reflect the runtime state of the NoC and periodically update the tables when link or router failures occur to aid in adapting the flit path [16–18]. However, these schemes have several drawbacks, including (1) the need for frequent global fault and route information updates that can take thousands of cycles at runtime, during which time the NoC is in an unstable state; (2) the lack of scalability—table sizes increase rapidly with NoC size, increasing router (or NI) area, energy, and latency; and (3) the strong possibility of deadlock, unless high overhead deadlock recovery mechanisms such as escape channels are used. *Partially adaptive* routing schemes enable a limited degree of adaptivity, placing various restrictions and requirements on routing around faulty nodes, primarily to avoid deadlocks. Turn model-based routing algorithms such as negative-first, north-last, and south-last [19–22] are examples of partially adaptive routing, where certain flit turns are forbidden to avoid deadlock. A recent work [23] combines the XY and YX schemes to achieve fault-resilient transfers. However, the degree of adaptivity provided by these routing algorithms is highly unbalanced across the network, which in some cases results in poor performance. A few works have proposed using convex or rectangular fault regions with turn models [24, 25]. Such region-based FT routing schemes that make use of fault regions are generally too conservative (e.g., disabling fully functional routers to meet region shape requirements), have restrictions on the locations of faults that can be bypassed, and also cannot adapt to runtime permanent faults.

For 3D NoCs, a few researchers have proposed routing algorithms other than the basic dimension order-routing approach (XYZ, ZYX, etc.). For instance, Ramanujam and Lin [27, 28] proposed a load-balancing routing scheme in which a packet is sent to a random layer, followed by XY routing to the same column and row as the destination node, and then finally traversing in the vertical direction to the destination. In Rusu et al. [29], a packet is first routed to the layer of its destination, and then from there to the destination using dimension-order routing. However, none of these routing algorithms possess fault tolerance. FT routing for 3D topologies in large-scale interconnection networks has been addressed in some works [30–33]. However, the overhead of these schemes makes them prohibitive for implementation on a chip, in the context of 3D NoCs. For instance, the planar adaptive scheme [31] uses three virtual channels (VCs) to avoid deadlocks, and requires building convex fault regions and creating routing rules that have significant implementation costs. The extended safety level scheme [32, 33] requires at least four VCs to avoid deadlock, and significant area and energy overhead to initialize and maintain the safety levels in every node.

In contrast to existing works described above, we propose a lightweight FT routing scheme for 3D NoCs that utilizes hybrid turn models. Our scheme introduces the 4N-First and the 4P-First 3D turn models, and then proceeds to combine these together to create the 4NP-First routing algorithm that achieves a higher packet arrival rate than the 4N-First and 4P-First schemes individually, as well as other FT routing schemes proposed in the literature, while maintaining a low implementation overhead.

10.3   PROPOSED 4NP-FIRST ROUTING SCHEME

10.3.1   3D Turn Models

The turn model for partially adaptive routing in large-scale 2D mesh interconnection networks was first proposed by Glass and Ni [34]. A turn in this context refers to a 90-degree change of traveling direction for a flit. In wormhole networks in particular, deadlock can occur because of flits waiting on each other in a cycle. Glass and Ni proposed three turn models: west-first, negative-first, and north-last, each prohibiting two out of eight allowed turns in 2D wormhole networks to break cyclic dependencies and ensure freedom from deadlock. For instance, the north-last turn model requires that if a packet's destination lies in the north (N) direction, then a turn to the N direction is the last turn that the packet should make. This essentially implies that the north–west and north–east turns are prohibited. We extend these 2D turn models into the third dimension in this work. The 2D west-first, negative-first, and north-last turn models are transformed into two negative-first (2N-First), three negative-first (3N-first), and four negative-first (4N-First) turn models in 3D, respectively. For the sake of discussion and because we use it in our approach, we focus on the 4N-First model.

In the 4N-First turn model, two out of the possible three positive directions (N, E, U) need to be selected as the last two directions before routing a packet to its destination. Thus at the beginning of the routing phase, we can adaptively choose to route the packet along the remaining one positive and three negative (S, W, D) directions. For our implementation, we chose the two positive last directions as N and E. Figure 10.1 shows the forbidden turns for such a 4N-First turn model. A packet coming from the W cannot be routed in the S, U, or D directions, while a packet coming from the south direction cannot be routed in the W, U, or D directions. To summarize, the forbidden turns in the 4N-First turn model are: E–U, E–D, E–S, N–U, N–D, and N–W. These turn

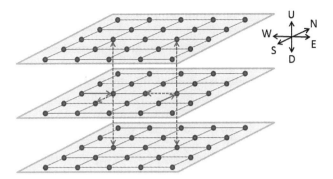

FIGURE 10.1. Forbidden turns in the 4N-First turn model.

restrictions allow deadlock-free routing with some measure of adaptivity in the 4N-First turn model. The analogous 4P-First turn model can be understood by inverting the 3D mesh by 180 degrees. In the 4P-First turn model, two out of the possible three negative directions must be selected as the last two directions before routing a packet to its destination. In our implementation of the 4P-First routing algorithm, we choose the two negative last directions as S and W.

10.3.2   4NP-First Overview

To ensure robustness against faults, redundancy is a necessary requirement, especially for environments with high fault rates and unpredictable fault distributions. As the level of redundancy is increased, system reliability improves as a general rule. However, redundancy also detrimentally impacts other design objectives such as power and performance. Therefore, in practice, it is important to limit redundancy to achieve a reasonable trade-off between reliability, power, and performance. Unlike directed and random probabilistic flooding algorithms that propagate multiple copies of a packet to achieve FT routing in NoCs, the proposed 4NP-First routing scheme sends only one redundant packet for each transmitted packet, and only if the fault rate is above a replication threshold δ. The original packet is sent using the 4N-First turn model, while the redundant packet is propagated using 4P-First turn model-based routing scheme. The packet replication happens only at the source. Two separate VCs, one dedicated to the 4N-First packets and the other for the 4P-First packets, ensure deadlock freedom. If the fault rate is low (i.e., below δ), replication is not utilized and only the original packet is sent using the 4N-First scheme while power and clock gating the 4P-First VC hardware in routers to save power. The value of δ is a designer-specified parameter that depends on several factors such as the application characteristics, routing complexity, and power reliability trade-off requirements.

The 4NP-First routing algorithm prioritizes minimal paths that have higher probabilities of reaching the destination even if faults are encountered downstream. Minimal paths and the replication threshold ensure low power dissipation under a diverse set of fault rates and distributions. No restriction on the number or location of faults is assumed, but the routers must know which of their adjacent (neighbor) links/nodes are faulty, which is accomplished using basic control signaling that already exists in any NoC architecture. The proposed routing approach can be combined with error control coding (ECC) techniques for transient fault resiliency and optimizations such as router buffer reordering [35] and router/NI backup paths [36] to create a comprehensive FT NoC fabric. In the following subsections, we describe the implementation of the proposed 4NP-First scheme. The implementation description focuses on the 4N-First turn model. The implementation for the analogous 4P-First turn model is similar, and therefore is excluded for brevity.

10.3.3   Turn Restriction Checks

Whenever a packet arrives at a router in the 4NP-First scheme, three scenarios are possible: (1) there is no fault in the adjacent links and the packet is routed to the output port based on the 4N-First (or 4P-First) scheme that selects a minimal path to the destination; (2) there are one or more faults on adjacent links that prevent the packet from propagating in a valid direction (an output port with a fault-free link that does not violate turn model routing rules), in which case the packet must be dropped, as a back turn is not allowed; and (3) one or more adjacent links have faults but a valid direction exists to route the packet toward, based on the 4N-First (or 4P-First) rules. A packet traversing an intermediate router must choose among six output directions (N, S, W, E, U, D). However, choosing some directions may lead to a violation of a basic 4N-First (or 4P-First) turn rule immediately or downstream based on the relative location of the destination.

Figure 10.2 shows the pseudocode of the turn restriction check phase for the 4N-First turn model to aid in detecting an invalid routing direction. The input to the turn restriction check phase is a potential output port direction for a header flit currently in the router input buffer, and the output of the phase indicates whether the output port direction is valid or not. Note that here our goal is to check not only for scenarios that lead to immediate violation of basic turn restriction rules for the 4N-First routing scheme, but also to identify special scenarios where even if a valid turn exists at the current router, taking that turn may ultimately not be a viable option, as it may lead to turn rule violations later along the communication path.

First, we check whether the output port direction has a fault in its attached adjacent link (steps 2 and 3) or is a back turn (steps 4 and 5), in which case this is an invalid direction. Next, we check for scenarios where choosing the given output port direction may lead to violations of the 4N-First turn model rules immediately or eventually at some point downstream. In steps 6 and 7

```
 1:  RESULT   check_neighbor(UI ip_dir, UI direct, ULL cur_id, ULL dest_id) {
     //ip_dir: direction where the packet comes from
     //direct: direction we want to check
     //cur_id: current node position
     //dest_id: destination node position.
     //dest_yco,dest_xco: y,x coordinates of destination node
     //cur_yco,cur_xco: y,x coordinates of current node
     //dif_yco=dest_yco-cur_yco
     //RESULT: LINK_FAULT, TURN_FAULT, BACK_TURN, OK
 2:     if (check_linkfault(direction)==LINK_FAULT)
 3:          return LINK_FAULT;
 4:     if (ip_dir==direct)
 5:           return BACK_TURN;
 6:     if(((ip_dir==S)||(ip_dir==W))&&((direct!=N)&&(direct!=E)))
 7:          return TURN_FAULT;
 8:     if(((dest_xco>cur_xco)&&((direct==N)||(direct==E)))
        ||((dest_yco<cur_yco)&&((direct==N)||(direct==E)))
        ||((dest_zco!=cur_zco)&&((direct==N)||(direct==E))))
 9:          return TURN_FAULT;
10:     if((dest_xco==cur_xco)&&(direct==N)){
11:          return TURN_FAULT;
12:     if((dest_yco==cur_yco)&&(direct==E)){
13:          return TURN_FAULT;
14:     if((borderD(cur_zco)&&borderS(cur_address.tile_id))
        &&(dest_zco==cur_zco)&&(dest_yco>cur_yco)
        &&(dest_xco==cur_xco)&&(direct==W))
15:          return TURN_FAULT;
16:     if((borderD(cur_zco)&&borderW(cur_address.tile_id))
        &&(dest_zco==cur_zco)&&(dest_yco==cur_yco)
        &&(dest_xco<cur_xco)&&(direct==S))
17:           return TURN_FAULT;
18:  return OK;
19: }
```

FIGURE 10.2. Pseudocode for 4N-First turn restriction checking.

we check for the basic 4N-First turn rules (Section 10.3.1) and prohibit a packet coming from S or W to go in any direction other than N or E. Steps 8 and 9 check for the following condition: if the destination is on the S, W, or another layer relative to the current node, then the packet cannot turn N or E, because once it does, it can only be routed N or E and will never reach the destination. Steps 10–13 check for two conditions. If a packet is in the same row as the destination, it should not be routed to the N direction as the packet will eventually need to be routed to the S direction that will break a basic turn model rule. Similarly, if a packet is on the same column as its destination, it should not be routed to the E direction as it eventually will have to turn W and this will break the basic turn model rule. Steps 14–17 check for two additional special situations. First, if a packet is on the bottom layer and on the S border, and the destination is on the same layer and to the E direction of the current node, then the packet should not be routed to the W direction. This is because the packet will eventually need to go N or U, then E, and then finally

S or D, which will break the basic turn model rules. Second, if a packet is on the bottom layer and on the W border, and the destination is on the same layer and to the S direction of the current node, then the packet should not be routed in the N direction, for reasons similar to the first scenario.

10.3.4   Prioritized Valid Path Selection

After the turn restriction checks, it is possible for a packet to have multiple available (valid) directions to choose from along minimal paths, or if a minimal path is not available, then we may need to select among directions that lead to nonminimal paths. In such a scenario, this section answers the question: *which out of the multiple available paths should be selected for the 4N-First routing algorithm to ensure maximum arrival rate?*

A packet in the process of being routed using the 4N-First scheme may encounter two situations:

> *Case 1*: The packet has already been routed along the N or E directions, and thus it can only choose the N or E directions to its destination.
>
> *Case 2*: The packet has not been routed along the N or E directions and therefore can have at most four choices (S, W, U, or D directions).

For each of these cases, we perform prioritized valid path selection in accordance with two general rules: (1) directions along the minimal path are always given the highest priority and (2) if a minimal path direction cannot be selected, we prefer a direction that has the largest next-hop node diversity number compared with other directions with the same distance to the destination. The *diversity number* of a node enumerates the number of available paths to the destination from that node. In the following sections, we describe the motivation behind these rules for prioritized valid path selection for the two cases identified above.

10.3.4.1   Prioritized Valid Path Selection for Case 1   Figure 10.3 shows the case with the destination (X) to the north and east of the current node (C), with each node annotated with its diversity number, determined recursively. For all minimal paths from C to X, it can be seen that the paths passing along the middle of the bounding submesh have the highest node diversity at each hop, and therefore the highest probability of reaching the destination in the presence of faults along the routed path from C to X. In other words, it would be preferable to choose a path where we toggle between the N and E directions as much as possible (as in the path highlighted in red in Fig. 10.3) than follow a path along the periphery (as, for instance, in XY routing). Thus, a path *along the middle of the mesh is preferable* than a path closer to the boundary.

10.3.4.2   Prioritized Valid Path Selection for Case 2   For Case 2, consider Figure 10.4 with the current node C to the E and U directions relative to

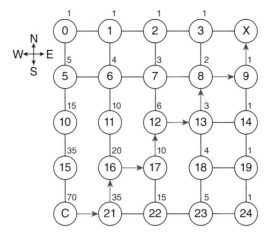

FIGURE 10.3. Node diversity and prioritized path for Case 1.

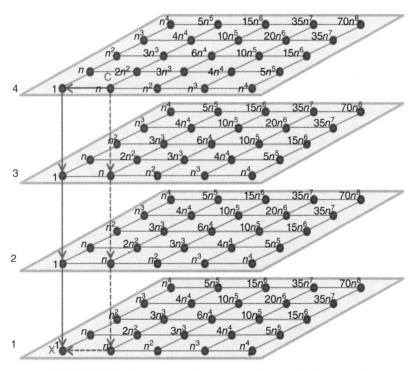

FIGURE 10.4. Node diversity and prioritized path for Case 2.

destination X. Each node in the mesh is annotated with its diversity number in terms of n, which is the number of layers (four in this case). It can be observed that sending the packet to the row coordinate of the destination before sending it in the D direction (depicted in the solid red line) leads to fewer path choices for the packet in case the path has a fault. In contrast, sending the packet down to the layer of the destination first before routing in the D direction (depicted in the dotted red line) is a path with greater diversity, and thus greater resiliency if faults are encountered. It should also be observed that nodes toward the middle of the mesh have greater diversity numbers, therefore paths along the middle of the mesh have a greater opportunity of arriving at the destination than paths along the periphery, which is a similar observation as what we described for Case 1 in the previous section. Thus, given a choice among S, W, U, and D directions, to ensure the highest arrival rate, the *packet should first be routed U or D to the same layer as the destination, and then routed in a direction along the middle of the mesh* as much as possible.

10.3.5 4NP-First Router Implementation

Figure 10.5 depicts our dual VC router architecture. Packets traversing the network have a header flit with a destination ID field that is used to determine

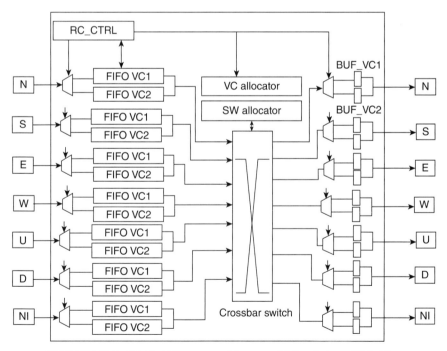

FIGURE 10.5. 3D NoC Router architecture. SW, switch; BUF, buffer.

the output port at each router. The router datapath consists of the buffers and the switch. The input FIFO (first in, first out) buffers store up to 16 flits waiting to be forwarded to the next hop. There are two input FIFO buffers each dedicated to a VC, with one VC for the 4N-First routed packets and the other for the 4P-First routed packets. When a flit is ready to move, the switch connects an input buffer to an appropriate output channel. To control the datapath, the router contains three major control modules: a route compute control unit (RC_CTRL), a virtual-channel allocator (VCA), and a switch allocator (SA). These control modules determine the next hop direction, the next VC, and when a switch is available for each packet/flit. The routing operation takes four phases: route computation (RC), VCA, SA, and switch traversal (ST). When a header flit (the first flit of a packet) arrives at an input channel, the router stores the flit in the buffer for the allocated VC and determines the next hop for the packet using the checks and priority rules described in the previous sections (RC phase). Given the next hop, the router then allocates a VC (VCA phase). Finally, the flit competes for a switch (SA phase) if the next hop can accept the flit, and moves to the output port (ST phase). In our scheme, a packet carries its VC (one of 4N-First or 4P-First) information in a bit in the header flit, therefore the output port and the destination VC can be determined early, at the end of the RC stage.

To reduce energy consumption, if the fault rate is below the replication threshold δ, the RC_CTRL unit shuts down the 4P-First VCs in the router. We assume that dedicated control signals connected to a lightweight fault detection unit (FDU) can allow estimation of the fault rate at runtime in the NoC, and the decision to change the replication status (initiate or cutoff) can be communicated to the nodes and implemented with a delay of at most a few hundred cycles. Typically, such a change in status would be a rare occurrence, as intermittent or permanent faults do not appear at runtime as frequently as transient faults.

An important requirement for any FT NoC fabric is a control network that can be used to send acknowledgment signals to inform the source whether the packet successfully arrived at the destination. We assume a lightweight and fault-free control network that is topologically similar to the data network for the purpose of routing ACK signals from the destination to the source node on successful packet arrival, or NACK signals from an intermediate node to the source node if the packet must be dropped due to faults that make it impossible to make progress toward the destination. In our implementation, a source can resend a packet at most two times after receiving NACK signals, before assuming that the packet can never reach its destination. While this can signal unavoidable failure for certain types of applications, other applications may still be able to operate and maintain graceful degradation. For instance, a multiple use-case CMP can shut down certain application use cases when their associated intercore communication paths encounter fatal faults, but there might be other use cases that can still continue to operate on fault-free paths. We assume that a higher level protocol (e.g., from the middleware or

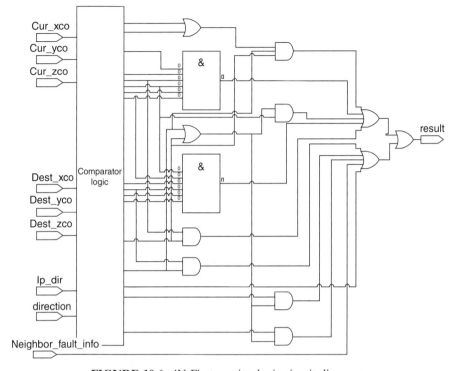

FIGURE 10.6. 4N-First routing logic circuit diagram.

operating system [OS] levels) can step in to initiate thread migration and other strategies to ensure useful operation in the presence of unavoidable system faults.

Figure 10.6 shows the circuit-level diagram for the 4N-First turn restriction check phase and prioritized path selection, implemented in the route compute unit of a NoC router. The circuit is replicated at each input port of the router. At the Taiwan Semiconductor Manufacturing Company (TSMC) 65 nm technology node, the circuit dissipates 0.98 μW of power on average, and has a critical path latency of 0.77 ns. This overhead is significantly lower than in the case of stochastic N-random walk or fully adaptive routing schemes.

10.4   EXPERIMENTS

10.4.1   Experimental Setup

We implemented and evaluated our proposed algorithm using a SystemC-based [37] cycle-accurate 3D mesh NoC simulator that was created by extending the open-source 2D NoC Nirgam simulator [38] by adding support for a

3D network, faults, and various FT routing schemes. In addition to communication performance estimates, we were interested in NoC energy estimation as well. For this purpose we incorporated dynamic and leakage power models for standard NoC components from Orion 2.0 [39] into the simulator. We also incorporated the power overhead of the circuits in the routers (obtained after synthesis) required to realize the FT routing algorithms. The NoC fabric was clocked at 1 GHz. The target implementation technology was 65 nm, and this node was used to determine parameters for delay and energy consumption. A $5 \times 5 \times 4$ (100 cores; 25 cores/layer in 4 layers) CMP with a 3D mesh NoC fabric was considered as the baseline system in our experiments. The stimulus for the simulations was generated for specific injection rates using uniform random, transpose, and hotspot traffic models. A fixed number of flits (3000 per core) were transmitted according to the specified traffic pattern in the simulations, to enable comparisons not only for successful arrival rates, but also for overall communication energy. We created a statistical fault model with random distributions for design time and runtime faults across the network fabric. Ten different distributions were considered for each evaluation point to obtain a more accurate estimate of fault resiliency for the routing schemes. To ensure fair comparison, the randomized fault locations were replicated across simulation runs for different FT routing schemes, to study the performance of the schemes under the same fault conditions.

10.4.2 Comparison with Existing FT Routing Schemes

Our first set of experiments compares the successful packet arrival rates for our 4NP-First FT routing scheme with FT routing schemes. Given the lack of any existing 3D NoC FT routing schemes, we extended FT routing schemes for 2D NoCs proposed in the literature to 3D NoCs. Additionally, we also extended variants of turn model-based partially adaptive routing schemes to 3D NoCs. The following schemes were considered in the comparison study:

1. XYZ dimension order routing (although not inherently fault tolerant, it is considered here because of its widespread use in 3D mesh NoCs proposed in literature)
2. *N*-random walk [15] extended to 3D NoCs for $N = 1, 2, 4, 8$
3. adaptive odd–even (OE) turn model [22] extended to 3D NoCs
4. 2N-First turn model
5. 3N-First turn model
6. 4N-First turn model
7. hybrid XYZ (XYZ and ZYX with replication on separate VCs)
8. 2NP-First (2N-First and 2P-First with replication on separate VCs)
9. 3NP-First (3N-First and 3P-First with replication on separate VCs)

10. 4NP-First (4N-First and 4P-First with replication on separate VCs)
11. hybrid OE + IOE (OE and inverted OE with replication on separate VCs)

Other FT routing schemes such as probabilistic flooding [14] and fully adaptive table-based routing [17] were considered, but their results are not presented because these schemes have major practical implementation concerns—their energy consumptions is an order of magnitude higher than other schemes and frequent deadlocks hamper overall performance. For our *4NP-First* scheme, we used prioritized valid path selection introduced in Section 10.3.4 and a replication threshold value of $\delta = 4\%$, based on our extensive simulation studies that showed a good trade-off between arrival rate and energy for $8\% \geq \delta \geq 2\%$. Ultimately, our goal with these experiments was to explore a variety of schemes with fault-tolerance capabilities, and compare the results for our proposed 4NP-First routing scheme with the other schemes for 3D NoCs.

Figure 10.7 shows the successful packet arrival rates for all the routing algorithms over a spectrum of fault rates from a low of 1% to a high of 20%, and with a flit injection rate of 20% (0.2 flits/node/cycle). On analyzing individual routing scheme performance, it can be immediately seen that the *N*-random walk schemes (*1_rand, 2_rand, 4_rand, 8_rand*) have a low arrival rate even for very low fault rates. This is due to frequent deadlocks during simulation, which the scheme is susceptible to, causing packets to be dropped. The *hybrid XYZ* scheme performs well for very low fault rates, but as the fault rate rises, the lack of adaptivity inherent to the XYZ and ZYX schemes results in its arrival rate dropping rapidly compared with turn model-based schemes. Overall, it can be clearly seen that the proposed *4NP-First* routing algorithm has the highest arrival rate compared with other schemes across the three traffic types. The reason for the high arrival rate with the *4NP-First* routing scheme is the greater path diversity due to the complementary combination of two analogous turn models, as well as smart path prioritization, which leads to better resiliency to faults in the network. For fault rates under 4%, replication in *4NP-First* is disabled to save energy, which explains the slight dip in successful arrival rate for low fault rates. However, with replication enabled, *4NP-First* outperforms every other scheme at low fault rates as well (i.e., 100% arrival rate at 1% fault rate). *4NP-First* scales particularly well under higher fault rates that are expected to be the norm in future CMP designs.

In addition to arrival rate, an increasingly important metric of concern for designers is the energy consumption in the on-chip communication network. Figure 10.8 shows the energy consumption for the different routing schemes under the same fault rates and types, and injection rates as in the previous experiment. The energy consumption accounts for circuit level implementation overheads as well as acknowledgments and retransmission in the schemes. It can be seen that the *N*-random walk FT routing scheme has significantly higher energy consumption compared with the other schemes. This is due to

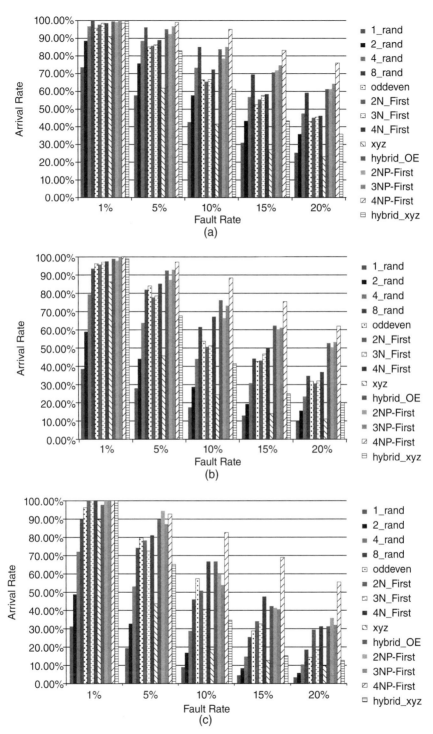

FIGURE 10.7. Packet arrival rate for $5 \times 5 \times 4$ NoC. (a) Uniform random; (b) transpose; (c) hotspot.

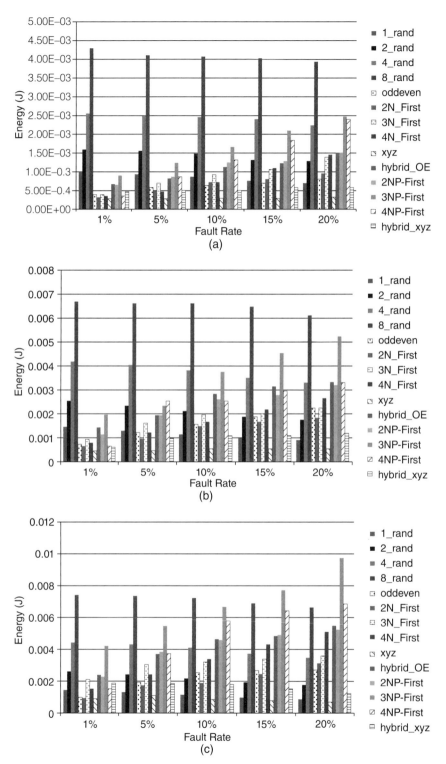

FIGURE 10.8. Energy consumption for $5 \times 5 \times 4$ NoC. (a) Uniform random; (b) transpose; (c) hotspot.

the high level of replication with increasing values of N, as well as because of the nonminimal paths chosen by default to route the packets on. Among single VC schemes (*oddeven, 2N-First, 3N-First, 4N-First, XYZ*), the *XYZ* scheme has the lowest energy dissipation, which explains its popularity in NoCs today. But as it is lacking in any fault tolerance, the scheme may not be a viable option for future CMPs.

Among the rest of the (hybrid) schemes that incorporate replication, the communication energy for the *hybrid XYZ* scheme is the lowest. However, in light of its extremely low successful arrival rates as fault rates increase, the energy savings become irrelevant. For low fault rates below 4%, the *4NP-First* scheme has the lowest energy among these schemes because it disables replication to trade-off energy with fault tolerance. As the fault rate increases, initially the energy consumption for the *4NP-First* scheme is competitive and generally lower than the other hybrid schemes. However, for higher fault rates, the energy consumption for 4NP-First becomes larger than all other hybrid schemes except *3NP-First*. This rise in energy for *4NP-First* happens primarily because the scheme uses more and more nonminimal paths as faults increase, to ensure high successful packet arrival rates. The other hybrid schemes have significantly lower packet arrival rates, because in a lot of cases, packets hit a fault soon after transmission and are dropped. Despite frequent retransmissions, the energy consumption of these hybrid schemes remains lower than that of *4NP-First*. In some cases, the replication threshold-based mechanism used in *4NP-First* can be extremely useful to trade off energy with packet arrival rate. For instance, for multimedia applications, occasionally dropped pixel data may not be perceivable by viewers, and in such scenarios the resulting lower energy consumption may lead to a much longer battery life and better overall user experience. Conversely, if fault resiliency is paramount, the value of δ can be reduced to close to 0, to enable the highest possible successful packet arrival rate. Given that ensuring high arrival rates is more important than saving energy in a majority of scenarios, the *4NP-First* scheme is a better choice for environments with both low and high fault rates. Figure 10.9 presents data to back this observation. We compare the various FT routing schemes for a hybrid energy $\times (1 - \text{arrival\_rate})^2$ metric. It can be seen that our proposed 4NP-First routing scheme has the lowest value for this metric out of all the schemes for low as well as high fault rates. These results demonstrate the promise of the proposed 4NP-First scheme in providing energy-efficient fault resilience during data transfers in 3D NoC architectures.

10.5 CONCLUSION

In this chapter, we explored FT routing schemes for 3D NoCs in detail. We proposed a novel FT routing scheme (4NP-First) for 3D NoCs that combines the 4N-First and 4P-First turn models to create a robust hybrid routing scheme. The proposed scheme is shown to have a low implementation overhead and

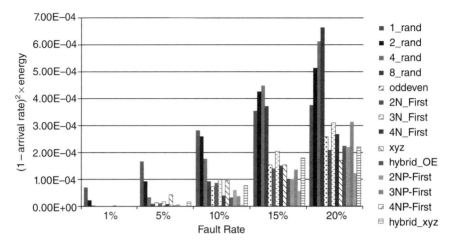

FIGURE 10.9. Hybrid energy arrival rate metric comparison.

adapts to design time and runtime faults better than existing turn model, stochastic random walk, and hybrid dual VC-based routing schemes. Our ongoing work is developing a theoretical formulation to characterize the coverage and FT capabilities of various NoC routing schemes.

REFERENCES

[1] ITRS Technology Working Groups, "International Technology Roadmap for Semiconductors (ITRS) 2007 Edition," 2007. Available: http://public.itrs.net

[2] K. Bernstein et al., "Interconnects in the third dimension: design challenges for 3D ICs," Proc. DAC, 2007, pp. 562–567.

[3] R.S. Patti, "Three-dimensional integrated circuits and the future of system-on-chip designs," Proceedings of the IEEE, 94(6), 2006, pp. 1214–1224.

[4] S. Vangal et al., "An 80-Tile 1.28 TFLOPS network-on-chip in 65 nm CMOS," Proc. IEEE ISSSC., Feb. 2007.

[5] C. Constantinescu, "Trends and challenges in VLSI circuit reliability," IEEE Micro, 23(4), July–Aug 2003, pp. 14–19.

[6] S. Nassif, "Modeling and analysis of manufacturing variations," Proc. CICC, May 2001.

[7] S. Pan et al., "IVF: characterizing the vulnerability of microprocessor structures to intermittent faults", Proc. DATE, 2010.

[8] C. Constantinescu, "Intermittent faults in VLSI circuits," Proc. IEEE Workshop on Silicon Errors in Logic (SELSE), 2007.

[9] L. Benini, G. De-Micheli, "Networks on chip: a new SoC paradigm," Proc. Computer, 49(1), 2002, pp. 70–71.

[10] W.J. Dally, B. Towles, "Route packets, not wires: on-chip interconnection networks," Proc. DAC, pp. 684–689, 2001.

[11] Picochip PC102, "Practical, Programmable Multi-Core DSP," 2007. Available: http://www.picochip.com/page/42/multi-core-dsp-architecture.

[12] Tilera Corporation, TILE64™ Processor. Product Brief. 2007.

[13] H. Zhu et al., "Performance evaluation of adaptive routing algorithms for achieving Fault Tolerance in NoC Fabrics," Proc. ASAP, 2007.

[14] T. Dumitras, R. Marculescu, "On-chip stochastic communication," Proc. DATE, 2003.

[15] M. Pirretti et al., "Fault tolerant algorithms for network-on-chip interconnect," Proc. ISVLSI, 2004.

[16] Y.B. Kim, Y.-B. Kim, "Fault tolerant source routing for network-on-chip," Proc. DFT, 2007.

[17] T. Schonwald et al., "Fully adaptive fault tolerant routing algorithm for network-on-chip architectures," Proc. DSD, Aug. 2007.

[18] T. Schonwald et al., "Region-based routing algorithm for network-on-chip architectures," Proc. Norchip, 2007.

[19] C.J. Glass, L.M. Ni, "The turn model for adaptive routing," Proc. ISCA, pp. 278–287, 1992.

[20] C.J. Glass, L.M. Ni, "Fault-tolerant wormhole routing in meshes without virtual channels," IEEE Transactions of Parallel and Distributed Systems, 7(6), 1996, pp. 620–635.

[21] C.M. Cunningham, D.R. Avresky, "Fault-tolerant adaptive routing for two-dimensional meshes," Proc. HPCA, 1995.

[22] G.-M. Chiu, "The odd-even turn model for adaptive routing," IEEE Transactions of Parallel and Distributed Systems, 11(7), 2000, pp. 729–738.

[23] A. Patooghy, S.G. Miremadi, "XYX: a power & performance efficient fault-tolerant routing algorithm for network on chip," Proc. ICPDNP, 2009.

[24] M. Andres et al., "Region-based routing: a mechanism to support efficient routing algorithms in NoCs," IEEE TVLSI, 17(3), 2009.

[25] J. Hu, R. Marculescu, "Dyad—smart routing for networks-on-chip," Proc. DAC, 2004.

[26] W.J. Dally, B. Towles, Principles and Practices of Interconnection Networks, Morgan Kauffman, San Francisco, CA, 2004.

[27] R.S. Ramanujam, B. Lin, "A layer-multiplexed 3D on-chip network architecture," IEEE Embedded Systems Letters, 1(2), 2009, pp. 50–55.

[28] R.S. Ramanujam, B. Lin, "Near-optimal oblivious routing on three-dimensional mesh networks," in IEEE Conference on Computer Design (ICCD), October 2008, pp. 134–141.

[29] C. Rusu et al., "Message routing in 3D networks-on-chip", Proc. NORCHIP, 2009.

[30] Y. Li et al., "Adaptive box-based efficient fault-tolerant routing in 3D torus," in 11th International Conference on Parallel and Distributed Systems, vol. 1, pp. 71–77, 2005.

[31] A.A. Chien, J.H. Kim, "Planar-adaptive routing: low-cost adaptive networks for multiprocessors," Proc. ISCA, 1992. pp. 268–277.

[32] J. Wu, "Fault-tolerant adaptive and minimal routing in mesh-connected multicomputers using extended safety levels," IEEE Transactions Parallel and Distributed Systems, 11(2), 2000, pp. 149–159.

[33] J. Wu, "A fault-tolerant adaptive and minimal routing approach in 3-D meshes," in Proc. of the 7th Int'l Conf. on Parallel and Distributed Systems (ICPADS), July 2000.

[34] C.J. Glass, L.M. Ni, "The turn model for adaptive routing," in Proceedings of the 19th International Symposium on Computer Architecture, May 1992, pp. 278–287.

[35] W.-C. Kwon et al., "In-network reorder buffer to improve overall NoC performance while resolving the in-order requirement problem," Proc. DATE, pp. 1058–1063, 2009.

[36] M. Koibuchi et al., "A lightweight fault-tolerant mechanism for network-on-chip," Proc. NOCS, 2008.

[37] "SystemC version 2.0 user's guide," 2005. Available: http://www.accellera.org/home/.

[38] "NIRGAM1: a simulator for NoC interconnect routing and application modeling version 1.1," 2007. Available: http://nirgam.ecs.soton.ac.uk/

[39] A. Kahng et al, "ORION 2.0: a fast and accurate NoC power and area model for early-stage design space exploration," Proc. DATE, 2009.

11 Interoperability in Electronic Systems

ANDREW LEONE

11.1 INTEROPERABILITY

Interoperability, or the idea of being "plug and play," has long been an implementation that engineers have looked for in the area of devices speaking the same language to each other. It takes the idea of open systems as opposed to closed systems. Multiple manufacturers can make different devices that all speak the same language and can communicate back to a single node within the system.

The basis for interoperability can be traced to the Open Systems Interconnection (OSI) Reference Model. Developed in 1974 by International Organization for Standardization (ISO) and American National Standards Institute (ANSI), it represents seven different layers within a network system and, in turn, an interoperable system. Since then, almost all interoperable networks can be related to the OSI Reference Model.

With the fundamental premise of the OSI model, many standards have been developed and implemented.

Traditionally products have communicated via proprietary, nonstandard communication structures. This means that a single entity controls the language and way of communication between two or more devices. In other words, the system is closed. The system cannot have other products enter the ecosystem without cooperation from a central entity that has created, set up, and maintains the interface and communication of the system.

Why is interoperability so important in electronic devices? Consumers, governments, and others are looking for ways for different devices to have their information collected and shared by a single aggregator. This is so important because one company may have a core competence of data acquisition

Embedded Systems: Hardware, Design, and Implementation, First Edition.
Edited by Krzysztof Iniewski.
© 2013 John Wiley & Sons, Inc. Published 2013 by John Wiley & Sons, Inc.

and algorithm, but that data algorithm is only a piece of the ecosystem measurement and analysis. I may need other company's products with their core competence to acquire and analyze data in order to create a more complete picture of what the ecosystem is trying to accomplish.

Later on in this chapter, I will discuss one of the more important interoperable ecosystems that are in development. The health-care ecosystem has the advantages I am talking about. The products are created by different companies with different competences and capabilities with the ability for all the information to be in a standard format so that the information is uniform for the whole ecosystem. Due to the openness of this standard, companies no longer need to worry about working with other products or worry about translators in the system to make sure all components work in unison. Also, it is much better for the user, either the hospital or home user. This allows for better competition and better product offerings with no need to wonder whether it will be useful in the ecosystem. The Continua Health Alliance is trying to fulfill this need.

Standards are critical in this implementation. Some document and framework common to all these interoperable devices must exist in order for each device to communicate with each other. This is similar to the way Italian speakers cannot communicate with Spanish speakers. They need to know the same language in order to communicate. In addition, this language needs to be decided upon ahead of time and with little room for interpretation. It needs to be complete. Standards that exist in the electronics market today are USB (Universal Serial Bus), Bluetooth, and Zigbee. In addition, Continua Health Alliance uses these established standards to add additional data structure information on top of them in order to be interoperable for medical devices. In order to understand further, we will need to review the OSI model in greater detail.

The IEEE (Institute of Electrical and Electronics Engineers) is responsible for many standards set in the market. Standards are published documents that establish specifications and procedures designed to ensure the reliability of the materials, products, methods, and/or services people use every day. Standards address a range of issues, including but not limited to various protocols that help ensure product functionality and compatibility, facilitate interoperability, and support consumer safety and public health.

Standards form the fundamental building blocks for product development by establishing consistent protocols that can be universally understood and adopted. This helps fuel compatibility and interoperability and simplifies product development and speeds time-to-market. Standards also make it easier to understand and compare competing products. As standards are globally adopted and applied in many markets, they also fuel international trade.

It is only through the use of standards that the requirements of interconnectivity and interoperability can be assured. It is only through the application of standards that the credibility of new products and new markets can be verified. In summary, standards fuel the development and implementation of technologies that influence and transform the way we live, work, and communicate.

Additionally, the ISO and International Electrotechnical Commission (IEC) are other standards organizations that work on interoperable standards and manage the upkeep of the standards and make sure they are in the best interest of the public and system.

The future of interoperability is endless. As more and more products enter the consumer and industrial market, the usefulness of those products will be gauged by their ability to communicate with other devices and systems. Healthcare, power consumption measurement, and consumer products are just a few to mention that do and will be best served by interoperability.

The remaining part of this chapter will discuss the technical aspects of interoperability, implementation, and how the ecosystems work in unison.

11.2 THE BASIS FOR INTEROPERABILITY: THE OSI MODEL

The OSI model was a product of the Open Systems Interconnection effort at the ISO. It is a way of subdividing a communications system into smaller parts called layers. Similar communication functions are grouped into logical layers. A layer provides services to its upper layer while receiving services from the layer below. On each layer, an instance provides service to the instances at the layer above and requests service from the layer below. It is also being referred that its name was proposed as the reverse of ISO.

For example, a layer that provides error-free communications across a network provides the path needed by applications above it, while it calls the next lower layer to send and receive packets that make up the contents of that path. Two instances at one layer are connected by a horizontal connection on that layer.

Virtually all networks in use today are based in some fashion on the OSI standard. OSI was developed in 1984 by the ISO, a global federation of national standards organizations representing approximately 130 countries.

The core of this standard is the OSI Reference Model, a set of seven layers that define the different stages that data must go through to travel from one device to another over a network (Figure 11.1).

Think of the seven layers as the assembly line in the computer. At each layer, certain things happen to the data that prepare it for the next layer. The seven layers, which separate into two sets, are the Application Set consisting of the upper layers of the OSI model and the Transport Set consisting of the lower layers of the OSI model.

The "upper part" of the stack consists of Layers 7 to 5 of the OSI model. Below is an explanation of the "upper layers" of the model.

Layer 7, the application layer, is the layer that actually interacts with the operating system or application whenever the user chooses to transfer files, read messages, or perform other network-related activities.

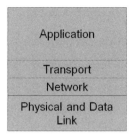

FIGURE 11.1. OSI model on the left and example of protocol stack in relation.

Layer 6, the presentation later, takes the data provided by the application layer and converts it into a standard format that the other layers can understand.

Layer 5, the session layer, establishes, maintains, and ends communication with the receiving device.

The "lower part" of the stack consists of Layer 4 down to Layer 1 of the OSI model. Below is an explanation of the "lower layers" of the model.

Layer 4, the transport layer, maintains flow control of data and provides for error checking and recovery of data between the devices. Flow control means that the transport layer looks to see if data are coming from more than one application and integrates each application's data into a single stream for the physical network.

Layer 3, the network layer, sets up the way that the data will be sent to the recipient device. Logical protocols, routing, and addressing are handled here.

Layer 2, the data link layer, is where the appropriate physical protocol is assigned to the data. Also, the type of network and the packet sequencing is defined.

Finally, Layer 1, the physical layer, is the level of the actual hardware. It defines the physical characteristics of the network such as connections (electrical, wireless, or optical interface), voltage levels, if applicable, and timing.

The OSI Reference Model is really just a guideline, though. Actual protocol stacks often combine one or more of the OSI layers into a single layer.

A protocol stack is a group of specific protocols that all work together to allow software or hardware to perform a function. Most protocol stacks, including Zigbee and Bluetooth, compress the OSI model into four basic layers.

Layer 1 and 2 Combined (Hardware Interface and Data Control): This combines the physical and data link layers and routes the data between devices on the interoperable network. This is generally referred to as the *transceiver* in the embedded world.

Layer 3 (Networking): This layer corresponds to the network layer and consists of network identification and a host identification, to determine the address of the device it is communicating with.

Layer 4 (Transport): This corresponds to the OSI transport layer and works by asking another device on the network if it is willing to accept information from an interoperable device on the network.

Layers 5–7 Combined: The protocol stack application layer combines the session, presentation, and application layers of the OSI model. The application layer is the part that the implementer uses as their code for data acquisition and processing. The application is fully customizable.

As you can see, it is not necessary to develop a separate layer for each and every function outlined in the OSI Reference Model. But developers are able to ensure that a certain level of compatibility is maintained by following the general guidelines provided by the model.

Now that the idea of the OSI Model and the protocol stack has been explained, we can now use this to better explain the two important parts of *interoperability*, hardware and firmware.

11.3 HARDWARE

Relating to what is considered Layer 1 in the protocol stack definition or the network interface, this is generally the physical interface or physical layer. Some in the electronics world might refer to it as the transceiver. This is the standard that sets up the electrical, wireless, or optical connection to the outside world. This creates an interface from the device system to the rest of the ecosystem.

The hardware for a particular interoperable embedded ecosystem or network is fairly straightforward. The physical layers for either electrical or wireless interfaces are usually, for most all cases, set by standards boards. For the purposes of this chapter, we will discuss standards-based physical layers.

The Universal Serial Bus, or better known as USB by everyone who owns a computer, is a wireline interface that has a specific physical layer interface to connect and communicate with other like devices. The physical layer or physical connection is the same for all USB devices with minor tweaks for HOST versus DEVICE versus On-The-Go DEVICES (these are different types of USB products that communicate). The physical layer doesn't dictate communication as much as the firmware stack does. The firmware portion will be discussed later on in the chapter. That being said, in its simplest form, wireline communication is implemented by varying voltages to represent the digital signal being transmitted. For instance, a digital 1 could be specified by +5 V and digital 0 could be specified by –5 V. The concept is sound and understood by many in the world of communications: digital implementation with regards to analog signals. And this can be facilitated by many off-the-shelf electronic components that exist. A USB device cannot communicate with non-USB devices. Since USB is a standard, in order to make compliant USB devices, one must follow the USB specification and reach interoperability compliance. When I say cannot communicate, I mean communicate with regards to the specification set by the USB standards association.

Moving on to the more complicated and somewhat cluttered space of wireless physical layers, standards organizations like IEEE have been integral in the development and adoption of physical layers in the wireless and now low-power wireless world.

As an example, Zigbee, ISA100.11a, and Wireless HART all use the 802.15.4 standard specification as their physical layer and data link layer. What makes these interoperable protocols different is just that, their protocols. The firmware layers that are on top of the physical and data link layers (Layers 1 and 2 of the OSI model) are where the differences lie.

To better illustrate this point, let us look at the differences between the wireless system schemes.

The 802.15 is a standards group within the IEEE. They have been conscientious enough to build specifications and standards for physical implementations that follow some semblance of order. The 802.15.1 is the basis for Classic Bluetooth. The 802.15.4 is the basis for Zigbee, ANT™, IS100.11a, Wireless HART®, and RF4CE protocols. That being said, what are the differences? Why can't a Bluetooth device communicate with a Zigbee device directly? What is holding this miracle back from happening?

Each time information is sent over the air wirelessly, those data need to be converted from digital information to analog information. The "real world" is analog. Signals are modulated and the digital signals are transformed into waves. All wireless schemes are not the same.

For example, 802.15.1 (Bluetooth's physical layer standard) and 802.15.4 (Zigbee's physical layer standard) are not the same in their modulation schemes, even though they transmit at the same 2.4 GHz spectrum.

The hardware becomes easier to implement if I know what type of network one needs to communicate with. The product created will either communicate via Zigbee, Bluetooth, USB, or one of many other standards. Selecting the protocol standard that is the best for the application is based on some critical criteria: What is the power consumption? What is the distance from device to device? What is the amount of data that need to be transmitted or received and how long will it take? Are there coexistence issues to be concerned about with regards to similar networks?

To look at this in more detail and drill down, what if my application needs to talk wirelessly and send and receive a lot of data, but there is only a limited power source (battery)? What if there is a device that is located in an area with many sensor networks sending data continuously? These are the types of questions that need to be assessed before making the selection of the network interface and protocol. Of course, an engineer may become limited if the network interface and protocol is already in place. If one needs to interoperably communicate on a Zigbee network, the choice has already been made. But this generally applies to certain types of devices: low power or low data rate.

These questions and more have been issues in network systems and embedded design for quite some time. The way the market responds is by petitioning the standard boards to modify the specifications to meet new requirements of the interoperable networks.

For instance, Classic Bluetooth is power hungry for battery-powered applications. But this is what was created years ago, before engineers, though of products and ideas that are limited by power consumption. Bluetooth had a particular market segment it was addressing, and as the cellphone became more and more pervasive, companies wanted to make different type of products to communicate with cellphones. Now that products have evolved and the shortcomings of Bluetooth have been discovered, corporations with their engineering teams need to figure out ways to improve the standards that exist. Bluetooth, a pervasive standard that is in virtually every cellphone, would be vastly improved if the power consumption could be reduced considerably. By reducing the power consumption, the battery life could be extended and smaller batteries could be used in application for a longer amount of time without changing them. This is an issue for the standards board to solve. How does the standard evolve without affecting the infrastructure that Bluetooth has created?

In turn, the Bluetooth standards board created Bluetooth Low Energy (LE). This new standard will address this low power consumption requirement and widen the types of applications that can communicate on a Bluetooth interoperable network.

Table 11.1 shows the differences between the Bluetooth standards and Zigbee.

The cellphone will remain the access point for the wireless network and soon cellphones will be able to handle both the Classic Bluetooth standard

TABLE 11.1. Differences between Wireless Standards (according to Standards Organizations)

	Classic Bluetooth	Bluetooth LE	Zigbee
IEEE physical layer standard	802.15.1	802.15.1	802.15.4
Network type	PAN	PAN	LAN
Frequency	2.4 GHz	2.4 GHz	2.4 GHz/915 MHZ
Data throughput	721 Kbps	1 Mbps	250/40 Kbps
Battery life (maximum based on application)	Months	Years	Years

PAN, personal area network; LAN, local area network.

and the new Bluetooth LE standard. This exemplifies the evolution that needs to take place in order for interoperability to ultimately take shape and be here to stay.

11.4   FIRMWARE

In my estimate, the most complex part of an interoperable system implementation is the firmware or protocol stack that takes the data from or sends data to the physical (PHY) and data link layers and arranges the information in the correct format. Depending on the complexity of the communication protocol, this will give insight into the various layers of the OSI model that will be used in the implementation. In other words, which layers are standard in the approach and which require customization.

Let's review the layers of the firmware important to the protocol stack (Figure 11.2).

The application layer that incorporates the application, presentation, and session layer is to be implemented or handled by the user. This, for example, could be the firmware code written by the design engineer/user in order for the information to be acquired or received and additionally processed, manipulated, or having a function or algorithm performed on the information.

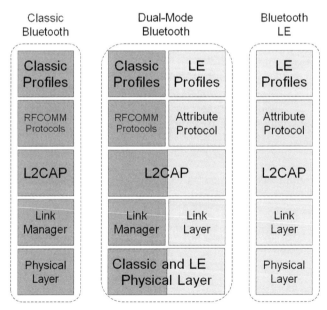

FIGURE 11.2. Classic Bluetooth, dual-mode Bluetooth, and Bluetooth LE stack layers. RFCOMM, radiofrequency communication; L2CAP, logical link control and adaptation protocol.

For the transport and network layers of the stack, there is generally no customization by the embedded engineer. It may be necessary for some firmware to be written to link the physical and data link layers to the network layer, but the communication flow control is in place. It is part of the implementation of the stack. Standards exist for interoperability, but also for ease of implementation.

This brings us to the application layer within the protocol stack. This is the part that needs customization and selections need to be made in order for functional interoperability. We need to make sure that the information acquired into the embedded system is formatted in such a way that every other device on the interoperable network can understand what is being broadcast.

If we take the protocol stack representation for the Continua Health Alliance's IEEE 11073 implementation, the transportation media for the information needs to be the same for interoperable devices on the network. The Bluetooth physical layer, the data link layer (previously mentioned as hardware), the healthcare-specific network, and transport layers all need to be the same for all interoperable devices on the network. These layers will also be the same for all devices on the network. The custom portion is in the application code that needs to be written for the device; in other words, the device receives data from the outside world and is put through particular process steps of the stack. Those data are then sent through the remaining upper layers of the protocol stack. In the case of Continua Health Alliance IEEE 11073, a particular type of device needs to be stipulated. What kind of health-care device is being created? What format does the information need to be in so that the system can understand? The device specification layer, which in the protocol stack layer would be present in the application layer, specifies which device we are receiving the information from. The choices could be a thermometer, a blood pressure monitor, or a glucose meter. This choice is something that the design engineer would obviously need to make and link the application code to the device specification layer that would then transfer down to the transport and network layers and out the data link and physical layers.

Speaking on the topic of core competency, a device manufacturer should more than likely concentrate their efforts on the product that they manufacture. If the manufacturer makes glucose meters, they should put most of their effort on the acquisition of glucose measurements. The embedded system engineer would not, in most cases, build a protocol stack from scratch. The time and effort involved would not be worth the return on investment. It is more cost-effective to purchase an industry-proven protocol stack from a provider in the market. This way, the integration/implementation can be the focus of the effort instead of reinventing what has already been invented and proven time and time again. It is by no means a simple task to integrate an established protocol stack into the embedded system and get it to work right without definitive development time, but one could imagine that if there is too much to worry about with regards to firmware, the debug and design issues could go on for a long time and end up costing more than the price of a proven, modular communication protocol stack.

11.5   PARTITIONING THE SYSTEM

Now that we have a cursory understanding of the separate components of the embedded interoperable system, let's take a look at how one might partition the system.

When considering the partitioning of the system, there is a definite trade-off between cost and time to market.

The simplest and most straightforward implementation is to have a host processor or microcontroller separate from the communication section of the system. This could mean that the application is handled in the host section and the communication stack is a resident on another microcontroller in the system. This has advantages from an integration standpoint, but a detriment with regards to cost and being fully integrated (full integration would save space in the design as well as cost).

Our discussion on the hardware and firmware looked at the system from its most integrated and most compact implementation. This is ideal, but in no way the easiest method.

Figure 11.3 shows the different ways to look at the implementation. The left side shows the most integrated and the right side shows what I call the modular approach.

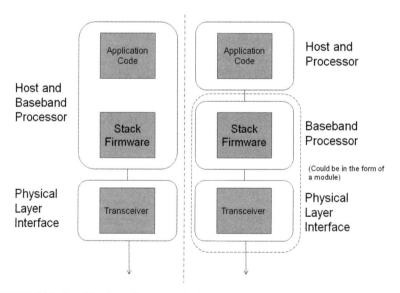

FIGURE 11.3. Partitioning. The left side is the most integrated and the right side is modularized.

As long as the designer of the system uses the prescribed physical layer interface and the prescribed firmware protocol stack, the device should meet the desired result of being interoperable in the system/ecosystem. But how is that verified? In almost all cases, the product that is created needs to go through interoperability testing in order to say that the product is interoperable with whatever standard the product was created to communicate with.

For instance, proposed Bluetooth products need to go through stringent interoperability testing that is overseen by the Bluetooth Standards Board. This testing and certification will maintain that the product is usable by other Bluetooth devices.

The standards board of the Continua Health Alliance is a little different in this regards. The Continua Health Alliance utilizes standard transport protocols such as USB and Bluetooth and specifies specific data structures for the different types of health-care devices that they will communicate with. In other words, a thermometer will gather certain types of data in a particular format. The system will know that this particular device is a thermometer. The relationship between Continua Health Alliance and the Bluetooth Special Interest Group (SIG) and USB standards organization is based on a petitioning by the Continua Health Alliance to the USB and Bluetooth working groups to form a group for new application layers that were specific to the Healthcare device field.

The IEEE 11073 specification, along with the USB and Bluetooth physical layer and stack, make up the full Continua Health Alliance interoperable product.

The created product must not only receive certification for the USB or Bluetooth body using their designated application layer for the particular device, but then subsequently must undergo Continua Health Alliance certification. It is drawn out, but will guarantee that the product will work in an interoperable ecosystem of health-care devices.

This is critical in order to drive down costs of implementation and create a "plug and play" environment for nontechnical users.

With the particular background in place, one can see that the choices need to be made for how the system is implemented.

Due to the fact that there is certification and compliancy that is needed, the ideal way would be to separate the application processor, where the system acquires and manipulates data from the communications section of the system. Why is this important? If the application code needs to be changed in any way, it could potentially force the need to change some of the interface into the protocol stack. I am not saying that the protocol stack would change; it is more about the interface to it. Then compound the fact that certification and compliancy *may* need to be redone because there have been changes to the code within the system. The communication stack portion may not have been changed, but rules are rules and this process may need to be restated. Not to mention in the world of Food and Drug Administration (FDA) certification and so forth, the combinations are endless as far as making changes to an established, existing device.

This compartmentalized approach is the best for those who have no experience with communication stack integration, but could potentially add tremendous cost to the system. There would need to be a serious discussion on what the trade-offs would be. Cost could be in the form of dollars and cents, but also in the form of power consumption and space considerations.

The trade-offs are difficult to assess. Imagine there is a device on the market that has no communication capabilities. The information acquired stays inside the device. Now imagine that an engineer needs to add USB, Bluetooth, and Zigbee capability. Those communication interfaces don't need to all be on the same device. The product will only have one. Is there a way to create a modular design that has "plug and play" capability for each communication protocol? Please see Figure 11.3, which shows the application with the possibility of adding different functions. It is not as easy to implement as it appears in the pictures, but the idea of keeping the communication protocols separate from the application processor adds value to the system.

Conversely, if a decision was made to use Bluetooth in the system, and cost and power consumption were main considerations, one would have to look at the design issue differently. Minimizing the number of components is ideal from a power and cost standpoint, but it creates some integration issues. The application code on the processor needs to be intertwined into the communication protocol firmware on the same processor. This is easier said than done, plus each time a change in made to the code in the processor, a recertification or compliancy may need to be completed. Those recertification steps can be as easy as documentation or as difficult and expensive as full-up hardware/firmware/network compliance testing.

I hope this section offers a better understanding of the partitioning of the interoperable system. There are many considerations to take into account and should be properly vetted and analyzed before moving forward with a particular design.

11.6   EXAMPLES OF INTEROPERABLE SYSTEMS

Interoperable systems have become more and more popular in the last coming years. Imagine if a company makes a really top-quality pressure transmitter for the oil and gas industry, but that pressure transmitter cannot speak the same language as other parts of the system. This particular pressure transmitter couldn't be used. How do we make sure that the best products in the market are all able to speak to each other? This is the fundamental idea expressed in the beginning of the chapter and we will now give examples of where this idea is most effective.

When I think of interoperable systems, two different ones come to mind in everyday life. The first is the headset that works with your cellphone. I can go to any store and purchase a headset and as long as it is Bluetooth compliant, I should be able to use it with my phone. The second is the USB mass storage drive. No matter which one I purchase, it should work with my computer, no matter how old the drive is or how old the computer is. These two application ideas are taken for granted, but over the years of working on standards and embedded systems, the consumer sees no difference and expects the products to work in unison within the system. I think this is a very important point. The ability for all engineers involved to create products that all understand each other is not a trivial task, but is necessary for the next generation of products to talk to each other.

The ideal situation would be that any engineer could create a product and make it so that it is agnostic to the transport protocol that is uses. More important is that the application is the core competency of the engineering team and the interoperable communication is a commodity that a buyer can pick and choose as they need, depending on the type of network they have.

Let's think about health-care systems and the implications of interoperable solutions. The Continua Health Alliance does not create any particular physical layer, transport layer, or upper layer stack layers. They decided to utilize already established standards that are known and well adopted in the market.

For wireline, their members decided to select USB as their first implementation. If the creation of healthcare-specific device class within USB, Continua supports could create an interoperable USB device network that could identify health-care devices and exchange information correctly so that all on the network understand.

For wireless, initially Continua selected Classic Bluetooth for the transport communication that all would follow. The Bluetooth SIG commissioned the creation of a healthcare-specific profile for Bluetooth devices. The Healthcare Device Profile (or HDP) was introduced and has started to be adopted by medical device manufacturers worldwide. Bluetooth is pervasive, similar to USB (all computers have USB ports), but Bluetooth is in almost all cellphones and there is no need for a cable. The Bluetooth-enabled cellphone acts as a conduit to the outside world. The information is gathered and forwarded to anywhere in the world. A glucose meter manufacturer can have their device talk to a cellphone and a pedometer manufacturer can have their device talk to the cellphone. The cellphone can run an application that could take the two pieces of information and analyze the data. Because the devices have set parameters for what they measure, all HDP-enabled, Bluetooth cellphones would understand the data and be able to do something with the data if the cellphone had an application on it for that purpose. The other cell phone could receive the information from various HDP-enabled Bluetooth devices and forward the information to a repository or a central computer that could do the analysis. The possibilities are endless and the applications are real and powerful. The healthcare-specific information, because it is interoperable, sets the ground rules for all future devices. As the devices become more and more complex in their function, but easier for the user, the possibility of measuring almost anything on the body in the comfort of your own home could help in the treatment and prevention of serious diseases and health problems.

As the Continua Health Alliance expands their reach, beyond USB and Bluetooth, they look to low power wireless, such as Bluetooth LE and Zigbee, and the adoption of additional established and proven standards make for more robust and application-specific devices. As products become lighter and need less power, Bluetooth LE and Zigbee will be there to serve the ideal purpose. In other words, as new devices are created and thought to be impossible for the system, the Continua Health Alliance will be there to provide the foresight and guidance to get new and once-thought impossible products to the market. Any medical device company can be part of the system, no matter how big or small, simple or complex, as long as they can communicate using the common language. The possibilities for innovation are then endless.

In conclusion, the concept of interoperable systems is becoming more and more important in the embedded systems space. The example of health-care devices and their need to become interoperable and communicate in a robust, complex ecosystem makes a compelling story on why interoperability is so important. The future will bring more and more devices to the market. The real task is how we make sure that these devices can effectively communicate with other devices and create value for the consumer and the world.

12 Software Modeling Approaches for Presilicon System Performance Analysis

KENNETH J. SCHULTZ and FREDERIC RISACHER

12.1 INTRODUCTION

Good-bye mainframes, and so long to server farms. Due to the economic driver of sheer market size, the smartphone is today's highest performing digital platform. Users have become conditioned to expect a single device to provide them with best in-class experiences for multiple functions that were formerly performed by single-function consumer electronics: computers, multimedia players, cameras, navigational devices, gaming consoles, and the list goes on. Moreover, consumers further expect the humble smartphone to provide all that exceptional functionality in the palms of their hands, running cool without fans, and without the need to plug into a power source, ideally for days. Welcome to the design problem faced by mobile handheld device architects.

The epicenter of this performance cyclone is the application processor (AP), the central processing unit (CPU) that is the hub to the various user and radio interfaces. The AP is sometimes integrated together with the baseband modem functionality into a combo chip referred to as a baseband processor (BP), though at the high end, these two functions are usually handled by separate chips. No matter the system partitioning, processing demands on the AP are driving architectures toward approaches like multicore, multimemory controller, heterogeneous graphics processing units (GPU), and other dedicated multimedia and gaming accelerators.

Surely many person-years of innovative intellectual property (IP) development goes into these various on-chip components, but in a sense the ubiquity of ARM architecture-based cores and the easy availability of third-party coprocessing elements combine to create commoditization at this level, and the true differentiation comes in the way the components are stitched together.

Embedded Systems: Hardware, Design, and Implementation, First Edition.
Edited by Krzysztof Iniewski.
© 2013 John Wiley & Sons, Inc. Published 2013 by John Wiley & Sons, Inc.

The questions faced by the integrator multiply. Does the cache coherency scheme work when various cores are disabled and re-enabled? Does the network-on-chip (NoC) provide adequate bandwidth to multiple components in complex aggregated user scenarios? Does it all fall apart at an insurmountable bottleneck at the memory controller? Or does the phone overheat at the frame rate the gaming enthusiast takes for granted? And even trickier, how should the software architects partition and schedule tasks across all those resources for optimum performance and battery life? The sheer complexity of interactions and the consequences of getting it wrong lead to two unfortunately opposed realities: presilicon performance investigation is essential, and it is very hard to do.

Another set of constraints is faced by phone manufacturers, like us at Research In Motion (RIM, Waterloo, Ontario, Canada), that purchases APs from silicon vendors, rather than making them ourselves. We do the following:

- provide the software that runs on someone else's hardware
- have *influence* over the AP architecture and design, but not control
- are dealing with the AP as a "gray box," which may be blacker or whiter in different cases
- do not get access to gate-level design, and consequently don't have the ability to alter it to explore what-if scenarios

The ultimate commercial success of a smartphone is heavily dependent on fast time to market. Being the first (or at least one of the first) to introduce a phone with a given performance class of AP is table stakes. If we wait until the chip is in our hands, and then take the time to analyze it, we lose market share because we're late. If we, on the other hand, blindly take the chip as is without doing any analysis, we run a high risk of missing performance requirements and losing market share for that reason.

Of course, we cannot influence fundamental architectural features and capabilities of an AP after it has begun volume production. Some form of presilicon analysis capability is required—the earlier, the better. Performance bottlenecks must be identified very early in the development cycle of the silicon—in the architecture definition phase, before design implementation. Then, if performance issues are found, architectural fixes are still possible.

Some of these fixes could be rather substantial in scope, but potential end-user performance improvements need not require major structural overhauls. Sometimes, by consideration of the overlay of our (perhaps unique) use-case scenarios on the tunable aspects of the design, we can get notable benefits from relatively minor changes—examples of such fine tuning include first in, first out (FIFO) size, arbitration policy, and scheduling schemes.

To perform presilicon analyses, we need a simulator (or a small set of them) and we need models. Simulators could be commercial or proprietary, and are not the topic of this chapter. If we focus on the models, instead, it is clear that we must model the key *hardware* components involved in the scenarios under investigation. For multimedia use cases, for example, this is usually the portion of the AP that comprises the processors, multimedia-related accelerators, system buses, and memory access path (including cache hierarchy and memory controllers). These components must be modeled with sufficient accuracy to provide confidence in the simulation results. Models of the hardware blocks can be derived from various sources: from previous designs, from register-transfer level (RTL), or they may be developed from scratch. Hardware models are also not the focus of this chapter.

Less obvious is the need to somehow represent *software*—the simulation stimulus. As mentioned earlier, we, the phone manufacturer, develop the software that runs on the AP, which is manufactured by the silicon vendor. If the AP has not yet been physically manufactured, we're not going to have software running on it yet. Reuse of legacy software as is would not result in stimulus that stresses the new silicon components under analysis. And any software porting exercise to explicitly take the new components into consideration would be too time consuming (recall our time-to-market imperative).

Therefore, rather than using actual software, we employ a model of that software. It is our focus in this chapter to describe how to go about the creation of such is a model.

As a primary requirement, our software model should stress the hardware in a way that mimics actual software as closely as possible. Further, a truly useful model is one that is easy and cheap to derive. Our success at meeting these criteria can be determined by test applications for which the software *does* exist. (Although a software model is only really useful in situations where the hardware is new and the actual software does not exist, we can evaluate various software modeling approaches using experimental scenarios in which we have both established hardware and software.)

In this chapter, we consider three approaches [1–2] and evaluate their relative suitability. One approach is to describe the software at a task level. The second option is to derive the models from captured traces of software execution, either on legacy hardware or on a virtual platform. A third approach is to replace the data processing element by a statistical description of the traffic it produces.

Section 12.2 of this chapter describes the methodology options in more detail, while Section 12.3 provides examples and results.

12.2 METHODOLOGIES

When performance analysis needs to be done before silicon, various factors influence the choice of the software modeling methodology. The three approaches below cover a wide range of needs and constraints.

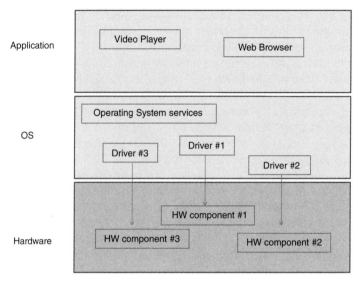

FIGURE 12.1. Simplified description of the software layers used for modeling purposes. Figure/Table © 2011 IEEE. Reprinted, with permission, from F. Risacher, K. Schultz, "Software agnostic approaches to explore pre-silicon system performance", in Proc. 2011 IEEE International High Level Design Validation and Test Workshop (HLDVT), pp. 116–120.

12.2.1 High-Level Software Description

The software architecture implemented on smartphones can be represented by a layered view, as illustrated in Figure 12.1. The operating system (OS) includes OS services, as well as the drivers that control the dedicated hardware blocks. The application layer resides on top of the OS and utilizes the various services provided by the lower layer.

To create a descriptive model of the software behavior, we need to expose greater detail in both the application and OS layers. Rather than expending the effort to model all aspects of the software, in our approach, we merge these two layers together and add details relevant only to the specific scenario we want to explore. The software interacts with the silicon in different ways for varied usage scenarios, such as video playback, Web browsing or three-dimensional (3D) gaming. For each of these use cases, we generate a unique representation of the software architecture. We have achieved good results by performing the necessary decomposition to the OS task level, which is the finest granularity at which the OS can schedule.

Such an abstraction is shown in Figure 12.2, where three different tasks are used to describe the simplified use case being considered. The execution of the flow is initiated by an interrupt from one of the hardware blocks. This could be generated by a FIFO that is part of a hardware accelerator (audio or video),

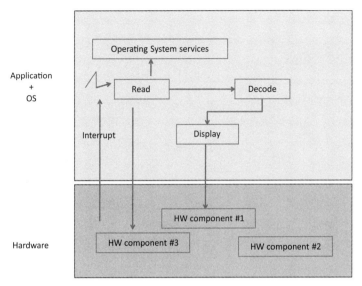

FIGURE 12.2. For scenario illustration purposes, the OS and application have been merged and abstracted as three tasks interacting together. In this configuration, the execution of the first task is triggered by an interrupt coming from one of the HW components. Figure/Table © 2011 IEEE. Reprinted, with permission, from F. Risacher, K. Schultz, "Software agnostic approaches to explore pre-silicon system performance", in Proc. 2011 IEEE International High Level Design Validation and Test Workshop (HLDVT), pp. 116–120.

for instance, and it may be providing notification that data are available for further processing. The first task ("Read") reads data, interacts with the OS services, and then hands off to a more specialized task, the decoder. Once data are decoded, the decoded frames are sent to the display. From a performance perspective, this representation, though conceptually simple, actually models most of the burden required for this scenario.

Each individual task in the flow ("Read," "Decode," and "Display," in this example) is then described using a modeling language such as SystemC. Though initial analyses can be performed with very simple descriptions, these can be refined as more information about the implementation (from both software and systems perspectives) becomes available, thus incrementally improving simulation accuracy. Therefore, at any point in the design process, the model is populated with the best available architecture description for the relevant elements.

When sufficient details become available, each task may be further decomposed into multiple sub-blocks having individual properties, as shown in Figure 12.3. A processing element is defined as a block that consumes resources. The model of such a processing element does not include any code representative of the actual processing, but instead is comprised only of a description of the

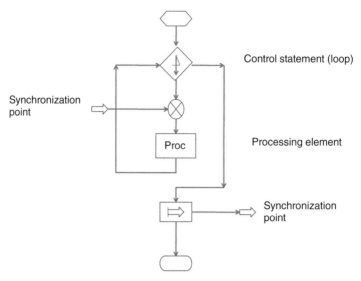

FIGURE 12.3. An OS task may be further broken down into processing elements, control statements, and synchronization points. Figure/Table © 2011 IEEE. Reprinted, with permission, from F. Risacher, K. Schultz, "Software agnostic approaches to explore pre-silicon system performance", in Proc. 2011 IEEE International High Level Design Validation and Test Workshop (HLDVT), pp. 116–120.

resource usage, which depends on the nature and amount of data that needs to be processed. Such a description could include parameters such as memory throughput, cycles, or any combination of descriptors for the load. The control statements are generally loops used to iterate over certain processing steps. Synchronization points are used to start or resume execution. They carry no payload; rather, they merely interact with other simulation components via a flag, or perhaps by signaling that a data threshold required for synchronization has been crossed.

12.2.2 Transaction Trace File

The above approach can be time consuming to implement, so for some performance investigations, we would select a quicker modeling methodology, even if it is less accurate. Such a method is described here, and it relies on measurements taken on existing hardware.

A trace file is a representation of activity measured on such hardware, recorded and thereafter available to be processed offline for further analysis. The data are gathered by either probing and monitoring signals exposed on the surface of a printed circuit board (PCB), or by relying on diagnostic modes implemented in the silicon that, when activated, stream otherwise internal buses to observable chip pins. Such modes can be proprietary to a given silicon vendor, or even specific and custom to a given AP. Alternatively, they may be available as a feature with broadly available IP (ARM Coresight [3], for example). Across this range of possibilities, the amount and usefulness of data that can be gathered and stored in trace files can vary greatly.

In cases where we have access to an AP's internal signals that can be directly logged, the resulting traces are immediately applicable as stimulus in our simulations, representative of the traffic generated by embedded components.

However, as AP customers (not AP designers and manufacturers), we often have to get data in situations where the proprietary means of gathering internal traces are not available to us. In such cases, we are constrained to the signals that otherwise pass through the PCB. Of these signals, the most useful by far are the memory buses. As straightforward as that seems, there are unfortunately other considerations working against us. Notably, these signals are routed on the PCB layers that best enable propagation of the very fast signals with optimum impedance and minimal delay, and these layers are buried below the surface of the board. As well, the security threat posed by exposing memory address and data lines during boot (or other supervisory modes) could leave the phone vulnerable to hacking, and this consideration also requires that these signals be routed below the surface.

The situation seems dire, indeed, except for the emergence of package-on-package technology that locates memory chips on top of APs, thus resulting in the exposure of the memory bus signals on balls on the top of the AP package. While these are certainly not accessible in production phones where the memory chips are securely bonded on top, we can, in prototype devices, insert an "interposer card." Such a card is effectively a vertical sandwich layer between the AP and the memories, and we can use signal routing on this card to break out a selection of the memory signals.

With unfettered access to this "fire hose" of useful information, how do we "drink from it"? (Note that this question also applies to the more fortunate case above, in which access to internal signals is possible.) To use rough numbers, tracing 64 bits of address and data at 300 MHz results in a data rate of approximately 2.4 GB/s. Given the challenges in routing this bandwidth a few centimeters on a board, it seems a leap indeed to get it into test instruments a meter or more away. And given that useful traces can be several seconds in duration, collecting the data in a dedicated high-speed memory on the interposer card is also not an option.

The solution is to drink only a portion of what's coming from the gushing hose, enough to adequately describe the pertinent aspects of the software behavior. Figure 12.4 shows a conceptual view of this approach, wherein the "custom capture board" has the capability, realized in a field-programmable gate array (FPGA), to analyze the exposed signals for the purpose of creating an accurate but compact representation of multicycle bursts. The analysis actually focuses on the control signals, rather than the address and data lines, and therefore requires in-depth understanding of the memory communication

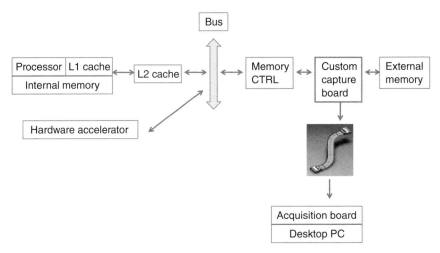

FIGURE 12.4. Description of the custom capture board used to record memory transactions. Figure/Table © 2011 IEEE. Reprinted, with permission, from F. Risacher, K. Schultz, "Software agnostic approaches to explore pre-silicon system performance", in Proc. 2011 IEEE International High Level Design Validation and Test Workshop (HLDVT), pp. 116–120.

protocol employed. This approach lowers bandwidth and trace storage requirements substantially, but still provides the information we need to create trace-driven stimulus for our simulations, effectively enabling reconstruction of the transaction.

The following line shows a simple example:

$$100: (8)\text{bwrite } 0xA0000000 \ 0xAA$$

The first number is the timestamp of the transaction (the time at which the transaction is active on the bus). The second parameter specifies the size of the transaction (burst of 8 in the example). The third parameter is the type of transaction, which can be a burst read or write. The two last parameters are the address and data. It turns out that logging the data is optional, and though the presence of data in a trace file can provide hints as to the nature of the operations taking place, it adds little value to the stimulus definition. In fact, omitting that data ensures that the trace files are virtually clear of proprietary information, and thus enables sharing of the resulting trace files between companies for cooperative investigations.

After the memory bus traces are filtered and processed by the FPGA on the interposer card, the resulting data are output to an acquisition card attached to a standard desktop personal computer (PC). The host application then translates these data to generate a text file in which transactions are represented using the Socket Transaction Language (STL) specified by the Open Core Protocol International Partnership Association (OCP-IP) [4]. STL is an easy-to-use standard transaction description language that is used to specify transactions exercising memory or other slave devices, and which enables different simulation environments to share stimulus and results.

Now that we have solved the access and bandwidth issues, we can manipulate the data gathering conditions to expose otherwise hidden information. For example, we can alter the software to selectively enable and disable various functions active in the AP, thus allowing us to decompose the traffic by source. Additionally, we can deactivate the level 2 (L2) cache, so that all transactions that would normally first attempt to access that cache would instead be forced off-chip and be fully exposed at the observable memory bus. Combining such manipulations allows us to perform cache hit rate analysis, as in the example we will describe in Section 12.3.2, among other interesting and useful studies.

There is an alternative method of gathering trace files that entirely avoids the messiness of wires and cables, as well as worries about signal accessibility and bandwidth limitations: use a virtual platform (VP) [5]. VPs are software emulations of hardware environments that have the primary purpose of enabling software development before silicon exists, thus accelerating development schedules, de-risking bring-up of new silicon, and ultimately enabling faster time to market. When a VP is available presilicon, it has the potential of solving all our problems—we have software running on hardware that does not yet exist, and which we therefore still have the opportunity to influence. Observability is perfect, and we can collect trace files at any internal node or bus and use them to drive that same internal point in our architectural simulations.

The practice is not quite as rosy as the potential, however. Software bring-up still takes time, even if that time is in the presilicon phase. By the time software with adequate application-like functionality is running on the VP, the silicon design may be almost complete, and the likelihood that the architecture is still influenceable is quite low. This cold reality leads to a modified approach: Instead of employing a VP of the AP under study, we employ the VP for an existing chip with established software—the same chip from which we were, a few paragraphs ago, physically gathering traces. We merely leverage the VP for ease of use and observability. But we caution that there are additional drawbacks to this approach, notably (1) the timing accuracy of VP simulations is only approximate, and not cycle accurate, and (2) depending on the scope of the presilicon software development efforts undertaken with the VP, the pertinent software functionality may never have been implemented on the VP, or support may be stuck at a relatively old and flakey build.

In this section, we have described an approach that allows us to capture traffic to and from memory on actual (or virtual) devices running production software. The resulting trace file can then be used as is in a modeling environment to exercise a specific hardware model. Although the trace file is a representation of the software of interest, more strictly speaking, it's actually a representation of that software *running on hardware* that is, in some significant way, *different* from the hardware we intend to simulate. This is a problem.

Maybe we can adjust either the software or the traces themselves to compensate for those hardware differences? But if we knew *exactly* how to make such adjustments, then that would imply that we have enough knowledge that we don't need to run the simulation at all! Fortunately, it is possible to compensate for some of the most obvious changes in the hardware by performing semiautomated processing on the trace files, such as upsampling or downsampling to represent a change in clock rate (that operation may sound complex, but practically it is implemented simply by modifying the time stamps of each line of the trace file).

An additional use for a trace file is as an input for a stochastic traffic generator, described in the next section.

12.2.3   Stochastic Traffic Generator

This third method, complementing the previous two, has the primary benefit of greater flexibility in trace file generation. Using this approach, the traffic injected into the system by any initiator can be represented using a statistical description.

Statistical traffic characterization is a well-established field of research, and the knowledge gained by such characterization can be used for various analysis purposes, in addition to (as in our usage) reproduction of the studied traffic in a simulation environment. Readers interested in detailed treatments of this field can review previous work on subject matter varying from Internet traffic [6, 7] to single-shooter and massively multiplayer online games [8, 9]. The beauty of this approach, as mentioned above, is flexibility. Depending on (1) the need for more accurate characterization, (2) the actual complexity of the traffic, (3) the statistics aptitude of the modeling team, and (4) the time available, one could undertake a wide variety of statistical modeling approaches, using Markov chains, Gaussian/Poisson/other probability density functions, or simple random distributions.

For memory investigations, the simplest traffic descriptions would comprise parameters like average bandwidth, address range, and read-to-write ratio. Although gathering data to support such a modeling methodology would be trivial, and the traffic generator would be simple to implement and efficient to run, such a representation may be too simple and the resulting traffic may not closely enough match what is observed on real hardware. More complex sets of parameters must then be used, including perhaps a characterization of the interburst latency, and testing of similarities between gathered data sets and various distribution models.

For longer transaction sequences, a single function is often not adequate to characterize the traffic. Sometimes, a use case can be decomposed into a number of well-understood phases or steps. Or the data content can vary in a predictable pattern, as with the different frame types in encoded video [10]. Or different statistical properties may be exhibited in short- and long-term characteristics, as with IP traffic on carrier networks [6]. A more refined

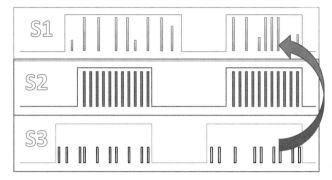

FIGURE 12.5. Three different segments are generated with their own parameters. The generator can loop through those segments to model a specific traffic pattern. Figure/Table © 2011 IEEE. Reprinted, with permission, from F. Risacher, K. Schultz, "Software agnostic approaches to explore pre-silicon system performance", in Proc. 2011 IEEE International High Level Design Validation and Test Workshop (HLDVT), pp. 116–120.

approach in such cases is to break down the traffic into small segments, each having different statistical properties [11]. These sequences are combined to form a longer pattern, as illustrated in Figure 12.5. We use an XML file to describe the statistical properties for each segment, as well as how the multiple segments are linked together.

The determination of the number of segments can make use of a semiautomated process based on k-*means* [12], an analysis tool used to group a large amount of data into clusters. The name is derived from the user's ability to choose the number of clusters (k) that are to be generated. The optimal choice of k is integral to the quality of results that are obtained from the clustering.

The determination of the correct parameters to represent the traffic is the challenging part of this method. Our approach is to first capture a transaction trace file using the methodology described above, and then analyze the trace results to estimate the parameters and the number of segments needed to appropriately represent the traffic.

12.3   RESULTS

The three approaches described above—the high level software description, transaction trace file capture, and the stochastic traffic generator—have been applied to three different projects: characterization of current consumption during MP3 playback, cache size impact on Web browsing, and GPU traffic characterization, respectively.

The application of a method to a specific project has been dictated by external constraints such as software architecture documentation or chipset availability. The more methodology choices there are available, the easier it is to adapt to such constraints and still generate useful results.

12.3.1 Audio Power Consumption: High-Level Software Description

One of the performance metrics of interest to consumers of handheld devices is battery life for audio scenarios. Modeling of such power consumption is a complex process that involves simulating the software–hardware interaction and then taking the results and performing postprocessing using milliwatts per megahertz (mW/MHz) scaling factors. This particular analysis was focused on MP3 audio playback. The MP3 application was modeled using the high-level software description method described in Section 12.2. For validation purposes, the current consumption results generated by the model were compared with measurements done on the actual device. As the software model was refined over time, the convergence of this comparison was tracked, as shown in Figure 12.6.

It can be seen that the first estimated data point is quite far off, in fact being half the measured value. The model was iteratively refined, adding more details and better performance estimation of the processing blocks. Two specific refinements in this case were:

- *The Dependency of Processing Duration on Phase-Locked Loop (PLL) and Clock Frequency Settings*: We initially assumed that this was a linear

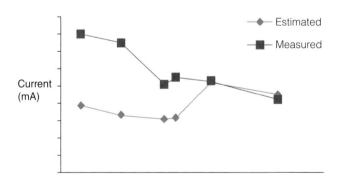

FIGURE 12.6. Estimated and measured MP3 current consumption. After some iterations, the model is accurate enough to give reliable predictions. Figure/Table © 2011 IEEE. Reprinted, with permission, from F. Risacher, K. Schultz, "Software agnostic approaches to explore pre-silicon system performance", in Proc. 2011 IEEE International High Level Design Validation and Test Workshop (HLDVT), pp. 116–120.

relationship, but after investigation realized that we needed to more accurately model the impacts of factors such as interrupts and memory access latencies. It turned out that, when combined with slower clocks rates, these secondary effects resulted in greater-than-linear increases in processing duration, and therefore current.

- *The Need to Model Interprocessor Communication*: For a given use case, some components of the system not initially modeled may have an indirect impact on the performance. In the case of MP3 decode, the communication mechanism between the modem processor (the source of the audio samples) and the AP doing the audio processing was not originally modeled, because we understood that this process was done in parallel and in a very short time. During our model refinement process, we performed specific profiling on this aspect, which revealed that the latency of the interprocessor communication was significantly higher than expected. This latency was impacting CPU utilization and power consumption in a way that was not accurately represented in earlier versions of the model.

Over time, as a result of our refinements, the current consumption values derived from our model and simulations converged to the experimentally measured value. As an aside, it should be noted that the experimental measurements done at early stages of software development also exhibited large fluctuations as time progressed, due to initial instability of the application, refinements in power management, and other optimizations that inevitably occur.

The model in this case has limited practical use because the silicon is already available. However, this kind of comparison is a necessary step to validate the accuracy of the software model. Now that the model has been shown to be accurate when run on "known hardware," it can then be easily retargeted to next-generation chipsets to predict power consumption before measurements are possible.

12.3.2   Cache Analysis: Transaction Trace File

The objective of this study was to better understand the impact of L2 cache size on Web browsing application performance. Our interposer capture board (described in Section 12.2.2) was used with a handheld device running the Sunspider [13] JavaScript benchmark. On this device, the L2 cache was disabled and transactions were directly sent from the level 1 (L1) cache to the external memory, thus enabling the external observation of memory traffic that would otherwise have been directed at the internal L2 cache. After acquisition, the timestamp of the trace file was changed to reflect the appropriate CPU clock frequency for internal transactions.

The resulting trace file was used in our modeling environment to generate transactions to an L2 cache model connected to a double data rate (DDR2)

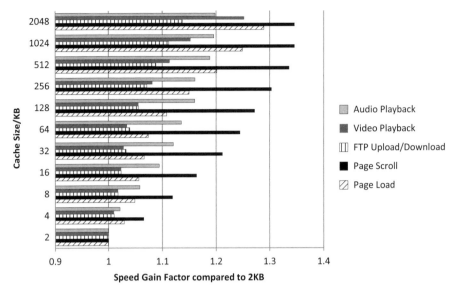

FIGURE 12.7. L2 cache analysis.

memory. Models were time-annotated with information from silicon vendors. The trace file accurately represented the combined impact of the Web browsing application running on the core processor and interacting with L1 cache, while the model represented the behavior of the L2 cache, the size of which was easily varied from 2 kB to 2 MB to perform the desired analysis. The duration of time to execute the trace file is a good indicator of the influence of the L2 cache size. When there is a cache hit, the round-trip latency is much shorter than for a miss, and we expect the hit rate to depend on the cache size. Results are presented in Figure 12.7; the speed gain is normalized to 2 kB.

Because the cache size must be selected to suit a range of applications, we gathered trace files running several different use cases for subsequent analysis, as shown in the figure. In all cases, we found that larger L2 cache sizes result in improved cache hit ratios and faster execution of the application.

Such an analysis provides insight into how changes in hardware can have different impacts depending on the intended (or most common) application. Among the examples, the one that derived the least benefit from a larger cache was File Transfer Protocol (FTP) upload and download. We only observed a 10% improvement of the speed for this use case for a 1000-fold increase in cache size. We can deduce that the limiting factor here is probably the network—a bigger cache does not speed up the network data rate. On the other extreme, the most significant benefit (30%) was observed for Web browsing, particularly during page scrolling. When a larger portion of the Web page content fits in a bigger cache, execution time is notably faster than when the page must be fetched from external memory.

12.3.3   GPU Traffic Characterization: Stochastic Traffic Generator

The traffic output by a GPU can be very difficult to approximate and model when it is used to accelerate processing tasks other than video, as its activity is then neither periodic nor frame based. We therefore chose it as a challenging test case for stochastic traffic modeling. Specifically, we employed the k-means clustering approach defined in Section 12.2.3 to create a statistical model representative of the GPU traffic, based on a hardware trace obtained from observation of the actual GPU activity. Our goal was to derive a model that, for a carefully selected set of parameters, accurately reproduced the behavior of the measured trace file. Having achieved that, we could then adjust the parameters to represent different GPUs performing different processing tasks.

After gathering the trace file and using k-means to generate an initial model, we assembled a simple test bench, comprised of a traffic generator connected to an ARM AXI bus model, in turn connected to a memory controller model. Two different traffic generators were alternatively inserted in the test bench: a simple STL trace file reader that read back the measured trace file (captured on the actual GPU, as mentioned above) and the derived stochastic traffic generator. The objective of the exercise was to verify the similarity of the results, and thus the accuracy of the representation of the stochastic model. The memory controller model reported various parameters that we used to compare the two traffic generators. Numerous trials were run with various values of k used to drive the k-means algorithm, and the best results were found for $k = 5$. The final results are presented in Table 12.1.

The first row gives the bandwidth and the page hit rate for the reference file. The second line shows the same metrics for the stochastic traffic generator. For total memory bandwidth, read bandwidth, and page hit rate, the difference is less than 10%. The biggest difference is for the write bandwidth, at 18%.

The accuracy achieved by the stochastic model is satisfactory for our performance investigation needs.

TABLE 12.1. Comparison between a Reference Transaction Trace File and the Output of the Stochastic Traffic Generator Configured Using the k-Means Analysis of the Reference Transaction Trace File

	Total Bandwidth (MB/s)	Read Bandwidth (MB/s)	Write Bandwidth (MB/s)	Page Hit Rate (%)
Reference file	1156.3	680.2	476.1	50.68
Stochastic traffic generator	1243.9	683.2	560.7	53.2

Figure/Table © 2011 IEEE. Reprinted, with permission, from F. Risacher, K. Schultz, "Software agnostic approaches to explore pre-silicon system performance", in Proc. 2011 IEEE International High Level Design Validation and Test Workshop (HLDVT), pp. 116–120.

12.4   CONCLUSION

Due to the increased complexity of a handheld system-on-chip (SoC), architecture exploration is an increasingly important part of the development cycle. Given the brief interval during which simulation can provide useful results to drive that exploration, software must be abstracted in some way. This chapter presented three different methods that have been successfully applied to three different projects. A high-level software description gives very good accuracy if the appropriate level of detail is available, at the cost of increased modeling complexity and effort. Transaction trace file capture is an alternative that makes use of bandwidth measurements on existing hardware and software, and uses these measurements as a means to exercise models; it can be fast, once the required infrastructure for measurement is in place. Stochastic traffic generators can be used early on, when only limited details of the software architecture are available, but practically need to be developed in advance, again by analyzing existing system behavior. These three methods each have their own advantages and disadvantages, but, when combined, constitute a powerful toolbox for software modeling. In general, the more methodology choices there are available, the easier it is to adapt to constraints and still generate useful results.

REFERENCES

[1] F. Risacher, "Pre-silicon system performance exploration using traffic generator based approach," presented at Ecole d'hiver Francophone sur les Technologies de Conception des Systèmes embarqués Hétérogènes (FETCH), January 2011, http://sites.google.com/site/fetch2011/programme.

[2] F. Risacher, K. Schultz, "Software agnostic approaches to explore pre-silicon system performance," in Proc. 2011 IEEE International High Level Design Validation and Test Workshop (HLDVT), pp. 116–120, 2011.

[3] ARM Coresight, "CoreSight on-chip debug & trace IP," 2012. Available: http://www.arm.com/products/system-ip/debug-trace/index.php.

[4] Open Core Protocol International Partnership (OCP-IP), "What is STL?" 2012. Available: http://www.ocpip.org/faqs_simulation_and_test.php#01.

[5] F. Schirrmeister et al., "Using virtual platforms for pre-silicon software development," 2008. Available: http://www.iqmagazineonline.com/IQ/IQ24/pdfs/IQ24_Using%20Virtual%20Platforms.pdf.

[6] T. Karagiannis et al., "A nonstrationary Poisson view of Internet traffic," in Proc. 23rd INFOCOM, pp. 1558–1569, 2004.

[7] U. Krieger et al., "Methods, techniques and tools for IP traffic characterization, measurements and statistical methods," Euro-NGI Public Report No. D.WP. JRA.5.1.2. Available: http://www.eurongi.org, 2006.

[8] P.A. Branch et al., "An ARMA(1,1) prediction model of first person shooter game traffic," in Proc. IEEE 10th Workshop on Multimedia Signal Processing, pp. 736–741, 2008.

[9] P. Svoboda et al., "Traffic analysis and modeling for World of Warcraft," in IEEE Communications Conference Digest, pp. 1612–1617, 2006.

[10] M. Dai et al., "A unified traffic model for MPEG-4 and H.264 video traces," IEEE Transactions on Multimedia, 11(5), August 2009.

[11] A. Scherrer et al., "Automatic phase detection for stochastic on-chip traffic generation," in Hardware/Software Codesign and System Synthesis, CODES+ISSS '06. Proceedings of the 4th International Conference, pp. 88–93, 2006.

[12] J. MacQueen, "Some methods for classification and analysis of multivariate observations," in Berkeley Symposium on Mathematical Statistics and Probability, pp. 281–297, 1967.

[13] The WebKit Open Source Project, "SunSpider JavaScript Benchmark," 2010. Available: http://www.webkit.org/perf/sunspider/sunspider.html.

13 Advanced Encryption Standard (AES) Implementation in Embedded Systems

ISSAM HAMMAD, KAMAL EL-SANKARY, and
EZZ EL-MASRY

13.1 INTRODUCTION

The word "cryptography" refers to the change of data representation from its original form into another different form in order to make it hidden and secured. Cryptography has two processes; the first process is the encryption, where the original data are converted into secured form using certain steps. The second process is the decryption, where the encrypted data are restored to the original form by applying the inverse to the steps applied in the encryption process. Classical cryptography started thousands of years ago. Historically, classical cryptography was used for secret communications between people. This kind of cryptography is usually applied by substituting the message letters by other letters using certain formula, for example substituting each letter in a message with the next letter in the alphabets so that the word "Test" would become "Uftu."

In modern ages, cryptography development has been a major concern in the fields of mathematics, computer science, and engineering. One of the main classes in cryptography today is the symmetric-key cryptography, where a shared key of a certain size will be used for the encryption and decryption processes.

Nowadays, cryptography has a main role in embedded systems design. As the number of devices and applications which send and receive data are increasing rapidly, the data transfer rates are becoming higher. In many applications, these data require a secured connection, which is usually achieved by cryptography.

Embedded Systems: Hardware, Design, and Implementation, First Edition.
Edited by Krzysztof Iniewski.
© 2013 John Wiley & Sons, Inc. Published 2013 by John Wiley & Sons, Inc.

Many cryptographic algorithms were proposed, such as the Data Encryption Standard (DES), the Elliptic Curve Cryptography (ECC), the Advanced Encryption Standard (AES), and other algorithms.

Many researchers and hackers are always trying to break these algorithms using brute force and side channel attacks. The DES algorithm starts to fail after several published brute force attacks. The linear cryptanalysis attack [1] could break the DES and make it an insecure algorithm. The National Institute of Standards and Technology (NIST) started to search for another algorithm to replace the DES, where the Rijndael cipher was selected as the new AES.

AES is considered nowadays as one of the most secured published cryptographic algorithms, where it was adopted by the NIST after the failing of the DES. Moreover, it is used in many applications such as in radiofrequency identification (RFID) cards, automated teller machines (ATM), cell phones, and large servers.

Due to the importance of the AES algorithm and the numerous applications that it has, the main concern of this chapter will be presenting new efficient hardware implementations for this algorithm.

13.2 FINITE FIELD

Most cryptographic algorithms are implemented using the finite field mathematics. In algebra, the field that has a finite number of elements is called finite field or Galois field (GF). Each finite field has a prime integer which represents its characteristic. For example, the finite field GF(p) represents a field with the range of integers $\{0, 1, \ldots, p - 1\}$. The total number of elements in the finite field is called the finite field order. Fields with prime integer orders has characteristic equal to their order.

Some finite fields use nonprime integer orders; in this case, the finite field will be represented using the prime number p, which represents the characteristic along with the positive integer power n. Equation (13.1) shows how to represent the finite field with order k and using the prime number p and the power n:

$$k = GF(p^n). \tag{13.1}$$

The finite field in Equation (13.1) has a range of integers that vary by $\{0, 1, \ldots, k - 1\}$.

The AES uses the finite field $GF(2^8)$, where each data byte represents a value in the range $(00–FF)_H$.

Each data byte can be represented as a polynomial over the $GF(2^8)$. Equation (13.2) shows the polynomial representations in $GF(2^8)$:

$$a(x) = b_7x^7 + b_6x^6 + b_5x^5 + b_4x^4 + b_3x^3 + b_2x^2 + b_1x + b_0. \tag{13.2}$$

Equation (13.2) can be also written as:

$$a(x) = \sum_{i=0}^{7} b_i x^i, \tag{13.3}$$

where $b_i \varepsilon \{0,1\}$.

The next subsections will explain the arithmetic in finite fields based on the characteristic $p = 2$.

13.2.1   Addition in Finite Field

Arithmetic in finite field is different from normal algebra arithmetic. In finite field with a characteristic of 2, an addition is obtained by applying bit-wise exclusive OR (XOR) operation between the operands. Equation (13.5) shows the result of the finite field addition in Equation (13.4):

$$a_3(x) = a_1(x) + a_2(x) = \sum_{i=0}^{7} b_{1i} x^i + \sum_{i=0}^{7} b_{2i} x^i, \tag{13.4}$$

$$a_3(x) = \sum_{i=0}^{7} (b_{1i} (\text{XOR}) b_{2i}) x^i. \tag{13.5}$$

13.2.2   Multiplication in Finite Field

In finite field, the multiplication product of two polynomials will be modulo an irreducible polynomial so that the result can be within the range of the applied finite field. Irreducible polynomial means it cannot be factorized and expressed as a product of two or more polynomials over the same field [2].

Equation (13.6) represents the multiplication operation of the polynomials $a_2(x)$ and $a_1(x)$ using the modulus $m(x)$:

$$a_3(x) = (a_2(x) \times a_1(x)) \bmod m(x). \tag{13.6}$$

More detailed explanation about finite field operations can be found in Daemen and Rijmen [3].

13.3   THE AES

In 1998, the Rijndael cipher, developed by the two Belgian cryptographers John Daemen and Vincent Rijmen, was published. This cipher was selected later on by the NIST as the AES to supersede the old DES. The NIST has

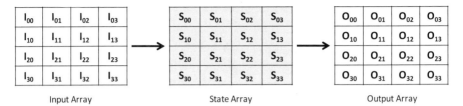

FIGURE 13.1. Input, state, and output arrays.

published full details of the AES under the Federal Information Processing Standards (FIPS) Publication 197 [4].

The AES, according to Reference [4], has a constant block size of 128 bits (16 bytes), with three different key sizes of 128, 192, and 256 bits, where 10, 12, and 14 encryption rounds will be applied for each key size, respectively. During the encryption and decryption processes, the 16 bytes of data will form a changeable (4 × 4) array called the state array. During the encryption process, the state array consists initially of the input data; this array will keep changing until reaching the final enciphered data. In the decryption process, the state array will start by the enciphered data and will keep changing until retrieving the original data. Figure 13.1 demonstrates the input, state, and output arrays.

Each encryption round has four main steps: Shift Rows, Byte Substitution using the Substitution Box (S-BOX), Mix Columns, and Add Round Key. The decryption process consists of the inverse steps, where each decryption round consists of: Inverse Shift Rows, Byte Substitution using Inverse S-BOX, Add Round Key, and Inverse Mix Columns. The round keys will be generated using a unit called the key expansion unit. This unit will be generating 176, 208 or 240 bytes of round keys depending on the size of the used key; more details about the key expansion unit will be explained later in Section 13.3.4. Figure 13.2 shows the AES encryption and decryption processes.

As can be seen from Figure 13.2, the encryption and decryption processes start by adding the round key to the data. This round key is called the initial round key and it consists of the first 16 bytes of round keys in case of encryption and the last 16 bytes in case of decryption. The encryption iteration starts with the Shift Rows step, and then the Bytes Substitution (Sub.) is applied, followed by the Mix Columns step, and finally the Round Key is added. In the decryption iteration, the Round Key is obtained before the Inverse (Inv.) Mix Columns step. These iterations are repeated 9, 11, and 13 times for the key sizes 128, 192, and 256 bits, respectively. The last encryption and decryption iterations exclude the Mix column and Inverse Mix column steps.

This section will explain the AES encryption and decryption steps. As most applications and designs also use the AES with 128 bits key size, all the used examples and algorithms in this section are based on this key size.

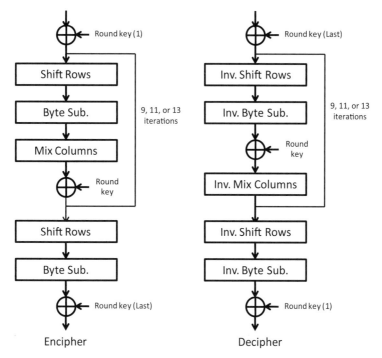

FIGURE 13.2. AES encryption and decryption processes.

13.3.1 Shift Rows/Inverse Shift Rows

In the Shift Rows step, the second, third, and fourth rows of the state array are shifted one, two, and three cyclic shifts to the left, respectively. Most references consider the shift rows step as the first step in the encryption iteration; however, it can be done after the Byte Substitution step without affecting the algorithm. The Inverse Shift Rows step is obtained during the decryption process by shifting the second, third, and fourth rows, one, two, and three cyclic shifts to the right, respectively.

13.3.2 Byte Substitution and Inverse Byte Substitution

Byte Substitution and Inverse Byte Substitution are the most complex steps in the encryption and decryption processes. In these steps, each byte of the state array will be replaced with its equivalent byte in the S-BOX or the Inverse S-BOX. As AES algorithm use elements within the $GF(2^8)$, each element in the state array represents a byte with a value that varies in the range 00_H–FF_H. The S-BOX has a fixed size of 256 bytes represented as a (16×16) bytes matrix. Figure 13.3 shows the AES S-BOX. In this figure, the

b2 \ b1	0	1	2	3	4	5	6	7	8	9	A	B	C	D	E	F
0	63	7C	77	7B	F2	6B	6F	C5	30	01	67	2B	FE	D7	AB	76
1	CA	82	C9	7D	FA	59	47	F0	AD	D4	A2	AF	9C	A4	72	C0
2	B7	FD	93	26	36	3F	F7	CC	34	A5	E5	F1	71	D8	31	15
3	04	C7	23	C3	18	96	05	9A	07	12	80	E2	EB	27	B2	75
4	09	83	2C	1A	1B	6E	5A	A0	52	3B	D6	B3	29	E3	2F	84
5	53	D1	00	ED	20	FC	B1	5B	6A	CB	BE	39	4A	4C	58	CF
6	D0	EF	AA	FB	43	4D	33	85	45	F9	02	7F	50	3C	9F	A8
7	51	A3	40	8F	92	9D	38	F5	BC	B6	DA	21	10	FF	F3	D2
8	CD	0C	13	EC	5F	97	44	17	C4	A7	7E	3D	64	5D	19	73
9	60	81	4F	DC	22	2A	90	88	46	EE	B8	14	DE	5E	0B	DB
A	E0	32	3A	0A	49	06	24	5C	C2	D3	AC	62	91	95	E4	79
B	E7	C8	37	6D	8D	D5	4E	A9	6C	56	F4	EA	65	7A	AE	08
C	BA	78	25	2E	1C	A6	B4	C6	E8	DD	74	1F	4B	BD	8B	8A
D	70	3E	B5	66	48	03	F6	0E	61	35	57	B9	86	C1	1D	9E
E	E1	F8	98	11	69	D9	8E	94	9B	1E	87	E9	CE	55	28	DF
F	8C	A1	89	0D	BF	E6	42	68	41	99	2D	0F	B0	54	BB	16

FIGURE 13.3. The S-BOX.

variable b2 represents the most significant nibble, while the variable b1 represents the least significant nibble.

The Inverse S-BOX, which is used during the decryption processes, will be retrieving the original byte that was substituted using the S-BOX during the encryption process. For example, from the S-BOX in Figure 13.3, we can see that the S-BOX will substitute the byte 00_H with the byte 63_H. Also the byte 63_H in the Inverse S-BOX will be substituted by 00_H.

The generation of S-BOX is done by two steps: first, by finding the multiplicative inverse for the numbers 00_H–FF_H in the $GF(2^8)$, and then applying the affine transformation on them. On the other hand, the generation of the Inverse S-BOX starts by applying the inverse affine transformation followed by finding the multiplicative inverse. The next subsections will explain these substeps in more detail.

13.3.2.1 Multiplicative Inverse Calculation The first step of S-BOX generation is finding the multiplicative inverse for the numbers 00_H–FF_H.

This requires using the irreducible polynomial p(x) defined in the Equation (13.7):

$$p(x) = x^8 + x^4 + x^3 + x + 1. \tag{13.7}$$

Since AES is dealing with numbers within the $GF(2^8)$, it uses the eighth-degree irreducible polynomial shown in Equation (13.7) as defined by Reference [4]. This polynomial is used as a reduction polynomial by applying it as a modulus for the multiplication result of two polynomials so that the final result can be within the finite field $GF(2^8)$.

Calculating the multiplicative inverse requires using the extended Euclidean algorithm [5], which state that for every polynomial a(x), there exists two polynomials b(x) and c(x), such that:

$$a(x).b(x) + p(x).c(x) = 1. \tag{13.8}$$

And since:

$$(a(x).b(x)) \bmod p(x) = 1, \tag{13.9}$$

we can rewrite Equation (13.9) as:

$$a(x)^{-1} = b(x) \bmod (p(x)). \tag{13.10}$$

13.3.2.2 Affine Transformation   The affine transformation is applied after the multiplicative inverse calculation in the Byte Substitution step, while it is applied first in the Inverse Byte Substitution step. The affine transformation and its inverse have two parts: the multiplication part, where a constant matrix will be multiplied with the data, and then the addition part, where a constant vector is added to the multiplication result. The matrix A1 and the vector C1 are used for the affine transformation as can be seen in Equation (13.11), while the matrix A2 and the vector C2 are used for the inverse affine transformation, as can be seen in Equation (13.12):

$$AT(q) = A1*q + C1 = \begin{bmatrix} 1 & 1 & 1 & 1 & 1 & 0 & 0 & 0 \\ 0 & 1 & 1 & 1 & 1 & 1 & 0 & 0 \\ 0 & 0 & 1 & 1 & 1 & 1 & 1 & 0 \\ 0 & 0 & 0 & 1 & 1 & 1 & 1 & 1 \\ 1 & 0 & 0 & 0 & 1 & 1 & 1 & 1 \\ 1 & 1 & 0 & 0 & 0 & 1 & 1 & 1 \\ 1 & 1 & 1 & 0 & 0 & 0 & 1 & 1 \\ 1 & 1 & 1 & 1 & 0 & 0 & 0 & 1 \end{bmatrix} * \begin{bmatrix} q_7 \\ q_6 \\ q_5 \\ q_4 \\ q_3 \\ q_2 \\ q_1 \\ q_0 \end{bmatrix} + \begin{bmatrix} 0 \\ 1 \\ 1 \\ 0 \\ 0 \\ 0 \\ 1 \\ 1 \end{bmatrix}, \tag{13.11}$$

$$AT(q)^{-1} = A2^*q + C2 = \begin{bmatrix} 0 & 1 & 0 & 1 & 0 & 0 & 1 & 0 \\ 0 & 0 & 1 & 0 & 1 & 0 & 0 & 1 \\ 1 & 0 & 0 & 1 & 0 & 1 & 0 & 0 \\ 0 & 1 & 0 & 0 & 1 & 0 & 1 & 0 \\ 0 & 0 & 1 & 0 & 0 & 1 & 0 & 1 \\ 1 & 0 & 0 & 1 & 0 & 0 & 1 & 0 \\ 0 & 1 & 0 & 0 & 1 & 0 & 0 & 1 \\ 1 & 0 & 1 & 0 & 0 & 1 & 0 & 0 \end{bmatrix} * \begin{bmatrix} q_7 \\ q_6 \\ q_5 \\ q_4 \\ q_3 \\ q_2 \\ q_1 \\ q_0 \end{bmatrix} + \begin{bmatrix} 0 \\ 0 \\ 0 \\ 0 \\ 0 \\ 1 \\ 0 \\ 1 \end{bmatrix}. \quad (13.12)$$

13.3.3 Mix Columns/Inverse Mix Columns Steps

After performing the Byte Substitution step during the encryption process, the Mix Columns step is applied. In the decryption process, the Inverse Mix Columns step is applied after adding the Round Key. The Mix Columns step and its inverse are not applied in the last encryption or decryption processes as described in Reference [4]. In these steps, each column of the state array will be processed using four polynomials. Each polynomial consists of four operands representing the old state array column elements and they will be used to obtain the new state array element.

According to Daemen and Rijmen [3], the polynomial c(x) given in Equation (13.13) is used to obtain the Mix Column step:

$$c(x) = \{03\}.x^3 + \{01\}.x^2 + \{01\}.x^2 + 02. \quad (13.13)$$

To obtain the Mix Column Step, each 4 bytes state array column is represented as polynomials over $GF(2^8)$, as shown in Equation (13.14). Each polynomial is multiplied by the fixed polynomial c(x) modulo the polynomial k(x) as described in Equation (13.15):

$$d(x) = S_{3,c}.x^3 + S_{2,c}.x^2 + S_{1,c}.x^2 + S_{0,c}, \quad (13.14)$$

$$k(x) = x^4 + 1. \quad (13.15)$$

According to Daemen and Rijmen [3], multiplication between the polynomials c(x) and b(x) modulo k(x) will result in the Matrix (13.16):

$$\begin{bmatrix} S'_{0C} \\ S'_{1C} \\ S'_{2C} \\ S'_{3C} \end{bmatrix} = \begin{bmatrix} \mathbf{02} & \mathbf{03} & \mathbf{01} & \mathbf{01} \\ \mathbf{01} & \mathbf{02} & \mathbf{03} & \mathbf{01} \\ \mathbf{01} & \mathbf{01} & \mathbf{02} & \mathbf{03} \\ \mathbf{03} & \mathbf{02} & \mathbf{01} & \mathbf{01} \end{bmatrix} * \begin{bmatrix} S_{0C} \\ S_{1C} \\ S_{2C} \\ S_{3C} \end{bmatrix}. \quad (13.16)$$

Matrix (13.16) can be written in the polynomials as shown in Equation (13.17):

$$S'_{0,c} = \{02\}.S_{0,c} + \{03\}.S_{1,c} + S_{2,c} + S_{3,c},$$
$$S'_{1,c} = \{02\}.S_{1,c} + \{03\}.S_{2,c} + S_{3,c} + S_{0,c},$$
$$S'_{2,c} = \{02\}.S_{2,c} + \{03\}.S_{3,c} + S_{0,c} + S_{1,c},$$
$$S'_{3,c} = \{02\}.S_{3,c} + \{03\}.S_{0,c} + S_{1,c} + S_{2,c}.$$
(13.17)

The Inverse Mix Column step is obtained by multiplying the 4 bytes state array column polynomial $d(x)$ given in Equation (13.14) by the Inverse Mix Columns polynomial $c(x)^{-1}$ in Equation (13.18) modulo $k(x)$ given by Equation (13.15):

$$c(x)^{-1} = \{0B\}.x^3 + \{0D\}.x^2 + \{09\}.x + 0E.$$
(13.18)

The latter multiplication can be represented using the matrix in Equation (13.19):

$$
\begin{bmatrix} S'_{0C} \\ S'_{1C} \\ S'_{2C} \\ S'_{3C} \end{bmatrix} =
\begin{bmatrix}
0E & 0B & 0D & 09 \\
09 & 0E & 0B & 0D \\
0D & 09 & 0E & 0B \\
0B & 0D & 09 & 0E
\end{bmatrix} *
\begin{bmatrix} S_{0C} \\ S_{1C} \\ S_{2C} \\ S_{3C} \end{bmatrix}.
$$
(13.19)

The matrix in Equation (13.19) can be written using the polynomials as shown in Equation (13.20):

$$S'_{0,c} = \{0E\}.S_{0,c} + \{0B\}.S_{1,c} + \{0D\}.S_{2,c} + \{09\}.S_{3,c},$$
$$S'_{1,c} = \{0E\}.S_{1,c} + \{0B\}.S_{2,c} + \{0D\}.S_{3,c} + \{09\}.S_{0,c},$$
$$S'_{2,c} = \{0E\}.S_{2,c} + \{0B\}.S_{3,c} + \{0D\}.S_{0,c} + \{09\}.S_{1,c},$$
$$S'_{3,c} = \{0E\}.S_{3,c} + \{0B\}.S_{0,c} + \{0D\}.S_{1,c} + \{09\}.S_{2,c}.$$
(13.20)

13.3.4 Key Expansion and Add Round Key Step

Add Round Key is applied one extra time compared with the other encryption and decryption steps. The first Add Round Key step is applied before starting the encryption or the decryption iterations, where the first 128 bits of the input key are added to the input data block. This round key is called the initial round key. For the decryption process, the initial round key is the last 128 bits of the generated keys as will be explained later.

In addition to the initial 16 bytes round keys, another 16 bytes of round keys will be required for each encryption or decryption iteration; this makes the total as 176, 224, and 240 bytes for the key sizes 128, 192, and 256 bits, respectively. These round keys are generated using an operation called the key expansion. In the key expansion, all the round keys will be generated from the original input key. The next subsection explains the round keys generation using the key expansion operation.

The key expansion term is used to describe the operation of generating all Round Keys from the original input key. The initial round key will be the original key in case of encryption and the last group of the generated expansion keys in case of decryption—the first and last 16 bytes in case of key sizes of 192 and 256 bits. As mentioned previously, the initial round key will be added to the input before starting the encryption or decryption iterations. Using the 128 bits key size, for example, 10 groups of round keys will be generated with a size of 16 bytes for each round.

The first 4 bytes column in each group will be generated as follows:

1. Taking the S-BOX equivalent to the last column of the previous group (one previous column).
2. Perform one cyclic permutation "rotate elements" from $[R_{0r}, R_{1r}, R_{2r}, R_{3r}]$ to $[R_{1r}, R_{2r}, R_{3r}, R_{0r}]$.
3. Add the round constant.
4. Add the result to the first column of the previous group (four previous columns).

The remaining second, third, and fourth columns of each group will be created by adding the direct previous column with the equivalent column in the previous group (four previous columns). This will create a total of 176 bytes of round keys.

The round constant matrix, known as "RCON," is a constant matrix used during the key expansion process. Each row of the round constant matrix will be added to the first row of each group during the key generation, as explained previously. The first column of this matrix is generated according to Equation (13.21), while the second, third, and fourth rows are zeros. The standard AES reduction polynomial will be used to keep the elements in the $GF(2^8)$:

$$RC[0] = x^0 = 1,$$
$$RC[1] = x,$$
$$RC[j] = x.RC[j-1] = x^j, j > 1.$$
(13.21)

13.4   HARDWARE IMPLEMENTATIONS FOR AES

Since the announcement of the AES algorithm in 2001, various hardware implementations were proposed for it. Most of these implementations have targeted the AES with a 128-bit key size. This key size is considered to be sufficient for most of the commercial applications, where using higher key sizes is considered a waste of resources, as it requires higher area implementations with longer processing time. Key sizes of 192 and 256 bits are used mainly in top-secret military applications to ensure the maximum level of security [6].

AES implementations can be divided into three main types depending on datapath width. The first type comes with an 8-bit datapath as implemented in Good and Benaissa [7], aiming for low-area architectures. The second type is the 32-bit datapath architectures which process each state array row or column together as implemented in Gaj and Chodowiec [8], Rouvroy et al. [9], Chang et al [10], and in Daemen and Rijmen [3], targeting medium-throughput applications. The last type of implementation is the 128-bit loop-unrolled architectures, which target very high-speed applications as presented in Hammad [11], Granado-Criado et al. [12], and in Hodjat and Verbauwhede [13].

Mainly, designs with 8 and 32-bit datapaths use looping architectures. Looping architectures use one stage of AES encryptor/decryptor with a feedback at the end as shown in Figure 13.4a. In this way, the data will go through this stage until completing the required number of iterations, which is determined according to the size of the used key. This AES stage could be an encryptor only or an encryptor with decryptor, and it includes the hardware implementation for the four AES steps: Shift Rows Step, Byte Substitution using the S-BOX, Mix Columns, and Add Round Key.

For very high-speed applications, which is implemented as a full 128-bit datapath, the throughput can be ideally doubled N times by applying the loop-unrolled architecture. In this architecture, replicates of the AES stages are

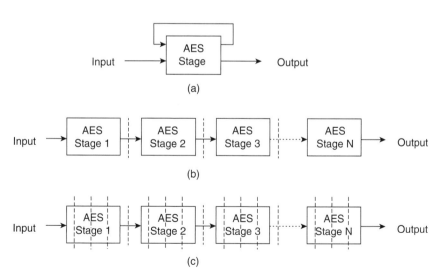

FIGURE 13.4. AES looping and loop-unrolled architectures. (a) Looping architectures; (b) loop-unrolled architecture with pipelining; (c) loop-unrolled architecture with subpipelining.

implemented in series, where N is the number of used stages. In an AES 128-bit key size architecture, N is 10, as 10 AES iterations are required to complete the encryption/decryption processes. Figure 13.4b shows the loop-unrolled architectures with pipelining technique.

In order to get benefits from the loop-unrolled architecture, a pipelining stage is implemented at the end of each AES stage, which allows entering new data at each clock cycle; therefore, all AES stages will be working in parallel. The design in Saggese et al. [14] presents a loop-unrolled AES implementation with pipelining techniques.

More advanced pipelining techniques were used in the designs in Jarvinen et al. [15], Zhang and Parhi [16], and in Hodjat and Verbauwhede [13], where the idea of using subpipelining stages is presented. In subpipelining, instead of applying pipelining stage at the end of each AES stage, the latter is divided into a certain number of pipelining stages. This method doubles the throughput a couple of times compared to what is achievable using normal pipelining. Figure 13.4c shows the loop-unrolled architectures with subpipelining techniques.

These loop-unrolled subpipelined AES designs, which achieves tens of gigabytes of throughput, are used in many applications such as high-traffic servers and in e-commerce servers [7].

S-BOX implementation is a main concern in the AES hardware design. Two main methods were proposed for the implementation of the S-BOX: prestoring the S-BOX elements or using composite field S-BOX. In prestoring the S-BOX elements, block random access memories (BRAMs), which are field-programmable gate array (FPGA) blocks of RAM, can be used to store data. The design in Granado-Criado et al. [12] has used the BRAMs to present high-speed loop-unrolled architecture. Using composite field S-BOX is proposed in Satoh et al. [17], and implemented as high-speed loop unrolled subpipelined AES design in Zhang and Parhi [16].

As pipelining cannot be applied to BRAM as it is a one-memory block, implementing S-BOX using BRAMs will limit the number of subpipelining stages in the design. Also, implementation using BRAMs usually requires a larger area than the composite field arithmetic designs, as will be shown later in this chapter.

The following subsections will present the implementation of the composite field S-BOX, in addition to different AES hardware designs, which used the FPGA technology in the implementation.

13.4.1 Composite Field Arithmetic S-BOX

The implementation of the composite field S-BOX is accomplished using combinational logic circuits rather than using prestored S-BOX values. S-BOX substitution starts by finding the multiplicative inverse of the number in $GF(2^8)$, and then applying the affine transformation. Implementing a circuit to find the multiplicative inverse in the finite field $GF(2^8)$ is very complex and costly, therefore Rijmen [18] has suggested using the finite field $GF(2^4)$ to find the multiplicative inverse of elements in the finite field $GF(2^8)$. The first detailed implementation for the composite field S-BOX was published in Satoh et al. [17].

Each element in a higher order field can be expressed using the polynomial $bx + c$, where b and c are elements in the lower order field. For example any element in $GF(2^8)$ can be expressed using the polynomial $(bx + c)$, where b and $c \in GF(2^4)$, and they represent the most and the least significant nibbles of that element.

After expressing the $GF(2^8)$ element as a polynomial over $GF(2^4)$, the multiplicative inverse can be found using the polynomial shown in Equation (13.22) [18]:

$$(bx+c)^{-1} = b(b^2\lambda + c(b+c))^{-1}x + (c+b)(b^2\lambda + c(b+c))^{-1}. \qquad (13.22)$$

Figure 13.5 shows the composite field S-BOX which was proposed by Satoh et al. [17]. This model applies Equation (13.22) in finding the multiplicative inverse for $GF(2^8)$ elements.

As can be seen from the figure, isomorphic mapping must be applied on the $GF(2^8)$ element before applying it as a polynomial over $GF(2^4)$. Also,

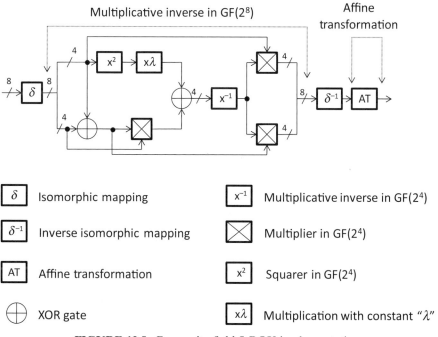

FIGURE 13.5. Composite field S-BOX implementation.

inverse isomorphic mapping is required after finding the multiplicative inverse for the number. According to Satoh et al. [17] and Zhang and Parhi [2], higher order fields can be built from the lower order field using the irreducible polynomials shown in Equation (13.23):

$$\begin{cases} GF(2) \rightarrow GF(2^2): x^2 + x + 1 \\ GF(2^2) \rightarrow GF((2^2)^2): x^2 + x + \phi \\ GF((2^2)^2) \rightarrow GF(((2^2)^2)^2): x^2 + x + \lambda \end{cases}. \tag{13.23}$$

The polynomials $x^2 + x + \lambda$ and $x^2 + x + \phi$ are used in the implementation of the composite field S-BOX. The constants λ and φ are chosen to ensure the polynomials' irreducibility. The used values for these constants according to those proposed in Satoh et al. [17] are $\lambda = \{1100\}$ and $\phi = \{10\}$. These polynomials are mainly used in the derivation of the isomorphic mapping and its inverse in addition to the design of the composite field S-BOX sub-blocks. Matrices (13.24) and (13.25) represent the isomorphic mapping and its inverse based on the values $\lambda = \{1100\}$ and $\phi = \{10\}$. Derivation for the isomorphic mapping matrix can be obtained using the algorithms presented in Paar [19] and in Zhang and Parhi [2]:

$$\delta * q = \begin{bmatrix} 1 & 0 & 1 & 0 & 0 & 0 & 0 & 0 \\ 1 & 1 & 0 & 1 & 1 & 1 & 1 & 0 \\ 1 & 0 & 1 & 0 & 1 & 1 & 0 & 0 \\ 1 & 0 & 1 & 0 & 1 & 1 & 1 & 0 \\ 1 & 1 & 0 & 0 & 0 & 1 & 1 & 0 \\ 1 & 0 & 0 & 1 & 1 & 1 & 1 & 0 \\ 0 & 1 & 0 & 1 & 0 & 0 & 1 & 0 \\ 0 & 1 & 0 & 0 & 0 & 0 & 1 & 1 \end{bmatrix} * \begin{bmatrix} q_7 \\ q_6 \\ q_5 \\ q_4 \\ q_3 \\ q_2 \\ q_1 \\ q_0 \end{bmatrix}, \tag{13.24}$$

$$\delta^{-1} * q = \begin{bmatrix} 1 & 1 & 1 & 0 & 0 & 0 & 1 & 0 \\ 0 & 1 & 0 & 0 & 0 & 1 & 0 & 0 \\ 0 & 1 & 1 & 0 & 0 & 0 & 1 & 0 \\ 0 & 1 & 1 & 1 & 0 & 1 & 1 & 0 \\ 0 & 0 & 1 & 1 & 1 & 1 & 1 & 0 \\ 1 & 0 & 0 & 1 & 1 & 1 & 1 & 0 \\ 0 & 0 & 1 & 1 & 0 & 0 & 0 & 0 \\ 0 & 1 & 1 & 1 & 0 & 1 & 0 & 1 \end{bmatrix} * \begin{bmatrix} q_7 \\ q_6 \\ q_5 \\ q_4 \\ q_3 \\ q_2 \\ q_1 \\ q_0 \end{bmatrix}. \tag{13.25}$$

Detailed explanations on how to use these polynomials in building the composite field S-BOX can be found in Mui [20] and in Zhang and Parhi [2].

13.4.2 Very High Speed AES Design

The design presented in Zhang and Parhi [16] is a loop-unrolled 128-bit bus architecture based on the composite field arithmetic S-BOX proposed in Satoh et al. [17]. This design presented a new $GF(2^4)$ inversion block, which is used as a part of the $GF(2^8)$ multiplicative inverse block, in addition to presenting a joint encryptor/decryptor architecture for loop-unrolled designs. In the encryptor design, each stage was divided into three and seven subpipelining stages. By using seven subpipelining stages, this design was able to achieve an efficiency of 1.956 Mbps/slice using XCV1000e-8, where a throughput of 21,556 Mpbs was achieved using 11,022 slices. Figure 13.6 shows the proposed encryption stage in Zhang and Parhi [16].

Another very high-speed loop-unrolled design was proposed in Hodjat and Verbauwhede [13], where each encryption stage was divided into four and seven pipelining stages. By using the composite field S-BOX design with seven pipelining stages, a throughput of 21.64 Gbps was achieved using 9446 slices. This led to an efficiency of 2.3 Mpbs/slice using the Xilinx XC2VP20-7 device. In the BRAM implementation using four pipelining stages, a throughput of 21.54 Gbits/s was achieved using 84 BRAMs and 5177 slices.

Improvements on high-speed loop-unrolled designs can be done by applying new methods to decrease the critical path of the AES stage and to allow a lower area implementation. For example, in the design of Hodjat and Verbauwhede [13], repeated blocks of isomorphic mapping, inverse isomorphic mapping, and affine transformation were used at each AES stage; also, repeated blocks of isomorphic mapping and merged affine transformation inverse isomorphic mapping was used in Zhang and Parhi [16]. The next section presents the proposed architecture in Hammad et al. [21], which merges these three blocks into one new block which allows the implementation with shorter critical path and lower area, therefore presenting a system with higher efficiency.

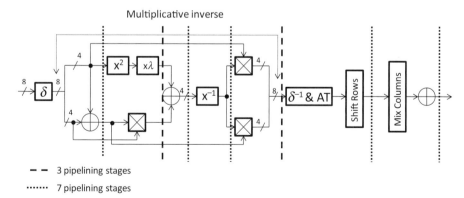

FIGURE 13.6. AES encryption stage in Zhang and Parhi [16].

13.5 HIGH-SPEED AES ENCRYPTOR WITH EFFICIENT MERGING TECHNIQUES

This section presents the high-speed AES encryptor proposed in Hammad et al. [21]. This architecture is implemented using the loop-unrolled technique with subpipelining. This encryptor simplifies the implementation of the composite field S-BOX by relocating the Mix Column step and the merge among the inverse isomorphic mapping, the affine transformation multiplication, and the isomorphic mapping of the next encryption stage. This merging combines these three operational blocks into one block. Moreover, the affine transformation vector "C" shown in Equation (13.11) is implemented in the key expansion unit instead of the main round unit. These improvements enabled the implementation to have higher efficiency by reducing the area and shortening the total path length, where a less number of subpipelining stages are required for achieving certain throughput. This architecture has shown higher efficiency in terms of FPGA (Throughput/Area) compared with previous loop-unrolled pipelined AES encryptors.

13.5.1 The Integrated-BOX

Applying the current composite field design in a loop-unrolled system resulted in repeated isomorphic mapping , inverse isomorphic mapping, and affine transformation operations in each encryption stage as what happened in Zhang and Parhi [16] and in Hodjat and Verbauwhede [13].

This section presents the new proposed composite field loop-unrolled design for the AES encryptor. In this design, the mix column step is relocated and performed before the inverse isomorphic mapping and the affine transformation operations, which are required in the byte substitution step, while the round keys are mapped with the isomorphic mapping. This relocation for the mix column and the mapping for the round keys allowed an efficient merging between the inverse isomorphic mapping, the affine transformation multiplication, and the required isomorphic mapping in the next encryption stage. By this merging, these three operations were substituted by a new operational block called the "ζ transformation."

Moreover, the affine transformation addition were placed in the key expansion unit instead of the encryption stages. This allowed the reduction of these addition blocks from 160 blocks in all encryption stages to only 16 blocks implemented in the key expansion unit.

With these merging techniques, a new block, which is called the I-BOX or "Integrated BOX," is introduced. This block generates one of the encryption iterations for each state matrix column. To derive the I-BOX equations and to explain the mathematical theory behind this proposed merging and rearrangements, we start by Equation (13.26), which represents the state array element after applying the inverse isomorphic mapping and the affine transformation to the multiplicative inverse result in Equation (13.22). Parameters r and c represents the row and column locations for the state matrix element, where it is assumed to be at the Ith encryption stage:

$$S'_{r,c}(i) = AT(\delta^{-1}(bx+c)_{r,c}^{-1}).$$ (13.26)

After byte substitution is performed using Equation (13.26), the mix column step is applied on the state array elements using the matrix in Equation (13.27):

$$\begin{bmatrix} S''_{0C} \\ S''_{1C} \\ S''_{2C} \\ S''_{3C} \end{bmatrix} = \begin{bmatrix} 2 & 3 & 1 & 1 \\ 1 & 2 & 3 & 1 \\ 1 & 1 & 2 & 3 \\ 3 & 2 & 1 & 1 \end{bmatrix} * \begin{bmatrix} S'_{0C} \\ S'_{1C} \\ S'_{2C} \\ S'_{3C} \end{bmatrix},$$ (13.27)

where the state array element after the mix columns step can be written as:

$$S''_{r,c}(i) = \{02\}.S_{r,c}(i)' + \{03\}_{16}.S_{\mathrm{mod}(r+1,4),c}(i)' + S_{\mathrm{mod}(r+2,4),c}(i)' + S_{\mathrm{mod}(r+3,4),c}(i)'.$$ (13.28)

By rewriting Equation (13.28) and adding the Round Key, the state array element will be written as:

$$S'''_{r,c}(i) = \{02\}.\left(S'_{r,c} + S'_{\mathrm{mod}(r+1,4),c}\right) + S'_{\mathrm{mod}(r+2,4),c} + S'_{\mathrm{mod}(r+3,4),c} + S'_{\mathrm{mod}(r+1,4),c} + RK_{r,c}.$$ (13.29)

By letting $\beta = (bx+c)^{-1}$ and $S_{r,'c'}(i+1) = \delta(S_{r,c}(i)''')$,, we obtain:

$$\begin{aligned} S_{r,'c'}(i+1) = \delta\Big(\{02\}.\Big(A\big(\delta^{-1}(\beta)_{r,c}\big) + A\big(\delta^{-1}(\beta)_{\mathrm{mod}(r+1,4),c}\big)\Big) \\ + A\big(\delta^{-1}(\beta)_{\mathrm{mod}(r+2,4),c}\big) + A\big(\delta^{-1}(\beta)_{\mathrm{mod}(r+3,4),c}\big) \\ + A\big(\delta^{-1}(\beta)_{\mathrm{mod}(r+1,4),c}\big) + C + RK_{r,c}\Big). \end{aligned}$$ (13.30)

The term $S_r, S_{r,'c'}(i+1)$ in Equation (13.30) represents the isomorphic mapped value of state element after shifting and before applying it to the byte substitution in the $(i+1)$ stage, where the affine transformation vector "C" in each operand is cancelled by XOR addition with the vector "C" in the next operand. By rearranging the location of the mix columns coefficient $\{02\}$ matrix, and replacing it with a new matrix $\{02\}'$, and by taking the affine transformation and the inverse isomorphic mapping as common factors, Equation (13.30) can be rewritten as:

$$\begin{aligned} S_{r,'c'}(i+1) = \delta\Big(A\big(\delta^{-1}\big(\{02\}'.\big((\beta)_{r,c} + (\beta)_{\mathrm{mod}(r+1,4),c}\big) \\ + (\beta)_{\mathrm{mod}(r+2,4),c} + (\beta)_{\mathrm{mod}(r+3,4),c} + (\beta)_{\mathrm{mod}(r+1,4),c}\big)\big)\Big) \\ + \delta(RK_{r,c} + C). \end{aligned}$$ (13.31)

From Equation (13.31) we can define the new transformation ζ as:

$$\zeta = (\delta)*(A)*(\delta^{-1}),\tag{13.32}$$

$$\zeta = \begin{bmatrix} 0 & 0 & 0 & 0 & 1 & 0 & 0 & 0 \\ 0 & 0 & 0 & 1 & 0 & 1 & 0 & 0 \\ 0 & 1 & 1 & 1 & 0 & 0 & 1 & 0 \\ 1 & 1 & 1 & 1 & 0 & 0 & 1 & 1 \\ 1 & 0 & 0 & 0 & 0 & 0 & 0 & 0 \\ 1 & 1 & 1 & 0 & 0 & 1 & 0 & 0 \\ 1 & 1 & 1 & 0 & 0 & 0 & 1 & 0 \\ 1 & 0 & 1 & 1 & 0 & 1 & 1 & 0 \end{bmatrix}$$

$$= \begin{bmatrix} 1 & 0 & 1 & 0 & 0 & 0 & 0 & 0 \\ 1 & 1 & 0 & 1 & 1 & 1 & 1 & 0 \\ 1 & 0 & 1 & 0 & 1 & 1 & 0 & 0 \\ 1 & 0 & 1 & 0 & 1 & 1 & 1 & 0 \\ 1 & 1 & 0 & 0 & 0 & 1 & 1 & 0 \\ 1 & 0 & 0 & 1 & 1 & 1 & 1 & 0 \\ 0 & 1 & 0 & 1 & 0 & 0 & 1 & 0 \\ 0 & 1 & 0 & 0 & 0 & 0 & 1 & 1 \end{bmatrix} * \begin{bmatrix} 1 & 1 & 1 & 1 & 1 & 0 & 0 & 0 \\ 0 & 1 & 1 & 1 & 1 & 1 & 0 & 0 \\ 0 & 0 & 1 & 1 & 1 & 1 & 1 & 0 \\ 0 & 0 & 0 & 1 & 1 & 1 & 1 & 1 \\ 1 & 0 & 0 & 0 & 1 & 1 & 1 & 1 \\ 1 & 1 & 0 & 0 & 0 & 1 & 1 & 1 \\ 1 & 1 & 1 & 0 & 0 & 0 & 1 & 1 \\ 1 & 1 & 1 & 1 & 0 & 0 & 0 & 1 \end{bmatrix}$$

$$* \begin{bmatrix} 1 & 1 & 1 & 0 & 0 & 0 & 1 & 0 \\ 0 & 1 & 0 & 0 & 0 & 1 & 0 & 0 \\ 0 & 1 & 1 & 0 & 0 & 0 & 1 & 0 \\ 0 & 1 & 1 & 1 & 0 & 1 & 1 & 0 \\ 0 & 0 & 1 & 1 & 1 & 1 & 1 & 0 \\ 1 & 0 & 0 & 1 & 1 & 1 & 1 & 0 \\ 0 & 0 & 1 & 1 & 0 & 0 & 0 & 0 \\ 0 & 1 & 1 & 1 & 0 & 1 & 0 & 1 \end{bmatrix},$$

$$\zeta * q = \begin{bmatrix} 0 & 0 & 0 & 0 & 1 & 0 & 0 & 0 \\ 0 & 0 & 0 & 1 & 0 & 1 & 0 & 0 \\ 0 & 1 & 1 & 1 & 0 & 0 & 1 & 0 \\ 1 & 1 & 1 & 1 & 0 & 0 & 1 & 1 \\ 1 & 0 & 0 & 0 & 0 & 0 & 0 & 0 \\ 1 & 1 & 1 & 0 & 0 & 1 & 0 & 0 \\ 1 & 1 & 1 & 0 & 0 & 0 & 1 & 0 \\ 1 & 0 & 1 & 1 & 0 & 1 & 1 & 0 \end{bmatrix} * \begin{bmatrix} q_7 \\ q_6 \\ q_5 \\ q_4 \\ q_3 \\ q_2 \\ q_1 \\ q_0 \end{bmatrix}.\tag{13.33}$$

Due to changing the Mix Columns Step location, the new $\{02\}'$ multiplication matrix is required because the multiplication is obtained before the inverse isomorphic mapping and the affine transformation. This matrix is derived by implementing a relation between the inputs 00_H–FF_H and the values which will yield to the $\{02\}$ multiplication after applying the inverse isomorphic mapping and the affine transformation. Equation (13.34) represents the multiplication of input byte q by the new matrix $\{02\}'$:

$$\{02\}'*q = \begin{bmatrix} 1 & 0 & 1 & 1 & 1 & 1 & 1 & 0 \\ 1 & 1 & 0 & 0 & 0 & 0 & 1 & 1 \\ 0 & 0 & 0 & 1 & 0 & 0 & 0 & 0 \\ 1 & 1 & 1 & 0 & 1 & 0 & 0 & 1 \\ 0 & 1 & 1 & 0 & 0 & 0 & 1 & 1 \\ 0 & 0 & 1 & 0 & 1 & 1 & 0 & 1 \\ 0 & 0 & 1 & 0 & 1 & 0 & 0 & 1 \\ 1 & 1 & 1 & 1 & 0 & 1 & 0 & 1 \end{bmatrix} * \begin{bmatrix} q_7 \\ q_6 \\ q_5 \\ q_4 \\ q_3 \\ q_2 \\ q_1 \\ q_0 \end{bmatrix}. \tag{13.34}$$

Using Equation (13.32), we can rewrite Equation (13.31) as:

$$S_{r,'c'}(i+1) = \zeta\left(\begin{array}{c} \{02\}'_{16} \cdot \left((\beta)_{r,c} + (\beta)_{\mathrm{mod}(r+1,4),c}\right) + \\ (\beta)_{\mathrm{mod}(r+3,4),c} + (\beta)_{\mathrm{mod}(r+1,4),c} \end{array} \right) + (\delta(RK_{r,c}) + \delta(C)). \tag{13.35}$$

Equation (13.35) represents one encryption stage calculation for state array element in the I-BOX, where each I-BOX will be calculating four state array elements. Figure 13.7 shows the state array element calculation inside the I-BOX based on Equation (13.35). As can be seen from this figure, each state element calculator in the I-BOX consists of the multiplicative inverse block $GF(2^8)$, followed by the mix column step which uses the new $\{02\}'$ multiplier. After the mix columns part, the new ζ transformation block is used. This block, as explained previously, is a result from the merging among the inverse isomorphic mapping, the affine transformation multiplication, and the isomorphic

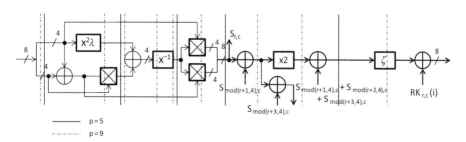

FIGURE 13.7. I-BOX state array element calculation.

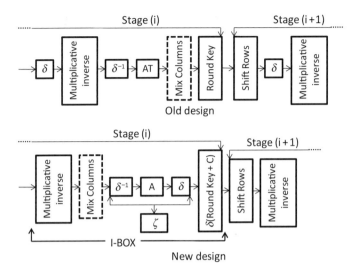

FIGURE 13.8. Rearrangements and merging steps.

mapping for the next stage. Finally, the Add Round Key XOR gate can be found.

By placing the mix column step before the inverse isomorphic mapping and the affine transformation and obtaining the Round Keys in the isomorphic mapping, the three transformation blocks were placed to be in sequence, therefore, their merging becomes feasible. Figure 13.8 depicts the implementation of this merging and rearrangement. Also, the affine transformation vector $\delta(C)$ has been added to each round key prior to adding it to the state array element, where the vector $\delta(C)$ is implemented in the key expansion unit. In this case it will be implemented 16 times, "one block for each byte," instead of 160 times in the main round units. Also, the I-BOX is merging between the λ Multiplier and the Square block.

Table 13.1 demonstrates the number of gates and the critical path for the isomorphic mapping, inverse isomorphic mapping, and the affine transformation compared to the ζ transformation.

13.5.2 Key Expansion Unit

The key expansion unit is responsible of generating the 176 bytes of round keys required during the encryption process. Figure 13.9 shows the proposed key expansion unit. This unit is using four S-BOXES in processing the first column of each group of the round keys, while the XOR gates are used in processing all the columns by adding the direct previous column with the equivalent column from the previous group. The four registers following the multiplexers are used to store the direct previous group of Round Keys to be

TABLE 13.1. Number of Gates and Critical Path for the Mappings and the Transformations

Operation	δ	$\delta - 1$	AT	Z
Total number of gates	12	14	20	15
Critical path	4	3	4	3

AT, affine transformation.

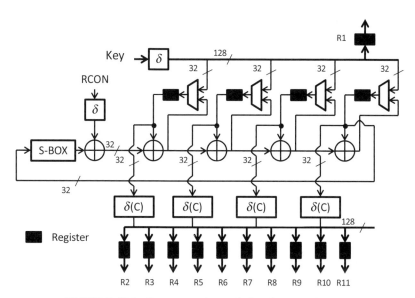

FIGURE 13.9. Key expansion unit for the 128 bits AES.

used in generation for the ongoing group, while the registers (R1–R11) will be storing all the 176 bytes of round keys.

To work in compatible with the I-BOX state array element calculation described in Equation (13.35), the original input key is transformed to the isomorphic mapping so that all generated round keys are also in the isomorphic transformation as required in Equation (13.35). The S-BOX in the key expansion unit is implemented using multiplicative inverse block with the ζ transformation block. The affine transformation constant $\delta(C)$ are implemented in the key expansion unit, where it is added to the Round Keys before being stored in the registers.

13.5.3 The AES Encryptor with the Merging Technique

The AES encryptor using the merging technique consists of 10 encryption stages. The first nine stages is implemented using four I-BOXES in each stage,

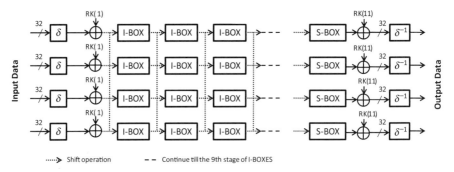

FIGURE 13.10. AES encryptor using I-BOX technique.

TABLE 13.2. Gates and the Critical Path for the GF(2^8) Multiplicative Inverse Sub-Blocks

Block	Critical Path	Total Number of Gates
GF(2^4) multiplier	5	30
GF(2^2) multiplier	3	7
(x^{-1}) inverse block	5	25
Merged (x^2) and ($x\lambda$) blocks	2	4
ζ transformation	3	15

while the last encryption stage is implemented using normal S-BOXES as the Mix Columns step is not required in the last encryption iteration. Each S-BOX in the last stage is similar to the S-BOX in the key expansion unit, where it will consist of the multiplicative inverse block with ζ transformation. As I-BOX uses the ζ transformation, isomorphic mapping and inverse isomorphic mapping will be required at the start and the end of the encryptor only. The Shift Rows step is obtained directly during the input/output operation between the I-BOX and S-BOX stages. Figure 13.10 shows the AES encryptor using the I-BOX technique.

13.5.4   Results and Comparison

The next two subsections present the subpipelining simulation results and the simulation results in comparison with previous proposed designs.

13.5.4.1   Subpipelining Simulation Results   The I-BOX has a total critical path of 26 gates, where each of its sub-blocks has a different critical path that plays a main role in determining the best number of used pipelining stages. Table 13.2 shows the number of gates and the critical path for each block for the composite field S-BOX [16].

TABLE 13.3. Simulation Result based on the Number of Subpipelining Stages

Subpipelining Stages	Frequency (MHz)	Throughput (Mpbs)	Slices	Critical Path	Efficiency (Mbps/area)
1	87.4	11,187	6,462	26	1.731
2	122.1	15,629	6,767	13	2.310
3	169.9	21,747	7,092	9	3.066
4	188.4	24,115	7,501	7	3.215
5	218.3	27,942	7,884	6	3.544
6	236.0	30,208	8,743	5	3.455
7	258.8	33,126	9,461	4	3.501
8	261.9	33,523	9,981	4	3.359
9	305.1	39,053	10,662	3	3.663

To determine the number of the used subpipelining stages, the AES encryptor was simulated with one to nine subpipelining stages and the best efficiency was achieved using nine subpipelining stages, while the second best efficiency was achieved with five subpipelining stages. Table 13.3 shows the simulation results for architectures with one to nine subpipelining stages using Xilinx XC2V6000-6.

As can be seen from Table 13.3, applying one pipelining stage achieves the lowest efficiency. This kind of pipelining is shown in Figure 13.4b, where a pipelining stage is applied between the AES stages without the usage of any subpipelining stages. In the case of two to five pipelining stages, the efficiency will increase as more pipelining stages are added. Applying six pipelining stages leads to less efficiency compared with what can be achieved using five pipelining stages. In the case of six pipelining stages, an internal pipelining must be applied in some of the sub-blocks of the composite field arithmetic S-BOX. This requires applying registers in each branch inside these sub-blocks, which result in higher usage of registers compared with the case of five pipelining stages, where the pipelining registers are needed only at the main buses in the composite field S-BOX.

Using eight pipelining stages also achieved less efficiency than using seven pipelining stages. The main reason for this is that in both cases the critical path of the system is four logic gates, which prevent a significant increase in the system throughput. Using nine pipelining stages achieves the best possible efficiency, while applying more than nine pipelining stages will cause a reduction in the efficiency as the critical path cannot be reduced to less than three gates until reaching a system with 13 pipelining stages. The key expansion unit has to be divided into the same number of subpipelining stages to maintain the synchronization between the main round units and the key expansion unit. Figure 13.7 shows the implementation of five and nine subpipelining stages for the I-BOX, where p in this figure represents the number of subpipelining stages in each case.

13.5.4.2 Comparison with Previous Designs Many designs rely on BRAMs in the implementation of AES encryptors, therefore Järvinen et al. [22] suggested using the metric megabits per second per area (Mbps/area) for a better performance–area relationship instead of the metric megabits per second per slice (Mbps/slice) to take into account the use of BRAMs in the efficiency calculation. The total area is obtained by adding the total slices and by considering each dual-port 256×8 bit BRAM as equivalent to 128 slices [22].

Owing to the proposed merging techniques, this design was able to achieve higher efficiencies than the previous loop-unrolled designs. By using five subpipelining stages, this design achieves almost the same throughput obtained by seven subpipelining stages as in Zhang and Parhi [16]. By comparing this design with that of Hodjat and Verbauwhede [13], a 37% improvement in efficiency was achieved based on the simulation results of the same FPGA device. The encryptor in Granado-Criado et al. [12] used BRAMs to implement the S-BOX. According to Granado-Criado et al. [12], these BRAMs are equivalent to 10,240 extra slices. Simulated on the same device, the proposed design was able to achieve higher efficiency and throughput. Also, the proposed design was able to achieve higher efficiency than the designs in Zambreno et al. [23] and in Jarvinen et al. [15]. Table 13.4 summarizes the obtained results and the comparison with previous different implementations.

TABLE 13.4. Results and Comparison of AES 128-bit Encryptors

Design	Device	Frequency (MHz)	Throughput (Mbps)	Slices	BRAMs	Efficiency (Mbps/ area)
Jarvinen et al. [15]	XCV1000e-8	129.2	16,500	11,719	0	1.408
Zhang and Parhi [16]	XCV1000e-8	168.4	21,556	11,022	0	1.956
This work (p = 5)	XCV1000e-8	168.3	21,542	9,104	0	2.366
Hodjat and Verbauwhede [13]	XC2VP20-7	169.1	21,645	9,446	0	2.291
This work (p = 5)	XC2VP20-7	220.7	28,250	9,028	0	3.129
Zambreno et al. [23].	XC2V4000	184.2	23,572	16,938	0	1.392
This work (p = 5)	XC2V4000-6	211.6	27,087	8,503	0	3.186
Granado et al. [12]	XC2V6000-6	194.7	24,920	3,576	80	1.804
This work (p = 5)	XC2V6000-6	218.3	27,942	7,884	0	3.544
This work (p = 9)	XC2V6000-6	305.1	39,053	10,662	0	3.663

13.6 CONCLUSION

AES hardware implementations vary according to the used bus size and technology. Several designs were proposed using ASIC technology while others uses FPGA designs. AES encryptor/decryptor is usually used as a coprocessor in the embedded systems; one example can be an MPEG2/H264 video encoder, which uses an AES coprocessor to encrypt the video stream. AES could be implemented on the software level; this requires using the central processing unit (CPU) processor for encryption, which usually results in low throughput. Comparing two different AES designs is challenging, as we have to take many aspects into consideration. In FPGA designs to apply a fair efficiency comparison between two designs, the same FPGA technology must be used; also, the usage of BRAMs should be taken into consideration while comparing the area. Furthermore, comparing AES designs based on different architectures—for example, comparing looping 32-bit bus design with 128-bit bus loop-unrolled design—cannot be fair and it is not logical, as a different architecture has its own application, in addition to the huge difference that the architecture will impose on the efficiency.

REFERENCES

[1] B. Eli, A. Shamir, Differential Cryptanalysis of the Data Encryption Standard, Springer-Verlag, Berlin, Germany, 1993.

[2] X. Zhang, K.K. Parhi, "On the optimum constructions of composite field for the AES algorithm," IEEE Transactions on Circuit and Systems-II, 53(10), 1153–1157, Oct. 2006.

[3] J. Daemen, V. Rijmen, The Design of Rijndael—The Advanced Encryption Standard, Springer, New York, 2002.

[4] "Advanced Encryption Standard (AES)," Federal Information Processing Standards Publication 197, Nov. 26, 2001.

[5] T.H. Cormen, C.E. Leiserson, R.L. Rivest, C. Stein, "Section 31.2: Greatest Common Divisor," in Introduction to Algorithms, 2nd ed., MIT Press and McGraw-Hill, 2001, pp. 859–861.

[6] Seagate—Technology Paper, "128 bit versus 256 bit AES encryption: practical business reasons why 128 bit solution provide comprehensive security for every need." Available: http://www.seagate.com/staticfiles/docs/pdf/whitepaper/tp596_128_bit_versus_256_bit.pdf.

[7] T. Good, M. Benaissa, "Very small FPGA application-specific instruction processor for AES," IEEE Transactions on Circuit and Systems-I, 53(7), 2006.

[8] K. Gaj, P. Chodowiec, "Very compact FPGA implementation of the AES algorithm," in Proceedings of CHES, Lecture Notes in Computer Science, Vol. 2779, Springer-Verlag, 2003, pp. 319–333.

[9] G. Rouvroy, F.-X. Standaert, J.-J. Quisquater, J.-D. Legat, "Compact and efficient encryption/decryption module for FPGA implementation of the AES Rijndael

very well suited for small embedded applications," Information Technology Coding and Computing 2004.

[10] C.-J. Chang, C.-W. Huang, K.-H. Chang, Y.-C. Chen, C.-C. Hsieh, "High Throughput 32-bit AES Implementation in FPGA," in APCCAS 2008. IEEE Asia Pacific Conference on Circuits and Systems, 2008.

[11] I. Hammad, "Efficient hardware implementations for the Advanced Encryption Standard algorithm," M.A.Sc Thesis, Dalhousie University, Halifax—NS , Canada, Oct. 2010.

[12] J.M. Granado-Criado, M.A. Vega-Rodríguez, J.M. Sanchez-Perez, J.A. G'Omez-Pulido, "A new methodology to implement the AES algorithm using partial and dynamic reconfiguration," Integration, the VLSI Journal, 43, 2010, pp. 72–80.

[13] A. Hodjat, I. Verbauwhede, "A 21.54 Gbits/s fully pipelined AES processor on FPGA," in Proc.12th Annual IEEE Symposium. Field Programmable Custom Computing Machines, FCCM'04, Napa, CA, USA, April 2004, pp. 308–309.

[14] G.P. Saggese, A. Mazzeo, N. Mazzocca, A.G.M Strollo, "An FPGA-based performance analysis of the unrolling, tiling, and pipelining of the AES algorithm," in Proc. 13th Int. Conf. Field Programmable Logic and Applications, FPL 2003, Lisbon, Portugal, September 2003, pp. 292–302.

[15] K. Jarvinen, M. Tommiska, J. Skytta, "A fully pipelined memoryless 17.8 Gbps AES-128 encryptor," in Proc. ACM/SIGDA 11th ACM Int. Symposium on Field-Programmable Gate Arrays, FPGA 2003, Monterey, CA, USA, February 2003, pp. 207–215.

[16] X. Zhang, K. K Parhi, "High-speed VLSI architecture for the AES algorithm," IEEE Transactions on Very Large Scale Integration (VLSI) System, 12(9), 957–967, Sep. 2004.

[17] A. Satoh, S. Morioka, K. Takano, S. Munetoh, "A compact Rijndael hardware architecture with S-Box optimization," in Proc. ASIACRYPT 2001, Gold Coast, Australia, Dec. 2000, pp. 239–254.

[18] V. Rijmen, "Efficient implementation of the Rijndael S-BOX," 2001. Katholieke Universiteit Leuven, Dept. ESAT. Belgium.

[19] C. Paar, "Efficient VLSI architecture for bit-parallel computations in Galois field," Ph.D. dissertation, Institute for Experimental Mathematics, University of Essen, Essen, Germany, 1994.

[20] E.-N. Mui, "Practical implementation of Rijndael S-BOX using combinational logic," 2007. Texco Enterprise Ptd. Ltd. Available: http://www.xess.com/projects/Rijndael_SBox.pdf.

[21] I. Hammad, K. El-Sankary, E. El-Masry, "High speed AES encryptor with efficient merging techniques," IEEE Embedded Systems Letters, 2(3), 2010, pp. 67–71.

[22] K. Järvinen, M. Tommiska, J. Skyttä, "Comparative survey of high performance cryptographic algorithm implementations on FPGAs," in IEE Proceedings—Information Security, 152(1), 2005, 3–12.

[23] J. Zambreno, D. Nguyen, A. Choudhary, "Exploring area/delay tradeoffs in an AES FPGA implementation," 2004. Department of Electrical and Computer Engineering, Northwestern University.

[24] "Data Encryption Standard (DES), FIPS PUB (46-3), Oct. 25, 1999, Federal Information Processing Standard 46-3," 1999.

14 Reconfigurable Architecture for Cryptography over Binary Finite Fields

SAMUEL ANTÃO, RICARDO CHAVES, and
LEONEL SOUSA

14.1 INTRODUCTION

Security is becoming mandatory for embedded systems in a wide range of
fields and applications. Not only fast, but scalable and flexible implementations
of cryptographic procedures are also demanded in these systems. Elliptic curve
(EC) cryptography and the Advanced Encryption Standard (AES) are two of
the main building blocks of the currently used cryptographic procedures on
embedded systems, enabling support for public-key and private-key protocols.
This chapter starts by introducing the mathematical details underlying the
addressed applications, highlighting possible directions for efficient implementations of the arithmetic units. Following this introduction, the developed
approach to exploit the programmability of a microcoded processor, the efficiency of hardwired finite-field arithmetic units, and the flexibility of the runtime hardware reconfiguration in an architecture targeting both EC and AES
is discussed. Furthermore, the development of other cryptographic relevant
functionalities, such as random number generators (RNGs), is also addressed.
A fully functional prototype supported by a field-programmable gate array
(FPGA) technology is designed and tested, tending an in-depth analysis of the
adopted approach.

This chapter is organized as follows. Section 14.2 presents the EC and the
AES properties, including different arithmetic approaches that can be used in
different implementations scenarios. Additionally in this section, a detailed
description of the RNG types and evaluation methods are provided. In Section
14.3, the best approaches are selected in order to design efficient arithmetic
units for the addressed applications as well as a fully functional microcoded

Embedded Systems: Hardware, Design, and Implementation, First Edition.
Edited by Krzysztof Iniewski.
© 2013 John Wiley & Sons, Inc. Published 2013 by John Wiley & Sons, Inc.

processor using the available reconfiguration capabilities. Section 14.4 presents and discusses the experimental results obtained for the developed processor, prototyped on an FPGA technology. Section 14.5 draws some conclusions about the developed methods and prototype.

14.2 BACKGROUND

Systems that implement security protocols typically require support for public-key and private-key ciphering as well as random number generation [1]. In this section, several algorithms suitable for efficient hardware implementations are introduced with this purpose. Additionally, known approaches to implement this kind of algorithms are briefly described and discussed. The preliminaries described in this section are further used in Section 14.3 to support an implementation of a flexible hardware architecture able to be embedded in any system that requires generic cryptographic support.

14.2.1 Elliptic Curve Cryptography

EC cryptography was proposed simultaneously by Neil Koblitz [2] and Victor Miller [3] in 1985 to support public-key protocols. Since then, EC cryptography has been confirmed as a competing alternative to the widely used Rivest–Shamir–Adleman (RSA) protocol due to its increased efficiency [4–6]. At the time of its proposals, EC cryptography suggested computing efficiency improvements of about 20% regarding the RSA protocol, which were expected to increase with the development of the computing capabilities of the target devices. More recently, Neil Koblitz et al. revisited the comparative efficiency of the EC cryptography regarding the RSA protocol [7], estimating that a 2048-bit public-key cipher for the RSA protocol can be tampered with 3×10^{20} million instructions per second (MIPS) years, whereas a 234-bit public-key cipher for an EC-based protocol requires 1.2×10^{23} MIPS years to be compromised. In other words, with EC cryptography it is possible to provide three orders of magnitude more security with eight times smaller keys regarding the RSA alternative. This analysis properly depicts the advantages of EC cryptography motivating the inclusion of this cryptosystem in the current standards for public-key ciphering published by recognized institutions such as the National Institute of Standards and Technology (NIST) [8], the Institute of Electrical and Electronics Engineers (IEEE) [9, 10], or the Certicom [11, 12]. These advantages and the wide acceptance of the EC cryptography as a secure and efficient solution for public-key protocols are the motivation for addressing this cryptosystem in this chapter.

The EC cryptography arithmetic relies on a hierarchical scheme that is depicted in Figure 14.1. The bottom layer of the hierarchy is the underlying finite field that supports addition, multiplication, and the respective additive and multiplicative inverses and identities. A reduction operation is also

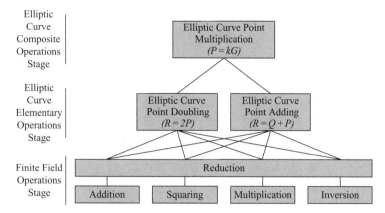

FIGURE 14.1. EC arithmetic hierarchy.

included in this layer to assure that the output of each of the aforementioned operations remains in the finite field. The middle layer of the hierarchy is composed of the EC group basic operations, namely the addition and doubling. Each of the operations in this layer is computed as a sequence of operations of the bottom layer. The top layer is a composite of the elementary operations in the middle layer and is the most important operation in EC cryptography, granting the security of the system. In the following sections, the relevant details of each layer are presented and discussed.

14.2.1.1 Finite Field Arithmetic   In an EC cryptosystem, all the operations are supported on an entity referred to as a finite field, the generic definition of which is the following:

Definition 14.1 (Finite Field). A finite field $GF(q)$ is an algebraic entity that consists of a finite set of q elements with the operations $+$ and $\times$ (this last operation symbol is often omitted), such that:

- $+$ and $\times$ have the property of closure: for any $f_1, f_2 \in GF(q), (f_1 + f_2) \in GF(q)$ and $(f_1 \, f_2) \in GF(q)$
- $+$ and $\times$ are associative: for any $f_1, f_2, f_3 \in GF(q), (f_1 + f_2) + f_3 = f_1 + (f_2 + f_3)$ and $(f_1 f_2) f_3 = f_1 (f_2 f_3)$
- $+$ and $\times$ are commutative: for any $f_1, f_2 \in GF(q), f_1 + f_2 = f_2 + f_1$ and $f_1 f_2 = f_2 f_1$
- $\times$ is distributive toward $+$: for any $f_1, f_2, f_3 \in GF(q), f_1(f_2 + f_3) = f_1 f_2 + f_1 f_3$
- there are identity elements f_{id+} and f_{idx} for the operations $+$ and $\times$, respectively: there are $f_{id+}, f_{idx} \in GF(q)$ such that for all $f_1 \in GF(q), f_1 + f_{id+} = f_1$ and $f_1 \times f_{idx} = f_1$
- all elements have an inverse f_{inv+} toward $+$: for any $f_1 \in GF(q)$ there is an element $f_{inv+} \in GF(q)$ such that $f_1 + f_{inv+} = f_{id+}$

- except for f_{id+}, all elements have an inverse f_{invx} toward $\times$: for any $f_1 \in GF(q)$, with $f_1 \neq f_{id+}$, there is an element $f_{inv+} \in GF(q)$ such that $f_1 \times f_{invx} = f_{idx}$

It is known that such a finite field exists for $q = p^m$, with p a prime and m a positive integer. There are different finite fields that have practical interest to implement an EC cryptosystem, each one tailored for different implementation approaches. The most common finite fields are the prime fields $GF(p)$ and the binary extension fields $GF(2^m)$. The notation $GF(.)$ is the acronym of Galois field, which is an alternative designation for a finite field that is named after the French mathematician Évariste Galois. These two different finite fields mainly differ in the sense that $GF(p)$ is more suitable for software implementation, whereas $GF(2^m)$ can prospect more advantages from hardware implementations. This is due to the carry-less operations over $GF(2^m)$ that can be accomplished with simple gates, such as the addition which corresponds to bitwise logic exclusive OR (XOR) operations. On the other hand, the operations over $GF(p)$ rely in the ordinary integer operations with carry for which efficient implementations already exist in many general-purpose processors, hence the reduced effort in implementing solutions based in this field in software. This chapter aims at the development of a hardware solution to support cryptographic protocols. Hence, given the aforementioned considerations, we focus the utilization of $GF(2^m)$. For details on the utilization of $GF(p)$ for the implementation of cryptographic accelerators, the reader is referred to Koç [13].

The finite field $GF(2^m)$ is constructed as an extension of the finite field $GF(2)$. Hence, it is mandatory to introduce the details of $GF(2)$ when discussing $GF(2^m)$.

Definition 14.2 ($GF(2)$). The finite field $GF(2)$ consists of the set $\{0, 1\}$, with the addition and multiplication modulo 2 operations. In this field, the addition results can be computed as the XOR of the inputs and the multiplication results from the logic AND.

Observing Definition 14.2 closely, it can be verified that the characteristics pointed out for $GF(2)$ fully comply with the statements in Definition 14.1. Now, let $P(x)$ be an irreducible polynomial of order m in $GF(2)$.

Definition 14.3 (Irreducible Polynomial). A polynomial $P(x)$ is irreducible if and only if it cannot be expressed as a product of two or more polynomials different from 1. In other words, $P(x)$ is irreducible if it cannot be factorized in two or more polynomials with smaller order.

Given $P(x)$, let α be a root of this polynomial, that is, $P(\alpha) = 0$. The first m powers of the value α can be used to construct a basis of $GF(2^m)$, often called polynomial or canonical:

$$\{1, \alpha, \alpha^2, \ldots, \alpha^{m-2}, \alpha^{m-1}\}. \tag{14.1}$$

Any element A of $GF(2^m)$ can be represented with m elements of $GF(2)$ $(a_{m-1} \ldots, a_1 a_0)$ (in practice, these elements are bits) designating a linear combination of the vectors in Equation (14.1):

$$A = \sum_{i=0}^{m-1} a_i \alpha^i, a_i \in GF(2). \tag{14.2}$$

Bases other than the polynomial in Equation (14.1) can be used for operating over $GF(2^m)$. Among these bases are the normal basis and optimal normal basis [14]. These bases enhance the performance of some of the operations in $GF(2^m)$ and are intended to be supported by specific dedicated hardware structures [15, 16]. Moreover, optimal normal bases do not exist for all fields $GF(2^m)$. Given that the aim of this chapter is the implementation of a generic and flexible accelerator for cryptography, the polynomial basis is used instead of the normal basis.

With the field basis already defined, it is now possible to define the operations in $GF(2^m)$. These operations are said to be modular in the sense that they are computed modulo $P(x)$, which is the irreducible polynomial used to define the basis. The modular addition is a straightforward operation in $GF(2^m)$, given that $a_i, b_i \in GF(2)$, corresponding to the bitwise XOR (addition in $GF(2)$) of the binary vector A and B.

Definition 14.4 (Modular Addition). For two elements A and B in $GF(2^m)$ defined by the irreducible polynomial $P(x)$, the addition modulo $P(x)$ is obtained as:

$$R = A + B \bmod P(x) = \sum_{i=0}^{m-1} a_i \alpha^i \oplus \sum_{i=0}^{m-1} b_i \alpha^i = \sum_{i=0}^{m-1} (a_i \oplus b_i) \alpha^i. \tag{14.3}$$

Note that the subtraction (additive inverse) corresponds to exactly the addition due to the XOR operand properties.

The other field operation is the modular multiplication. Prior to introducing the multiplication of generic elements of $GF(2^m)$, let us consider the multiplication of an element A by the root α of the irreducible polynomial $P(x)$:

$$\alpha A \bmod P(x) = \alpha \sum_{i=0}^{m-1} a_i \alpha^i \bmod P(x) = \sum_{i=1}^{m} a_{i-1} \alpha^i \bmod P(x)$$

$$= a_{m-1} \alpha^m + \sum_{i=1}^{m-1} a_{i-1} \alpha^i \bmod P(x). \tag{14.4}$$

In Equation (14.4), the value α^m is not defined in the basis, and thus needs to be rewritten as a linear combination of the vectors in the basis, or in other words, needs to be reduced. By definition, the coefficients p_m and p_0 of the irreducible polynomial $P(x)$ are different from zero because $P(x)$ has order m and cannot be divided by x (it is irreducible). Therefore:

$$P(\alpha) = 0 \Leftrightarrow \sum_{i=0}^{m} p_i \alpha^i = 0 \Leftrightarrow \alpha^m + \sum_{i=1}^{m-1} p_i \alpha^i + 1 = 0 \Leftrightarrow \alpha^m = \sum_{i=1}^{m-1} p_i \alpha^i + 1. \quad (14.5)$$

The result in Equation (14.5) allows rewriting α^m in Equation (14.4) with the basis in Equation (14.1). Hence it is now possible to define the multiplication by α over $GF(2^m)$.

Definition 14.5 (Modular Multiplication by α). For an element A in $GF(2^m)$ defined by the irreducible polynomial $P(x)$ the multiplication by α modulo $P(x)$ is obtained as:

$$\alpha A \bmod P(x) = a_{m-1} \left(\sum_{i=1}^{m-1} p_i \alpha^i + 1 \right) + \sum_{i=1}^{m-1} a_{i-1} \alpha^i \bmod P(x)$$

$$= \sum_{i=1}^{m-1} (a_{m-1} p_i \oplus a_{i-1}) \alpha^i + a_{m-1}. \quad (14.6)$$

This expression corresponds to a left shift of the input A followed by a conditional modular addition with the terms of $P(x)$ with order lower than m. This condition depends on the content of a_{m-1}.

Definition 14.6 provides evidence on how to perform the other two important operations over $GF(2^m)$, namely the reduction and general multiplication. Using the property in Equation (14.5) it is possible to rewrite α^k for any value $k \geq m$ as terms of the basis in Equation (14.1), that is, α^k can be obtained from α^m by applying $(k - m)$ times the Definition 14.6. This property allows rewriting any result $B = \sum_{i=0}^{k} b_i \alpha^i$ as a linear combination of the elements in Equation (14.1), or in other words, allows reducing B modulo $P(x)$. Given that the reduction is provided, the multiplication can be obtained using the traditional schoolbook algorithm or any of the subquadratic complexity multiplication algorithms, such as the Karatsuba–Ofman multiplication algorithm [17]. The multiplication of two elements A and B in $GF(2^k)$ following the schoolbook algorithm is obtained as:

$$AB \bmod P(x) = \sum_{i=1}^{m-1} a_i \alpha^i \sum_{j=1}^{m-1} b_j \alpha^j = \sum_{i=1}^{m-1} \sum_{j=1}^{m-1} a_i b_j \alpha^{i+j}. \quad (14.7)$$

The reduction should be obtained afterward by rewriting the powers of α, for which $i + j \geq m$. Concerning the Karatsuba–Ofman multiplication, the inputs A and B should be rewritten as:

$$A = A_H \alpha^{\lfloor m/2 \rfloor} + A_L = \sum_{i=\lfloor m/2 \rfloor}^{m-1} a_i \alpha^i + \sum_{i=0}^{\lfloor m/2 \rfloor - 1} a_i \alpha^i, \qquad (14.8)$$

$$B = B_H \alpha^{\lfloor m/2 \rfloor} + B_L = \sum_{i=\lfloor m/2 \rfloor}^{m-1} b_i \alpha^i + \sum_{i=0}^{\lfloor m/2 \rfloor - 1} b_i \alpha^i.$$

Using the approach in Equation (14.8), it is possible to obtain the product over the binary field with only three multiplications of size $[m/2]$ by partitioning inputs in 2 as:

$$AB = A_H B_H \alpha^m + ((A_H + A_L)(B_H + B_L) - A_H B_H - A_L B_L) \alpha^{\lfloor m/2 \rfloor} + A_L B_L. \quad (14.9)$$

Concerning the modular inversion, two main approaches are often employed. One of them follows the extended Euclidean algorithm (EEA) and can be implemented with dedicated hardware [18]. In this chapter, the other approach is employed. This other approach follows Fermat's little theorem (FLT) [19] and allows reusing the hardware structures for modular multiplication and for squaring, enhancing the compactness of the design. Following the FLT, the inversion of an element A is obtained as:

$$A^{2^m - 1} \equiv 1 \bmod P(x) \Leftrightarrow A^{2^m - 2} \equiv A^{-1} \bmod P(x) \Leftrightarrow A^{2^m - 1} \dots A^2 \equiv A^{-1} \bmod P(x). \tag{14.10}$$

There are optimizations on Equation (14.10) that allow reducing the number of required multiplications and squaring operations. A method known as Itoh–Tsujii inversion [20] proposes rewriting Equation (14.10) as:

$$A^{-1} \bmod P(x) = \begin{cases} \left(A^{2^{m/2} - 1} \right)^2 A \bmod P(x) & \text{if } m \text{ even} \\ \left[\left(A^{2^{(m-1)/2} - 1} \right)^{2^{(m-1)/2} + 1} \right]^2 A \bmod P(x) & \text{if } m \text{ odd} \end{cases}, \quad (14.11)$$

which reduces the number of required multiplication to $[\log_2(m - 1) + h(m - 1) - 1]$ [16], where $h(.)$ stands for the Hamming weight. Given that the inversion is equivalent to several multiplications and squarings, one can conclude that inversion is the most demanding of the field operations discussed above, and thus should be avoided, unless strictly necessary.

14.2.1.2 Elliptic Curve Arithmetic   There are several ECs suiting different underlying finite fields. Concerning $GF(2^m)$ curves, the often called Edwards curves and Koblitz curves can be used, each one having different properties. A different generic nonsupersingular curve is herein addressed, which is also defined with the standard curves presented in NIST [8].

Definition 14.6 (Elliptic Curve over GF(2^m)). A nonsupersingular EC over $GF(2^m)$, referred to as $E(a, b, GF(2^m))$, consists of the set of two coordinated points $P_i = (x_i, y_i) \in GF(2^m) \times GF(2^m)$, complying with:

$$y_i^2 + x_i y_i = x_i^3 + ax_i^2 + b, a, b \in GF(2^m), \tag{14.12}$$

together with a point at infinity $\mathcal{O} = (0, 1)$.

The points in an EC form an additive group where the operation addition is defined and where $\mathcal{O}$ is the addition identity. This EC operation is constructed over the underlying finite field, in the sense that it can be constructed by a sequence of finite field operations on the coordinates of the EC points.

Definition 14.7 (Elliptic Curve Addition). Consider two points $P_1 = (x_1, y_1)$ and $P_2 = (x_2, y_2)$ in $E(a, b, GF(2^m))$ and the irreducible polynomial $P(x)$ that defines the underlying finite field. The addition in $E(a, b, GF(2^m))$ is commutative and consists of the field operations modulo $P(x)$ specified in the following rules [21]:

- For $P_1 \neq \mathcal{O}, P_2 \neq \mathcal{O}$, and $P_1 \neq P_2$, the point $P_3 = (x_3, y_3) = P_1 + P_2$ (addition) is obtained as:

$$x_3 = \lambda^2 + \lambda + x_1 + x_2 + a; \; y_3 = \lambda(x_1 + x_3) + x_3 + y_1; \lambda = \frac{y_1 + y_2}{x_1 + x_2}. \tag{14.13}$$

- For $P_1 \neq \mathcal{O}$, the point $P_3 = (x_3, y_3) = P_1 + P_1 = 2P_1$ (doubling) is obtained as:

$$x_3 = \lambda^2 + \lambda + a; \; y_3 = x_1^2 + (\lambda + 1)x_2; \lambda = x_1 + \frac{y_1}{x_1}. \tag{14.14}$$

- The point $P_3 = (x_3, y_3) = -P_1 = \mathcal{O} - P_1$ (additive inverse) is obtained as:

$$x_3 = x_1; \; y_3 = x_1 + y_1. \tag{14.15}$$

- For $P_1 = \mathcal{O}$, the point $P_3 = P_1 + P_2 = P_2$ and $P_3 = 2P_1 = \mathcal{O}$.

As suggested in Figure 14.1, there is an upper stage in the EC arithmetic that is constructed based on the operations in Definition 14.7. This upper stage consists of the EC point multiplication by a scalar, or simply EC point multiplication. For a scalar integer s and a point P_1 in $E(a, b, GF(2^m))$, the EC point multiplication P_3 is obtained as:

$$P_3 = sP_1 = \underbrace{P_1 + \cdots + P_1}_{s \text{ times}}. \tag{14.16}$$

This operation grants the security of the EC cryptography given that knowing P_3 and P_1, it is not possible to compute s in polynomial time, or in other words, it is possible to select reasonable large values of s that allow computing P_3 efficiently but preclude the computation of s from P_3 and P_1 in time short enough to compromise the security of the data being secured within the cryptosystem. This is known as the elliptic curve discrete logarithm problem (ECDLP). Consider the following example concerning the utilization of the EC point multiplication for secure data transfer, usually called the El-Gammal protocol for EC [22]:

- Alice wants to send a message M to Bob, which content is somehow mapped to an EC point P_M.
- Alice generates a random secret integer k_A that only she knows. Bob does the same, generating an integer k_B. These two integers are known as Alice's and Bob's private keys.
- There is an EC point P_G that is defined in the protocol, thus both Alice and Bob know about it.
- Alice and Bob compute their public keys, which correspond to EC points, by performing the EC point multiplications $P_A = k_A P_G$ and $P_B = k_B P_G$, respectively. Thereafter, Alice and Bob publish their public keys so that everyone knows them. Because of the ECDLP, Alice and Bob are sure that no one will be able to get their k_A and k_B from their public keys.
- After receiving Bob's public key, Alice is able to compute an EC point containing a ciphertext of the message M as $P_C = P_M + k_A P_B$.
- Alice can now send the cryptogram P_C to Bob. Given that no one knows k_A, no one intercepting P_C will be able to compute P_M, except for Bob. Bob can compute $P_C - k_B P_A = (P_M + k_A P_B) - k_B P_A = (P_M + k_A k_B P_G) - k_B k_A P_G = P_M$.
- Knowing P_M, Bob can retrieve the secret message M.

The computation of the EC point multiplication can be done with several different methods. One of them, and perhaps the simplest one, is the double-and-add method that is presented in Algorithm 14.1. In this algorithm it is possible to identify that $n - 1$ EC point doublings and $h(s) - 1$ EC point additions need to be performed in order to compute the final result P_3. Thus, considering Definition 14.7, $n + h(s) - 2$ field inversions are required to compute P_3. Regarding the higher computational demands of the field inversion regarding the other field operations, a different representation of the EC points can be adopted using three coordinates instead of two (see Definition 14.6). These coordinates are called *Projective Coordinates* while the coordinates originally introduced in Definition 14.6 are called *Affine Coordinates*. Several approaches using projective coordinates exist and have been employed in hardware implementations. These approaches are designated Standard [23], Jacobi [24], and López-Dahab [25] projective coordinates. In this chapter, the

---

ALGORITHM 14.1. Double-and-add EC point multiplication.

Require: n-bit scalar $s = (s_{n-1} \ldots s_0)$ and P_1 in $E(a, b, GF(2^m))$;
Ensure: $P_3 = sP_1$;
   $P_3 = P_1$;
   {Assume $s_{n-1} = 1$}
   for $i = n - 2$ down to 0 **do**
      $P_3 = 2P_3$;
      if $s_i = 1$ **then**
         $P_3 = P_3 + P_1$;
      end if
   end for
   return P_3;

---

standard projective coordinates are addressed not only because they avoid several inversions in the EC point multiplication, but also because they underlie an EC point multiplication algorithm whose execution time does not depend on the scalar s. As discussed ahead, this latter property is important to avoid leaking information on the secret scalar to attackers tracking the execution time [6].

Definition 14.8 (Standard Projective Coordinates). The standard projective coordinates refer to the triple $(\widehat{x}_i, \widehat{y}_i, \widehat{z}_i)$ in $GF(2^m) \times GF(2^m) \times GF(2^m)$, except for the triple $(0, 0, 0)$ in the surface of equivalence classes, where $(\widehat{x}_1, \widehat{y}_1, \widehat{z}_1)$ is said to be equivalent to $(\widehat{x}_2, \widehat{y}_2, \widehat{z}_2)$ if there exists a $\lambda \in GF(2^m)$ such that $\lambda \neq 0$ and $(\widehat{x}_1, \widehat{y}_1, \widehat{z}_1) = (\lambda \widehat{x}_2, \lambda \widehat{y}_2, \lambda \widehat{z}_2)$. In this surface, an EC point with affine coordinates (x_i, y_i) can be represented in projective coordinates by setting $\lambda \widehat{z}_i = 1 \Leftrightarrow \lambda = 1/\widehat{z}_i$. Therefore, applying $x_i = \widehat{x}_i / \widehat{z}_i$ and $y_i = \widehat{y}_i / \widehat{z}_i$ to Equation (14.12) results in the following equation to describe the EC:

$$\widehat{y}_i^2 \widehat{z}_i + \widehat{x}_i \widehat{y}_i \widehat{z}_i = \widehat{x}_i^3 + a\widehat{x}_i^2 \widehat{z}_i + b\widehat{z}_i^3, \quad a, b \in GF(2^m). \tag{14.17}$$

The projective representation of the points in Definition 14.8 is particularly interesting for a method to perform the EC point multiplication, the Montgomery EC point multiplication, presented in Montgomery [26] and adapted to $GF(2^m)$ in López and Dahab [27].

Definition 14.9 (Montgomery Point Multiplication). Consider the EC points $P_1 = (x_1, y_1)$, $P_2 = (x_2, y_2)$, $P_{add} = (x_{add}, y_{add}) = P_1 + P_2$, and $P_{sub} = (x_{sub}, y_{sub}) = P_1 - P_2$. Then, the following holds [27]:

$$x_{\text{add}} = \begin{cases} x_{\text{sub}} + \dfrac{x_1}{x_1 + x_2} + \left(\dfrac{x_1}{x_1 + x_2}\right)^2 & \text{if } P_1 \neq P_2, \\[3mm] x_1^2 + \dfrac{b}{x_1^2} & \text{if } P_1 = P_2, \end{cases} \qquad (14.18)$$

which suggests that it is possible to compute the x-coordinate of an addition, knowing only the x-coordinates of the input points and of the subtraction of the inputs, without requiring the y-coordinate of any of the points. This suggests that it is possible to compute the EC point multiplication using the EC point addition and doubling such as in Equation (14.18) without handling the y-coordinate at all.

As proposed in López and Dahab [27], it is possible to obtain a version of Equation (14.18) using standard projective coordinates. The resulting addition and doubling operations can thus be obtained with:

$$\widehat{x_{\text{add}}} = \begin{cases} x_{\text{sub}}\,\widehat{z_{\text{add}}} + \widehat{x_1}\,\widehat{z_2}\,\widehat{x_2}\,\widehat{z_1} & \text{if } P_1 \neq P_2, \\[2mm] \widehat{x_1}^4 + b\widehat{z_1}^4 & \text{if } P_1 = P_2. \end{cases} \qquad \widehat{z_{\text{add}}} = \begin{cases} \left(\widehat{x_1}\,\widehat{z_2} + \widehat{x_2}\,\widehat{z_1}\right)^2 & \text{if } P_1 \neq P_2, \\[2mm] \widehat{x_1}^2\,\widehat{z_1}^2 & \text{if } P_1 = P_2. \end{cases}$$

$$(14.19)$$

Note that the operations in Equation (14.19) depend on the affine coordinate x_{sub}. That is the reason for using the operations in Equation (14.19) to support an EC point multiplication algorithm called Montgomery ladder that is presented in Algorithm 14.2, where `Madd` and `Mdouble` stand for the EC point addition and doubling, respectively. Algorithm 14.2 has the property that x_{sub} is invariant and corresponds to the affine x-coordinate of the EC point multiplication input point, that is, $x_{\text{sub}} = x_1$. Algorithm 14.2 has a final step, `Mxy`, that aims to compute the affine coordinates of $P_3 = (x_3, y_3)$. This can be obtained from the projective coordinates as [21]:

$$x_3 = \widehat{x_1}/\widehat{z_1}, \; y_3 = (x_1 + \widehat{x_1}/\widehat{z_1})(x\,\widehat{z_1}\,\widehat{z_2})^{-1}\Big[(\widehat{x_1} + x_1\widehat{z_1})(\widehat{x_2} + x_1\widehat{x_2}) + (x_1^2 + y_1)\widehat{z_1}\,\widehat{z_2})\Big] + y_1.$$

$$(14.20)$$

If the application underlined by the EC arithmetic does not need the y-coordinate, the expression for y_3 does not have to be implemented at all.

14.2.2 Advanced Encryption Standard

The AES is a block cipher algorithm that was standardized to support symmetric encryption as of 2001 [28]. The AES supersedes the Data Encryption Standard (DES) [29], which has known successful methods for its cryptanalysis [30]. The establishment of the AES was accomplished with the selection of the Rijndael block cipher, among several candidates [31, 32]. In contrast to the

---

ALGORITHM 14.2. Montgomery ladder for computing the EC point multiplication.

Require: n-bit scalar $s = (s_{n-1} \ldots s_0)$ and P_1 in $E(a, b, GF(2^m))$;
Ensure: $P_3 = sP_1$;

$$\widehat{x_1} = x_1; \widehat{z_1} = 1; \widehat{x_2} = x_1^4 + b; \widehat{z_2} = x_1^2.$$

  for $i = n - 2$ down to 0 **do**
    if $s_i = 1$ **then**

$$(\widehat{x_1}, \widehat{z_1}) = \mathtt{Madd}(\widehat{x_1}, \widehat{z_1}, \widehat{x_2}, \widehat{z_2}), (\widehat{x_2}, \widehat{z_2}) = \mathtt{Mdouble}(\widehat{x_2}, \widehat{z_2});$$

  else

$$(\widehat{x_2}, \widehat{z_2}) = \mathtt{Madd}(\widehat{x_2}, \widehat{z_2}, \widehat{x_1}, \widehat{z_1}), (\widehat{x_1}, \widehat{z_1}) = \mathtt{Mdouble}(\widehat{x_1}, \widehat{z_1});$$

  end if
  end for
  return $P_3 = \mathtt{Mxy}\left(\widehat{x_1}, \widehat{z_1}, \widehat{x_2}, \widehat{z_2}\right)$;

EC cryptography, AES is a symmetric encryption protocol, which requires that both parties involved in a communication know the encryption key. This means that the encryption key needs to be somehow distributed, hopefully using a secured mechanism. Between these mechanisms are the public-key protocols such as the one described in Section 14.2.1.

The arithmetic underlying the AES has similarities to the one for the ECs in the sense that it is over $GF(2^m)$. However, while for an EC the order m of the irreducible polynomial supporting the field is around hundreds for the AES, it has an order $m = 8$. Moreover, the AES also contains in its constructs nonlinear operations that are the security recipient of the algorithm. In the AES, the data to be ciphered/deciphered are organized in a 4×4 matrix D of bytes called *state*, that is, $4 \times 4 = 16$ bytes (128 bits) is the size of the basic block of the data to be processed. The bytes in matrix D are addressed as $d_{(i,j)}$, where elements with the same value i configure a row and elements with the same value j configure a column, with $0 \le i < 4$ and $0 \le j < 4$. Algorithm 14.3 presents the ciphering of the AES.

Several entries and routines in Algorithm 14.3 require a definition. Between the entries are the values N_{key} and N_{rounds}. The AES comprises three variations of the algorithm for different security levels which are designated as AES-128, AES-192, and AES-256 (in increasing security order). As the designations suggest, the security key size for these variations are 128 (16 bytes), 192 (24 bytes), and 256 (32 bytes), respectively. The value N_{key} stands for the size of

---

ALGORITHM 14.3. AES ciphering algorithm.

Require: D, 4×4 matrix with 16-byte input data (state);
Require: N_{key}, size of the security key;
Require: N_{rounds}, number of rounds of the algorithm;
Require: K, $4 \times (S_{rounds} + 1)$ matrix with the expanded key;
Ensure: D contains a secure ciphertext for the input data.
  D =AddRoundKey(D, $k_{(i,0...3)}$);
  for $r = 1$ to $N_{rounds} - 1$ **do**
    D =SubBytes(D);
    D =ShiftRows(D);
    D =MixColumns(D);
    D =AddRoundKey(D, $k_{(i,4r...3(r+1))}$);
  end for
    D =SubBytes(D);
    D =ShiftRows(D);
    D =AddRoundKey(D, $k_{(i,4N_{rounds}...3(N_{rounds}+1))}$);
  return D;

---

the key in bytes and the value N_{rounds} is directly inferred from the key size and is 10, 12, and 14 for increasing levels of security. The routines which require definition are the SubBytes, ShiftRows, and MixColumns.

Definition 14.10 (AES SubBytes Routine). The SubBytes routine applies the same transformation to each byte $d_{(i,j)}$ of the state. Each byte is addressed as an element of the field $GF(2^8)$ with irreducible polynomial $P(x) = x^8 + x^4 + x^3 + x + 1$; thus, $d_{(i,j)} = \sum_{k=0}^{7} d_{(i,j)k}\alpha^k$, where α is a root of $P(x)$. The first step in the computation of the SubBytes routine is the computation of the inverse of each byte $d_{(i,j)} = d_{(i,j)}^{-1} \mod P(x)$, except to $d_{(i,j)} = 0$, which is mapped to 0. The second step allows obtaining the final result of the state performing the following operation in $GF(2)$:

$$
\begin{bmatrix} d_{(i,j)0} \\ d_{(i,j)1} \\ d_{(i,j)2} \\ d_{(i,j)3} \\ d_{(i,j)4} \\ d_{(i,j)5} \\ d_{(i,j)6} \\ d_{(i,j)7} \end{bmatrix} = \begin{bmatrix} 1 & 0 & 0 & 0 & 1 & 1 & 1 & 1 \\ 1 & 1 & 0 & 0 & 0 & 1 & 1 & 1 \\ 1 & 1 & 1 & 0 & 0 & 0 & 1 & 1 \\ 1 & 1 & 1 & 1 & 0 & 0 & 0 & 1 \\ 1 & 1 & 1 & 1 & 1 & 0 & 0 & 0 \\ 0 & 1 & 1 & 1 & 1 & 1 & 0 & 0 \\ 0 & 0 & 1 & 1 & 1 & 1 & 1 & 0 \\ 0 & 0 & 0 & 1 & 1 & 1 & 1 & 1 \end{bmatrix} \begin{bmatrix} d_{(i,j)0} \\ d_{(i,j)1} \\ d_{(i,j)2} \\ d_{(i,j)3} \\ d_{(i,j)4} \\ d_{(i,j)5} \\ d_{(i,j)6} \\ d_{(i,j)7} \end{bmatrix} + \begin{bmatrix} 1 \\ 1 \\ 0 \\ 0 \\ 0 \\ 1 \\ 1 \\ 0 \end{bmatrix}. \tag{14.21}
$$

The conjunction of both steps of the SubBytes routine can also be efficiently performed by a table lookup.

Definition 14.11 (AES ShiftRows Routine). The ShiftRows routine is accomplished by circularly shifting to the left the bytes $d_{(i,j)}$ in the same row. The number of positions to shift depends on the row: for the row i, the number of positions to shift is i.

Definition 14.12 (AES MixColumns Routine). For the MixColumns routine the bytes of a column of the state $d_{(0...3,j)}$ refer to elements constructed using the basis $\{1, \alpha, \alpha^2, \alpha^3\}$:

$$C_j = \sum_{k=0}^{3} d_{(k,j)} \alpha^k, \tag{14.22}$$

where α is a root of the polynomial $P(x) = x^4 + 1$ in $GF(2^8)$. The transformation being held for the MixColumns routines consists of the multiplication of each of the elements c_j by:

$$\bar{C} = 3\alpha^3 + \alpha^2 + \alpha + 2, \tag{14.23}$$

modulo $P(x) = x^4 + 1$. The reduction modulo $P(x)$ is computed in a similar way as described in Section 14.2.1.1 for $GF(2^m)$: given that $P(\alpha) = 0 \Leftrightarrow \alpha^4 = 1$, hence $\alpha^i \equiv \alpha^{i \bmod 4} \bmod P(x)$. The operation computed in the MixColumns routine can be represented using matrix operations over $GF(2^8)$ for each column j:

$$\begin{bmatrix} d_{(0,j)} \\ d_{(1,j)} \\ d_{(2,j)} \\ d_{(3,j)} \end{bmatrix} = \begin{bmatrix} 2 & 3 & 1 & 1 \\ 1 & 2 & 3 & 1 \\ 1 & 1 & 2 & 3 \\ 3 & 1 & 1 & 2 \end{bmatrix} \begin{bmatrix} d_{(0,j)} \\ d_{(1,j)} \\ d_{(2,j)} \\ d_{(3,j)} \end{bmatrix}. \tag{14.24}$$

Note that the inverse modulo $P(x)$ for the polynomial $\bar{C}$ exists, thus an inverse operation to the MixColumns routine also exists.

The remaining two definitions required to Algorithm 14.3 are the matrix K and the routine AddRoundKey. The matrix K is built from the algorithm's input key and is split in 4×4 submatrices for the AddRoundKey routine, which only computes the addition in $GF(2)$ between these submatrices and the state. The matrix K is built only using operations over $GF(2^8)$ and remains unchanged for the same input key and operation mode [28].

All the operations introduced for the AES ciphering are all invertible, which means that the deciphering is accomplished by a proper sequence of the inverses of the routines involved in the ciphering, including the substitution box (S-box) operations.

14.2.3   Random Number Generators

The RNGs can be divided in two main classes: the pseudo RNGs (PRNGs) and true RNGs (TRNGs). The PRNGs include simple and low-security generators such as the linear feedback shift registers (LFSRs) [33] as well as more elaborated and secure generators such as the Mersenne Twister generator [34] that is employed as the default generator of widely used tools such as MATLAB. These generators produce a deterministic sequence departing from a given seed and are considered safe if the period of the sequence is sufficiently large to avoid prediction and if patterns in a period cannot be efficiently identified. The other class of generators, the TRNGs, uses physical sources of randomness to produce a sequence that has no period. These randomness sources have evolved to sophisticated sources that take advantage of the quantum phenomena and prototypes of generators based in this approach have already been proposed [35]. However, commercially available devices able to support cryptographic applications do not yet include such sophisticated sources of randomness. Hence, TRNGs to be integrated in such devices have to get their randomness from more conventional sources such as the jitter of oscillators [36].

Whatever chosen RNGs, methods to verify their randomness properties are required. The most popular methods used for this purpose consist of a battery of statistical tests. These methods formulate the hypothesis (null hypothesis) that a sequence is random and obtain a reference/theoretical probability distribution for such hypothesis by using any mathematical method. Given the null hypothesis probability distribution, the probability of accepting that a sequence is random when it is not is evaluated and set to have a negligible value (say 1%). The statistic that separates the 99% probability of accepting a sequence that is random from the 1% probability of accepting a sequence that is not is referred to as the critical value. Then, the test is performed as follows: (1) a sequence is sampled from the RNG under test, (2) a statistic is computed from that sequence, (3) the obtained statistic is compared with the critical value, and (4) a decision is made and the test fails if the statistic is beyond the critical value (within the 1% probability range). Two main batteries of tests are widely used to test RNGs: the diehard battery of tests of randomness proposed by G. Marsaglia [37] and the NIST Statistical Test Suite for Random and Pseudorandom Number Generators for Cryptographic Applications [38].

14.3   RECONFIGURABLE PROCESSOR

A fully functional processor able to supply the cryptographic support to a host system is described in this section addressing the algorithms specified in Section 14.2. The main key feature of this processor is flexibility. Flexibility, because the workload of a cryptographic processor varies and the processor should be adaptable to these different workloads for efficiency sake. Moreover, given the constant evolution of the attacks, a more flexible processor can better adapt and include countermeasures. The flexibility is also important to face budget constraints in the sense that a single design can fit the demands of several applications with different demands in terms of performance and/or resources. The flexibility in the design is prospected in two different dimensions: programmability and reconfigurability. The programmability of the processor herein addressed is expressed in the utilization of a microcoded program that controls the behavior of the processing units (PUs). The reconfigurability of the processor is obtained from the underlying hardware platform for which the processor is designed to, that is, an FPGA. Following the aforementioned features, an overview of the proposed processor is presented in Figure 14.2. The processor contains a customizable number n of PUs which can be dynamically changed through runtime reconfiguration. These PUs contain a random access memory (RAM) for storing input data as well as the intermediate and final results. Each PU is intended to individually support potentially different operations in order for the arithmetic units available in each PU to match the requirements of the applications being supported. In this chapter, the focused applications are the EC cryptography and the AES, but the concept introduced with this processor allows embedding any other application by designing a different PU. Each PU contains a simple finite state machine (FSM) to control

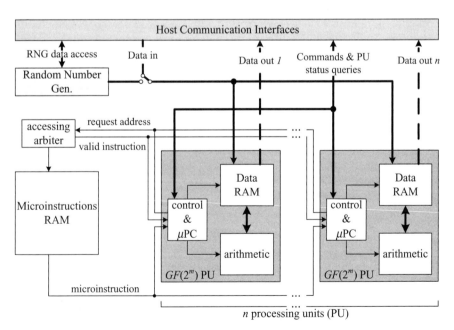

FIGURE 14.2. Programmable and reconfigurable processor overview.

the arithmetic units and data transfers inside the PU. This FSM receives a microinstruction and a valid instruction bit signaling that the input microinstruction is valid. After decoding it, the FSM behaves accordingly and increments a local μ program counter (μPC) requesting the following microinstruction. This FSM also receives simple commands (input data, write enable, etc.) and answers to queries (e.g., is the PU idle?) from the processor's host. In order to save resources, and given that each PU operates on multiclock cycle instructions, a RAM centralizes the microprogram required by each of the PUs, even if the programs to be executed in each of them are different. An arbiter controls the access to the microinstructions RAM so as to solve the conflicts due to simultaneous requests from different PUs. The arbiter is the only component in the processor design that should be dimensioned for the maximum number of PUs. The input and output are directly written/read to/from the PUs and controlled by direct queries from the processor's host. Furthermore, the input data origin is controlled so that data can be obtained from a built-in RNG, allowing the generation of secrets inside the processor for maximum privacy/security. In the following subsections, the individual components of the processor, including the different PUs, are described.

14.3.1 Processing Unit for Elliptic Curve Cryptography

The EC arithmetic over $GF(2^m)$ is supported by operands that usually have from 163 to 521 bits [8] or, in other words, $163 \leq m \leq 521$. This means that directly using these operands would require a datapath of the same size, which would be inefficient due to the large resulting datapath, difficult data handling, and resources constraint. Hence, splitting these operands in smaller limbs is the best approach, given that smaller limbs are easier to access and handle in the available memory resources. Moreover, this approach enables the resources reusing, given that a unit can operate on each limb of an operand. Therefore, for an m-bit operand, $l = \lceil m/w \rceil$ limbs of w bits are set. For an m-bit operand A, a limb of w bits is referred to as $\langle A \rangle_{w,i}$ such that:

$$\langle A \rangle_{w,i} = (a_{(w+1)i-1}, \dots, a_{wi}), \ A = \sum_{i=0}^{l-1} \langle A \rangle_{w,i} \, \alpha^{wi}. \tag{14.25}$$

A PU, such as the one depicted in Figure 14.3a, can be used to operate on these limbs for EC arithmetic [6]. Note that, for simplicity's sake, the control circuitry is omitted from Figure 14.3a. Four main hardware structures can be identified in Figure 14.3a: (1) storage logic, (2) addition logic, (3) multiplication logic, and (4) reduction/squaring logic. Furthermore, the PU contains a register that is able to store a field element that is directly provided by the host or read from the RNG. This register can be browsed and its bits used to control the program flow in the PU which enables the implementation of dependent branches and, consequently, loops. This register can be used, for example, to store the scalar employed in the EC point multiplication (see Algorithm 14.2),

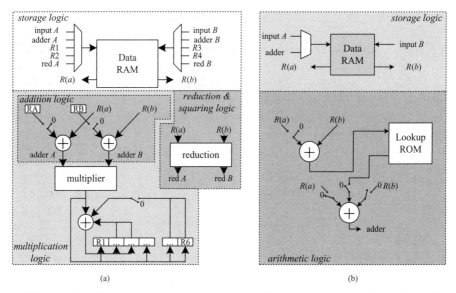

FIGURE 14.3. PUs to accelerate the cryptographic procedures based on (a) an EC and (b) the AES.

supporting the required branches in this algorithm. The following subsection details the remaining logic in the PU.

14.3.3.1 Storage Logic This logic consists of a dual-port RAM that is used for storing the operands in limbs. The contents of the RAM are forwarded to the arithmetic units inside the PU and the results retrieved from these units after processing, allowing for two simultaneous read or write operations. The input data are multiplexed to the RAM from the circuitry in the PU and also from outside the PU, that is, from the processor's host or RNG.

14.3.1.2 Addition Logic The addition is the simplest operation supported inside the PU. As discussed in Section 14.2, the $GF(2^m)$ addition corresponds to inexpensive bitwise XOR operations. This is an advantage for splitting the computation by limbs, given that applying the bitwise XOR operation to the m-bit operand is equivalent to performing the same operation on each limb. In order to take advantage of the dual-port RAM, the addition is implemented with two adders, adding two limbs simultaneously. The addition of two operands $C = A + B$ is accomplished with two steps, that is, two clock cycles, for each two limbs as Table 14.1 suggests. Note that each of the bit adders can be efficiently implemented with a lookup table (LUT) with at least four inputs which are available in the commercial FPGA [39].

14.3.1.3 Multiplication Logic The multiplication is performed as a 2-limb × 2-limb multiplication so that full advantage of the dual-port RAM

TABLE 14.1. Addition Schedule for the EC PU for Each Two Limbs of the Operands

	RAM Out		Registers		RAM In	
Step	A	B	A	B	A	B
0	–	–	–	–	–	–
1	$\langle A \rangle_{w,i}$	$\langle A \rangle_{w,i+1}$	–	–	–	–
2	$\langle B \rangle_{w,i}$	$\langle B \rangle_{w,i+1}$	$\langle A \rangle_{w,i}$	$\langle A \rangle_{w,i+1}$	$\langle C \rangle_{w,i}$	$\langle C \rangle_{w,i+1}$

can be exploited. Hence, the result obtained from the multiplication logic will be a four-limb operand. To obtain the full multiplication in $GF(2^m)$, these four-limb operands can be combined by using any algorithm for multiple-precision multiplication, including the subquadratic ones such as the Karatsuba–Ofman's introduced in Section 14.2.1.1. This can be accomplished by introducing a suitable microcoded program to control the PU. In fact, the Karatsuba–Ofman approach is also used in the multiplication logic to obtain the resulting four-limb operand. Thus, the hardwired multiplier in the PU only supports a $w \times w$, and the $2w \times 2w$ multiplications of the two-limb operands is obtained by using Equation (14.9). This approach allows reducing the multiplication delay, enhancing the maximum operating frequency of the PU. Consider the following identities so that a proper operation schedule can be defined for the multiplication of two limbs from each operand A and B, $C = AB$:

$$
\begin{aligned}
L &= \langle L \rangle_{w,1} \alpha^w + \langle L \rangle_{w,0} = \langle A \rangle_{w,i} \langle B \rangle_{w,i}, \\
H &= \langle H \rangle_{w,1} \alpha^w + \langle H \rangle_{w,0} = \langle A \rangle_{w,i+1} \langle B \rangle_{w,i+1}, \\
M' &= \langle M' \rangle_{w,1} \alpha^w + \langle M' \rangle_{w,0} = \left(\langle A \rangle_{w,i+1} + \langle A \rangle_{w,i} \right) \left(\langle B \rangle_{w,i+1} + \langle B \rangle_{w,i} \right), \\
M'' &= \langle M'' \rangle_{w,1} \alpha^w + \langle M'' \rangle_{w,0} = H + L, \\
M &= \langle M \rangle_{w,1} \alpha^w + \langle M \rangle_{w,0} = M' + M'' + \langle H \rangle_{w,0} \alpha^w + \langle L \rangle_{w,1}, \\
C &= \langle H \rangle_{w,1} \alpha^{3w} + \langle M \rangle_{w,1} \alpha^{2w} + \langle M \rangle_{w,0} \alpha^w + \langle L \rangle_{w,0}.
\end{aligned}
\tag{14.26}
$$

Given the identities in Equation (14.26), it is possible to obtain the final value of C in three steps (three clock cycles) with the schedule in Table 14.2. The resulting limbs are stored in the last two steps, with the most and less significant limbs stored in step 2 and the mid-significant limbs in step 3.

14.3.1.4 Reduction and Squaring Logic   Reduction is one of the most complex operations to be computed in the sense that each limb of the result potentially depends on several limbs of the inputs. The largest operand that is expected as input of a reduction operation in $GF(2^m)$ has up to $2m - 1$ bits, which is equivalent to the output of a multiplication or squaring. Consider the reduction of the operand $A = H\alpha^m + L$. The m-bit term L is already reduced

TABLE 14.2. Two-Limb Multiplication Schedule for the EC PU

	RAM Out		Registers								RAM In	
Step	A	B	A	B	1	2	3	4	5	6	A	B
0	–	–	–	–	0	0	0	0	0	0	–	–
1	$\langle A \rangle_{w,i}$	$\langle B \rangle_{w,i}$	$\langle A \rangle_{w,i}$	$\langle B \rangle_{w,i}$	$\langle L \rangle_{w,0}$	$\langle L \rangle_{w,0}$	$\langle L \rangle_{w,1}$	0	$\langle L \rangle_{w,1}$	0	–	–
2	$\langle A \rangle_{w,i+1}$	$\langle B \rangle_{w,i+1}$	$\langle A \rangle_{w,i}$	$\langle B \rangle_{w,i}$	$\langle L \rangle_{w,0}$	$\langle M'' \rangle_{w,0}$	$\langle M'' \rangle_{w,1}$	$\langle H \rangle_{w,1}$	$\langle L \rangle_{w,1}$	$\langle H \rangle_{w,0}$	$\langle L \rangle_{w,0}$	$\langle H \rangle_{w,1}$
3	$\langle A \rangle_{w,i+1}$	$\langle B \rangle_{w,i+1}$	$\langle A \rangle_{w,i}$	$\langle B \rangle_{w,i}$	$\langle L \rangle_{w,0}$	$\langle M \rangle_{w,0}$	$\langle M \rangle_{w,1}$	$\langle H \rangle_{w,1}$	$\langle L \rangle_{w,1}$	$\langle H \rangle_{w,0}$	$\langle M \rangle_{w,0}$	$\langle M \rangle_{w,1}$

in $GF(2^m)$, thus there is no need to reduce it. The $(m-1)$-bit term H is written in terms of powers of α^i with $i \geq m$, thus it should be reduced in order to be generated from the field basis specified in Equation (14.1). Given the irreducible polynomial $P(x) = \sum_{i=0}^{m} p_i x_i$, whose root generates $GF(2^m)$, this task can be accomplished using the property in Equation (14.5). Given Equation (14.5), the following identities hold:

$$\alpha^m = \sum_{i=0}^{m-1} p_i \alpha^i, \tag{14.27}$$

$$\alpha^{m+1} = \alpha^m \alpha = p_{m-1} \alpha^m + \sum_{i=0}^{m-2} p_i \alpha^{i+1} = p_{m-1} \left(\sum_{i=0}^{m-1} p_i \alpha^i \right) + \sum_{i=0}^{m-2} p_i \alpha^{i+1},$$

$$\vdots$$

Hence, given the coefficients p_i of $P(x)$, it is possible to obtain all the powers α^i for $i \geq m$ written as powers of α^j for $j < m$. Considering that the operands H and L are column arrays in $GF(2)$, the reduction result $C = A \bmod P(x)$ can be obtained by computing:

$$C = RH + L \Leftrightarrow \begin{bmatrix} c_0 \\ \vdots \\ c_{m-1} \end{bmatrix} = \begin{bmatrix} r_{0,0} & \cdots & r_{0,m-2} \\ \vdots & \ddots & \vdots \\ r_{m-1,0} & \cdots & r_{m-1,m-2} \end{bmatrix} \begin{bmatrix} h_0 \\ \vdots \\ h_{m-2} \end{bmatrix} + \begin{bmatrix} l_0 \\ \vdots \\ l_{m-1} \end{bmatrix}, \tag{14.28}$$

where R is a $m \times (m-1)$ matrix which operates the transformation of H such that RH uses the canonical basis to specify H. For this purpose, R is constructed directly from the identities in Equation (14.27) such that the column j is a column matrix containing α^{j+m} written with terms of the canonical basis, or in other words $r_{i,j} = \alpha_i^{j+m}$. Given the matrix R, the final reduction result is obtained by computing the multiplications and additions of Equation (14.28) over $GF(2)$, that is, with the appropriate AND and XOR operations. The complexity of the reduction operand is thus strongly related with the number of entries in the matrix R different from 0. Furthermore, the entries of matrix R are strongly dependent of the coefficients of $P(x)$. Fortunately, *good* polynomials can be selected so as to obtain an improved reduction operation. In particular, the standards available for EC-based protocols [8] suggest the utilization of polynomials with very few coefficients different from zero, more precisely pentanomials and trinomials (with five and three coefficients different from zero). The following example presents the reduction logic based on the matrix R using practical values for the relevant parameters:

Example 14.1 (Reduction over $GF(2^{163})$). Consider the irreducible polynomial $P(x) = x^{163} + x^7 + x^6 + x^3 + 1$ defined in the EC-based protocols standards in NIST [8], which states that root α can be used to build a canonical basis which generates the field $GF(2^{163})$. Given $P(x)$, the value of α^m in terms of the canonical basis is given by $(00 \ldots 011001001)$. The reduction matrix R can be obtained by organizing in its columns the values α^i with $m \leq i < 2m - 1$:

$$
R =
\begin{bmatrix}
1 & 0 & 0 & 0 & 0 & 0 & 0 & 0 & 0 & 0 & 0 & 0 & 0 & \cdots & 0 & 0 & 0 & 0 & 0 & 0 & 0 & 1 & 1 & 0 & 0 & 1 & 0 \\
0 & 1 & 0 & 0 & 0 & 0 & 0 & 0 & 0 & 0 & 0 & 0 & 0 & \cdots & 0 & 0 & 0 & 0 & 0 & 0 & 0 & 0 & 1 & 1 & 0 & 0 & 1 \\
0 & 0 & 1 & 0 & 0 & 0 & 0 & 0 & 0 & 0 & 0 & 0 & 0 & \cdots & 0 & 0 & 0 & 0 & 0 & 0 & 0 & 0 & 0 & 1 & 1 & 0 & 0 \\
1 & 0 & 0 & 1 & 0 & 0 & 0 & 0 & 0 & 0 & 0 & 0 & 0 & \cdots & 0 & 0 & 0 & 0 & 0 & 0 & 0 & 1 & 1 & 0 & 1 & 0 & 0 \\
0 & 1 & 0 & 0 & 1 & 0 & 0 & 0 & 0 & 0 & 0 & 0 & 0 & \cdots & 0 & 0 & 0 & 0 & 0 & 0 & 0 & 0 & 1 & 1 & 0 & 1 & 0 \\
0 & 0 & 1 & 0 & 0 & 1 & 0 & 0 & 0 & 0 & 0 & 0 & 0 & \cdots & 0 & 0 & 0 & 0 & 0 & 0 & 0 & 0 & 0 & 1 & 1 & 0 & 1 \\
1 & 0 & 0 & 1 & 0 & 0 & 1 & 0 & 0 & 0 & 0 & 0 & 0 & \cdots & 0 & 0 & 0 & 0 & 0 & 0 & 1 & 1 & 0 & 1 & 0 & 0 \\
1 & 1 & 0 & 0 & 1 & 0 & 0 & 1 & 0 & 0 & 0 & 0 & 0 & \cdots & 0 & 0 & 0 & 0 & 0 & 0 & 1 & 0 & 1 & 0 & 0 & 0 \\
0 & 1 & 1 & 0 & 0 & 1 & 0 & 0 & 1 & 0 & 0 & 0 & 0 & \cdots & 0 & 0 & 0 & 0 & 0 & 0 & 0 & 1 & 0 & 1 & 0 & 0 \\
0 & 0 & 1 & 1 & 0 & 0 & 1 & 0 & 0 & 1 & 0 & 0 & 0 & \cdots & 0 & 0 & 0 & 0 & 0 & 0 & 0 & 0 & 1 & 0 & 1 & 0 \\
0 & 0 & 0 & 1 & 1 & 0 & 0 & 1 & 0 & 0 & 1 & 0 & 0 & \cdots & 0 & 0 & 0 & 0 & 0 & 0 & 0 & 0 & 0 & 1 & 0 & 1 \\
0 & 0 & 0 & 0 & 1 & 1 & 0 & 0 & 1 & 0 & 0 & 1 & 0 & \cdots & 0 & 0 & 0 & 0 & 0 & 0 & 0 & 0 & 0 & 0 & 1 & 0 \\
0 & 0 & 0 & 0 & 0 & 1 & 1 & 0 & 0 & 1 & 0 & 0 & 1 & \cdots & 0 & 0 & 0 & 0 & 0 & 0 & 0 & 0 & 0 & 0 & 0 & 1 \\
\vdots & \vdots & \vdots & \vdots & \vdots & \vdots & \vdots & \vdots & \vdots & \vdots & \vdots & \vdots & \vdots & \ddots & \vdots & \vdots & \vdots & \vdots & \vdots & \vdots & \vdots & \vdots & \vdots & \vdots & \vdots & \vdots & \vdots \\
0 & 0 & 0 & 0 & 0 & 0 & 0 & 0 & 0 & 0 & 0 & 0 & 0 & \cdots & 0 & 1 & 0 & 0 & 0 & 0 & 0 & 0 & 0 & 0 & 0 & 0 & 0 \\
0 & 0 & 0 & 0 & 0 & 0 & 0 & 0 & 0 & 0 & 0 & 0 & 0 & \cdots & 0 & 0 & 1 & 0 & 0 & 0 & 0 & 0 & 0 & 0 & 0 & 0 & 0 \\
0 & 0 & 0 & 0 & 0 & 0 & 0 & 0 & 0 & 0 & 0 & 0 & 0 & \cdots & 1 & 0 & 0 & 1 & 0 & 0 & 0 & 0 & 0 & 0 & 0 & 0 & 0 \\
0 & 0 & 0 & 0 & 0 & 0 & 0 & 0 & 0 & 0 & 0 & 0 & 0 & \cdots & 0 & 1 & 0 & 0 & 1 & 0 & 0 & 0 & 0 & 0 & 0 & 0 & 0 \\
0 & 0 & 0 & 0 & 0 & 0 & 0 & 0 & 0 & 0 & 0 & 0 & 0 & \cdots & 0 & 0 & 1 & 0 & 0 & 1 & 0 & 0 & 0 & 0 & 0 & 0 & 0 \\
0 & 0 & 0 & 0 & 0 & 0 & 0 & 0 & 0 & 0 & 0 & 0 & 0 & \cdots & 1 & 0 & 0 & 1 & 0 & 0 & 1 & 0 & 0 & 0 & 0 & 0 & 0 \\
0 & 0 & 0 & 0 & 0 & 0 & 0 & 0 & 0 & 0 & 0 & 0 & 0 & \cdots & 1 & 1 & 0 & 0 & 1 & 0 & 0 & 1 & 0 & 0 & 0 & 0 & 0 \\
0 & 0 & 0 & 0 & 0 & 0 & 0 & 0 & 0 & 0 & 0 & 0 & 0 & \cdots & 0 & 1 & 1 & 0 & 0 & 1 & 0 & 0 & 1 & 0 & 0 & 0 & 0 \\
0 & 0 & 0 & 0 & 0 & 0 & 0 & 0 & 0 & 0 & 0 & 0 & 0 & \cdots & 0 & 0 & 1 & 1 & 0 & 0 & 1 & 0 & 0 & 1 & 0 & 0 & 0 \\
0 & 0 & 0 & 0 & 0 & 0 & 0 & 0 & 0 & 0 & 0 & 0 & 0 & \cdots & 0 & 0 & 0 & 1 & 1 & 0 & 0 & 1 & 0 & 0 & 1 & 0 & 0 \\
0 & 0 & 0 & 0 & 0 & 0 & 0 & 0 & 0 & 0 & 0 & 0 & 0 & \cdots & 0 & 0 & 0 & 0 & 1 & 1 & 0 & 0 & 1 & 0 & 0 & 1 & 0 \\
0 & 0 & 0 & 0 & 0 & 0 & 0 & 0 & 0 & 0 & 0 & 0 & 0 & \cdots & 0 & 0 & 0 & 0 & 0 & 1 & 1 & 0 & 0 & 1 & 0 & 0 & 1 \\
0 & 0 & 0 & 0 & 0 & 0 & 0 & 0 & 0 & 0 & 0 & 0 & 0 & \cdots & 0 & 0 & 0 & 0 & 0 & 0 & 1 & 1 & 0 & 0 & 1 & 0 & 0 \\
\end{bmatrix}.
$$

The first column of R directly consists of α^m, whereas the following columns are obtained by multiplying the previous column by α, that is, by shifting the columns downward. Only the very last columns have a bit 1 moving out from the columns while shifting, hence the value of α^m is added to these columns after shifting. The reduction result C of an operand $A = H\alpha^m + L$ can now be easily accomplished from matrix R:

$$c_0 = h_{160} + h_{157} + h_{156} + h_0 + l_0,$$

$$\vdots$$

$$c_{162} = h_{159} + h_{156} + h_{155} + l_{162}.$$

Matrix R only depends on the irreducible polynomial, which is usually set for a given application, thus the operations based on matrix R can be efficiently hardwired. Two main characteristics observed in matrix R contribute for an efficient implementation: (1) very few entries are different from 0 in each row (up to 6), thus each bit of the result has few dependencies on the input operand's bits, and (2) except for the rightmost columns, a pattern can be identified while browsing the matrix R diagonal, suggesting that hardware reutilization in the computation of C is possible, enhancing the compactness of the design.

Let us now consider the circuit depicted in Figure 14.4, which allows computing the reduction based on matrix R for a limb size of $w = 21$ bits. This circuit receives two simultaneous input limbs and also outputs two simultaneous limbs and therefore can be embedded in the PU in Figure 14.3a. Three main building blocks can be identified in this circuit: the Reduction High, Middle, and Low. Each one of these blocks refer to different submatrices of

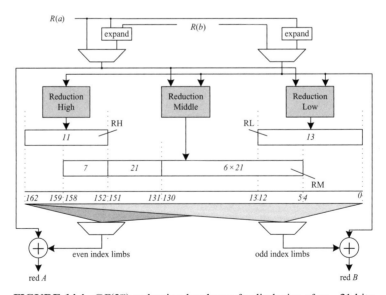

FIGURE 14.4. $GF(2^m)$ reduction hardware for limb size of $w = 21$ bits.

TABLE 14.3. Alignment of H in the Limbs of $A = H\alpha^m + L$

Limbs of A	$\langle A \rangle_{21,15}$	$\langle A \rangle_{21,14}$	$\langle A \rangle_{21,13}$	$\langle A \rangle_{21,12}$	$\langle A \rangle_{21,11}$	$\langle A \rangle_{21,10}$	$\langle A \rangle_{21,9}$	$\langle A \rangle_{21,8}$	$\langle A \rangle_{21,7}$
Bits of H	161	152 151	131 130	110 109	89 88	68 67	47 46	26 25	5 4 0

R and overcome alignment issues between the limbs themselves and the term H of A. Concerning the submatrices, three different ones should be considered: (1) an 11×6 matrix R_{TR} of the top-right corner entries of R, (2) a $(42 + 7) \times 42$ matrix R_{TL} of the top-left corner of R that has the property of being equal to any other submatrix with the same size identified while browsing the diagonal of R, except for (3) the 7×6 matrix R_{BR} of the bottom-right corner of R. With matrices R_{TR}, R_{BR} and the diagonal browsing with R_{TL}, all entries different from zero in matrix R are comprised. Concerning the alignment issues, the alignment of H with the limbs of A presented in Table 14.3 should be regarded. The Reduction Low block is responsible for computing the reduction due to R_{TR} as well as the reduction of the 5 least significant bits (LSBs) of H that are not aligned with the limbs of A. The Reduction Middle computes the reduction of two 21-bit limbs at once by using matrix R_{TL} and can be reused to several pairs of limbs since the matrix R_{TL} contains the pattern in the diagonal of R. Note that there is not an even number of complete limbs of H in this example, hence the last reduction of this unit is accomplished with only one limb instead of a pair. The reduction with the matrix R_{TL} produces, for each two input limbs, two output limbs plus 7 bits, hence the final result obtained from Reduction Middle have 7 extra bits that require to be properly aligned. The 10 most significant bits (MSBs) of H (note that the bit index 162 is 0) that are in $\langle A \rangle_{21,15}$ are reduced with the Reduction High block by using the information of the bottom-right corner of R, including R_{BR}. Once the computation of RH is accomplished with these blocks, the limbs of L are fed to the reduction unit in pairs and the final result's limbs are directly computed with the limb adders as depicted in Figure 14.4.

Example 14.1 presents and explains how to build a reduction unit from the parameters given in the standards. Nonetheless, the reduction unit presented in Example 14.1 allows for an extra feature using the blocks *expand*: the reduced squaring in $GF(2^m)$. Squaring is a particular case of Equation (14.7) such that:

$$A^2 \bmod P(x) = \sum_{i=1}^{m-1}\sum_{j=1}^{m-1} a_i a_j \alpha^{i+j} = \sum_{i=1}^{m-1} a_i \alpha^{2i}. \tag{14.29}$$

Note that Equation (14.29) is true given that the addition in $GF(2^m)$ is an XOR so that for $i \neq j$ the terms $a_i a_j \alpha^{i+j}$ cancel with the terms $a_j a_i \alpha^{j+i}$. From Equation

(14.29) results that the squaring $C = A^2$ can be accomplished by setting $c_i = 0$ for i odd and $c_i = a_{i/2}$ for i even, or in other words, C is obtained by expanding A to $(2m - 1)$ bits and filling the odd index spaces with zeros. The *expand* blocks in Figure 14.4 are included with this purpose. Therefore, the already reduced squaring operation is obtained exactly as in Example 14.1, except that the input limbs are expanded in the *expand* blocks prior to the computation by activating the multiplexing logic accordingly.

14.3.2   Processing Unit for the AES

A PU with support to the AES must handle the key expansion, data ciphering (as introduced in Section 14.2.2), and data deciphering. All these operations are supported by routines that operate on the bytes of the data and key, including nonlinear operations. Two main approaches for computing these nonlinear operations exist: (1) arithmetic and (2) LUT approach. Authors adopting the arithmetic approach (see Definition 14.10) claim that this approach allows better maximizing of the ciphering throughput due to deeper pipelines, while a lookup approach limits the pipeline operating frequency with the delay of getting the lookup from memory [40]. On the other hand, authors adopting a lookup approach overcome this limitation by using more than one LUT, allowing the parallelization of the processing, masking a lower operating frequency. In fact, the most efficient implementations in the related state of the art use the lookup approach, namely the ones supported on FPGA, given that these devices have built-in RAM blocks distributed throughout the device that can be efficiently used for this purpose [41]. Given that in this chapter a reconfigurable architecture based on FPGA is addressed, the PU depicted in Figure 14.3b used to support the AES also adopts a lookup approach. The PU in Figure 14.3b also contains a data memory to store the data and key bytes as well as the lookup memory. The remaining logic consists of XOR gates to perform the $GF(2^8)$ additions. Similarly to the EC PU, the control of this unit is assured by a simple local FSM that responds to the incoming microinstructions from the centralized microinstruction RAM of the processor.

The computation of the AES ciphering relies on the sequence of several routines (see Algorithm 14.3). These routines rely on the nonlinear operations often called *S-box*, matrix multiplication, and byte permutation and addition over $GF(2^8)$. The addition is assured by the XOR gates in the PU while the permutations are accomplished addressing the data RAM accordingly, which is specified in the microinstructions used to program the device. Concerning the *S-box* and matrix multiplication and following the suggestion in Daemen and Rijmen [31], these two operations can be efficiently chained into a single one by combining the multiplication with the constant coefficients of the matrix with the *S-box* itself, storing the possible results in the LUT. The rationale of the AES ciphering can also be used for deciphering, and the proper lookup values precomputed and stored in the LUT. Therefore, the LUT contains for all bytes b the values $S(b)$, $2S(b)$, and $3S(b)$ for ciphering, and the values $\overline{S(b)}$, $9b$, $11b$, $13\underline{b}$, and $14b$ for deciphering, where $S(.)$ stands for the value of the *S-box* and $\overline{S(.)}$ for its inverse. The constants adopted for the LUT are slightly different from the ones proposed in Daemen and Rijmen [31], namely for deciphering for which $\overline{S(.)}$ can also be combined with the matrix coefficients. However, the adopted PU does not require this combination, given that the addition of the round keys can be performed in the same instruction as the lookup of $\overline{S(.)}$. Summarizing, the number of entries in the LUT is $8 \times 2^8 = 2048$ bytes, which fits the size of a single block RAM available nowadays in FPGAs such as the Xilinx Virtex 4 onward [39].

Given the LUT approach, the computation of the AES algorithms is accomplished by the appropriate selection and addition over $GF(2^8)$. This PU is prepared to compute the operations: $R(c) = R(a) + R(b)$, $R(c) = R(a) + L(R(b))$, $R(c) = L(R(a) + R(b))$, and $R(c) = L(R(b))$, where $L(.)$ stands for an LUT. A last component is required to compute the addressed algorithms: a counter which value can be added to the data addresses, allowing browsing data in the RAM of this PU. Furthermore, this counter is used to control branch instructions, allowing the implementation of loops. The initialization, incrementation, and decrementation of this counter are directly controlled by the microinstructions that compose the program.

The efficiency of the PU in terms of area resources and throughput can be optimized, taking into account two features. The built-in RAM blocks in the FPGA devices are usually true dual port. This allows sharing the same RAM block between two PUs by implementing the lookup outside the reconfigurable confinement reserved for the PUs, that is, as part of the static resources of the processor. This approach is depicted in Figure 14.5. Furthermore, the

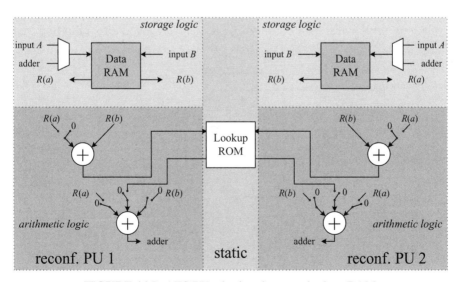

FIGURE 14.5. AES PUs sharing the same lookup RAM.

required resources of the AES PU are considerably less than the resources for the EC PU. Also, the same reconfigurable confinement, that is, the same resources, is reserved for both PUs to allow a dynamic assignment of those resources for both PUs through runtime configuration. Therefore, the datapath of the AES PU as well as the data and lookup RAM can be reproduced several times, allowing parallel ciphering/deciphering of different data of the input stream, increasing the throughput of the PU. The control of the PU can be the same for different datapaths, given that it is not data dependent.

14.3.3 Random Number Generator

In hardware implementations of RNGs, it is possible to prospect physical randomness sources in hardware parts. The utilization of these randomness sources avoids the usage of complex arithmetic and filters required to obtain long-period PRNG suitable for cryptographic applications with increased security demands. Thus, once the randomness is harvested from the physical components, compact yet highly secure RNG can be obtained. A source of randomness that has shown to be easily harvested while providing good results is the oscillators jitter, with proven results for FPGA devices as well [42, 43]. A theoretical analysis of such kind of RNG is presented in Sunar et al. [36] and a hardware structure is proposed. This structure was further improved in Antão et al. [42] and in Wold and Tan [43] and is depicted in Figure 14.6. This RNG structure consists of several oscillators based on rings of inverter gates. Each oscillator also has a NAND gate that is controlled by a reset signal that allows turning the oscillator on and off once it is set to 0 or 1, respectively. The number of inverters in the ring determines the ring frequency, and given that it is an unoptimized oscillator, there is uncertainty in the frequency itself and in the transaction edges, which is known as clock jitter. This uncertainty is the source of randomness. The output of each oscillator is fed into a flip-flop that

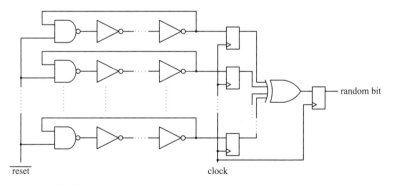

FIGURE 14.6. RNG based on several oscillator rings.

is activated by a common clock *clk*, of which frequency f_{clk} is lower than any of the frequencies in the oscillator. Due to the uncertainty in each oscillator, the values sampled in each *clk* rising edge are unpredictable, being that any bias is masked by a parity filter built with an XOR gate that combines the bits in each flip-flop to produce a final random bit. Three main factors contribute for the quality of the RNG: (1) the number of inverters in each oscillator, (2) the number of oscillators, and (3) the difference between the oscillators' frequency and the sampling clock *clk*. An increasing number of inverters in each oscillator results in a decrease of the oscillators' frequency but increases the uncertainty around the waveform transitions, that is, the clock jitter becomes more pronounced. The number of oscillators controls the number of random samples which contributes to increased randomness after the parity filter. The increase in the difference between f_{clk} and the oscillators' frequency results in the increase of the transitions in the oscillators between samples, which makes the effect of the clock jitter even more pronounced in the result of the sampling.

Concerning the implementation issues, the main difference between the original approach in Sunar et al. [36] and the improvement in Wold and Tan [43] relies on the introduction of the flip-flops after each of the oscillators. The authors in Reference [43] verified that feeding the parity filter directly with the oscillators imply very fast transitions in the filter output that might be faster than the time response of the filter itself, which would result in memory issues in the filter that would decrease the randomness of the output bitstream. The number of inverters in the different oscillators may vary. Nevertheless, experiments in Wold and Tan [43] suggest that this parameter does not have a strong impact in the quality of the RNG, as it is possible to build an RNG that passes the statistical tests in Marsaglia [37] and NIST [38], with 25 oscillators with three inverters (three LUTs) in each one, sampled at a frequency of $f_{clk} = 100$ MHz for an Altera Cyclone II FPGA.

The implementation of oscillator rings is not a standard application of FPGAs, hence special attention must be paid to some particularities of these devices and the tools supporting them (synthesis, place, and route). The design flow for FPGAs usually departs from a hardware description language (HDL) and produces a final configuration bitstream. Both synthesis and place and route steps introduce optimizations in the design logic; for example, the synthesis tool can turn a sequence of an odd number of inverters into a single inverter. Also, the clock signals inside an FPGA are usually set by dedicated hardware blocks that can be specified in the HDL description of the design, which limit the clock frequencies for which the FPGA was dimensioned to. Therefore, designing oscillators from logic gates can result in frequencies higher than the supported ones, which may result in abusive power dissipation, and thus potential device damage may occur if measures concerning the heat dissipation of the device are not taken. The following example illustrates how these concerns can be addressed at the design stage of such RNG.

Example 14.2 (RNG Design). Consider that one wants to implement an oscillator ring on a Xilinx FPGA using very high speed integrated circuit (VHSIC) HDL (VHDL). Several issues arise from the implementation of an oscillator with a ring of inverters in an FPGA. These issues mostly concern the tools used in the design of an HDL-based flow that mostly optimize the gates in the ring because these tools consider the ring a logic function instead of an oscillator. In order to avoid these problems, some procedures can be taken. The first one is the direct instantiation of the FPGA components using an appropriate library instead of letting the tools infer them. Hence, an inverter should be directly implemented with an LUT:

```
Library UNISIM;
use UNISIM.vcomponents.all;
inverter_instantiation : LUT1
generic map (
  INIT => "01") -- this entry specify the true table of
                -- the inverter
port map (
  O => inv_out, -- inverter output
  I0 => inv_in  -- inverter input
);
```

instead of

```
inv_out <= not inv_in;
```

The next issue to be addressed is to avoid the tools to optimize away the sequence of inverters that are instantiated. This can be done by using proper attributes. For example, for the Synopsys Synplify synthesis tool, the attribute

```
attribute syn_hier : string;
attribute syn_hier of behavioral: architecture is
"hard";
```

avoids optimizations across the lower hierarchies, or in other words, the tool will not look inside the instantiated LUTs for optimization purposes of the entity where the attribute is set. Another attribute that is useful is:

```
attribute syn_keep : boolean;
attribute syn_keep of inv_connections : signal is
true;
```

where `inv_connections` is the signal that links the consecutive inverters in the ring. This attribute forces the tool to include that signal in the final netlist,

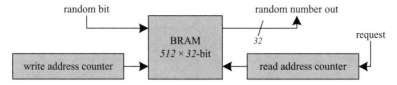

FIGURE 14.7. Conversion from a random bitstream to random words.

and thus avoids merging consecutive inverters. An equivalent attribute should also be used at the place and route stage by setting the entry

```
NET oscillator_instance/inv_connections KEEP;
```

in the user constraints file (usually referred to as the UCF file).

The final stage of the design of the RNG includes a component that turns a random bitstream into a sequence of limbs that can be easily handled by the other blocks in the processor. This can be easily accomplished by using the FPGA built-in RAMs, as Figure 14.7 depicts. This component benefits from the true dual-port implementation of these RAMs that allows configuring one port with a single bit width and the other port with the width of the limbs in the processor. A counter controls the addresses in both ports that allows the RNG to continue writing the random bitstream to the RAM and the processor to read the required random limbs whenever needed.

14.3.4 Microinstructions and Access Arbiter

All the PUs in the processor are controlled by a microcoded program stored in a centralized RAM. Together with the reconfiguration capabilities, the microcode enhances the flexibility of the processor, allowing the user to easily change the behavior of the processor by loading a different microcoded program into the central RAM. Furthermore, the utilization of a microcoded program allows simplifying the PUs' local control (say FSM) since the complexity of the control can be divided by the microcode and local FSMs. This latter observation is particularly interesting for the EC PU addressed in Section 14.3.1. In the EC PU, there are two granularities in the computation: the finite field and the limbs. Therefore, the microinstructions for the EC address two granularities which are the following:

- *Finite Field (Coarse) Granularity*: This addresses a finite field element directly, that is, all the limbs that compose a finite field element, being that the management of the limbs are controlled by the local FSM of the PU. The microinstructions supported in this class are:
  - ○ ADD: This computes $C = A + B \mod P(x)$, where $A, B, C \in GF(2^m)$ are stored in the data RAM

- ○ SQR: This computes $C = A^2 \bmod P(x)$, where $A, C \in GF(2^m)$ are stored in the data RAM
- ○ RED: This computes $C = A \bmod P(x)$, where A is a $(2m - 1)$-bit operand and $C \in GF(2^m)$, and are both stored in the data RAM
- ○ READ: This outputs the limbs of $C \in GF(2^m)$ stored in the data RAM to the host
- ○ WRITE: This stores the limbs of $C \in GF(2^m)$ in the data RAM from the host
- ○ KEY: This stores the limbs of $C \in GF(2^m)$ in the jump control register. The bits of this register are browsed by the jump instruction in order to perform a jump depending on their content
- *Limb (Fine) Granularity*: This directly addresses the limbs involved in the operations with two input limbs addressed simultaneously. The micro-instructions supported in this class are:
  - ○ LADD: This computes $C = \langle A \rangle_{w,i_2} \alpha + \langle A \rangle_{w,i_1} + \langle B \rangle_{w,j_2} \alpha + \langle B \rangle_{w,j_1}$, where $A, B \in GF(2^m)$ are stored in the RAM
  - ○ LMUL: This computes $C = \left(\langle A \rangle_{w,i_2} \alpha + \langle A \rangle_{w,i_1} \right)\left(\langle B \rangle_{w,j_2} \alpha + \langle B \rangle_{w,j_1} \right)$, where $A, B \in GF(2^m)$ are stored in the RAM
- *Control*: These instructions control the program flow, including branches. The supported microinstructions are:
  - ○ COST: This states that the following sequence of microinstructions is to be decoded as fine grained. The COST microinstruction also comprises specified data addresses which are stored in local registers to allow indirect addressing of the data RAM in the following sequence of microinstructions
  - ○ COSTE: This indicates that a sequence of fine-grained microinstructions has finished, that is, the following microinstructions are to be decoded as coarse grained
  - ○ JMP: This jumps to the microinstruction with address $addr_{\text{true}}$ if the bit of the jump control register under test is set to 1, jumps to $addr_{\text{false}}$ if it is set to 0, and jumps to $addr_{\text{finished}}$ if all bits in the jump control register have already been browsed
  - ○ END: This indicates the end of the program, that is, the PU is turned idle and the host is signaled

Figure 14.8 depicts how the aforementioned microinstructions can be coded for a 32-bit size microinstruction example. In these microinstructions, *Op.i* stands for direct addresses of data RAM and *Wj* stands either for direct addresses or indirect address, depending on the content of *Cj*. The indirect addresses are generated by adding the values *Wj* to one of the addresses *Op.i* that was provided in the last COST microinstruction.

For the AES PU, no instruction granularity is identified, given that all the operations handle bytes. Therefore, the microinstructions that allow operating the AES PU are the following:

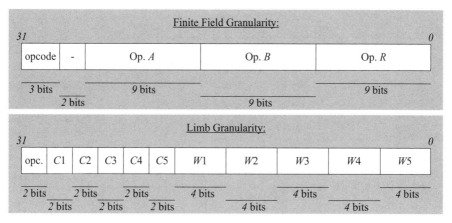

FIGURE 14.8. EC PU microinstruction coding.

- *Data Operations*: These microinstructions operate on the data stored in the data RAM. Considering $R(.)$, a data RAM byte, and $L(.)$, the lookup result of the LUT RAM, the supported data microinstructions are defined as:
  - AB: This computes $R(c) = R(a) + R(b)$
  - ALB: This computes $R(c) = R(a) + L(R(b))$
  - LAB: This computes $R(c) = L(R(a) + R(b))$
  - LB: This computes $R(c) = L(R(b))$
  - READ: This outputs 2 bytes stored in the data RAM to the host
  - WRITE: This stores 2 bytes in the data RAM from the host
- *Control*: These instructions control the program flow:
  - JMPSET: This sets an indexing counter that is used to control the jump and generate indirect addresses for the data
  - JMPINC: This means to jump if a given value matches the value in the indexing counter; the indexing counter is incremented
  - JMPDEC: This is similar to JMPINC, but the indexing counter is decremented
  - END: This indicates the end of the program, that is, the PU is turned idle and the host is signaled

In order to operate the aforementioned microinstructions, the microinstruction coding depicted in Figure 14.9 can be used. In the microinstruction, *Op.i* stands for a data address. This address can be added to the value of the indexing counter if I_i is set, enabling indirect addressing. The microinstruction also comprises an entry to select which of the lookup values is to be used among the eight available for a given byte. Note that the width of the AES PU microinstructions supersedes the width of the EC PU, thus the microinstruction

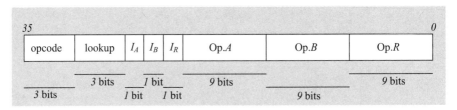

FIGURE 14.9. AES PU microinstruction coding.

RAM must be able to accommodate the maximum instruction width of the processor.

Since the several PUs in the processor require instructions from the same RAM, an accessing arbiter is required to solve conflicts due to simultaneous requests. The arbiter must be able to store all the requests and decide in a priority basis which request to issue at each moment so as to obtain the maximum performance of the processor. Each time the arbiter issues a request, it signals the correspondent PU, informing that a valid instruction is ready. Concerning the priority scheme underlying the decisions, its main property is to minimize the average time a PU is idle. In this processor, this is done by identifying the average time of the execution of a microinstruction in each PU. A PU with a larger average execution time will contribute less to the conflicts since it will request less microinstructions in the same time window than a PU with a lower average microinstruction execution time. Summarizing, a PU with a larger average microinstruction execution time should be prioritized. Thus, for PUs with an identical average microinstruction execution time, the priority scheme can be static, always providing the higher priority to always the same PU. A priority scheme such as this allows the user to define the highest priority depending on the application a PU is supporting. Concerning the discussed EC and AES PUs, the complexity of the operations in the EC PU is higher, hence these PUs should be prioritized over the AES PUs, following the afore-mentioned criteria.

14.4 RESULTS

In this section, experimental results for the processor detailed in Section 14.3 are presented and discussed. The experimental setup uses parameters and algorithms that are used nowadays in applications concerning EC and the AES arithmetic and prototyping devices available for commercial use. Hence, the experimental setup comprises the following considerations:

1. Algorithm 14.2 is used to support EC point multiplication in the EC PU using a limb size of $w = 21$.
2. The AES ciphering/deciphering are based on Algorithm 14.3 for a 128-bit key.

3. The EC implementation is based on the standard curve *B-163* defined in NIST [8], supported on the finite field $GF(2^{163})$ with irreducible polynomial $P(x) = x^{163} + x^7 + x^6 + x^3 + 1$ so that Example 14.1 applies.

4. The RNG topology is based on oscillator rings as presented in Section 14.3.3 and depicted in Figure 14.6. The considerations of Example 14.2 are used in this implementation.

5. The implementation was optimized and thoroughly tested based on a Xilinx Virtex 4 FPGA (part *XC4VSX35*) installed in the platform in Annapolis Micro Systems, Inc. [44].

6. The FPGA design tools *Synplify Premier C-2009-06* and *Xilinx ISE 9.2.04i_PR14* are used for synthesis and place and route, respectively, enabling the production of the configuration bitstreams (complete and partial) that can be uploaded in runtime.

14.4.1 Individual Components

In this subsection, the particular characteristics and performance figures of the main individual components of the processor are addressed, namely both EC and AES PUs, and the RNG. First of all, the applications for the PUs were programmed using the supported microinstructions. Microinstructions have different execution times, as stated in Table 14.4 for the aforementioned experimental setup. The read and write microinstructions are the fastest since the PUs support a write/read burst mechanism that allows the host to write/read two limbs per clock cycle. The finite field operations ADD, SQR, and RED are the most demanding in terms of latency, given that these microinstructions involve the manipulation of several limbs so that a complete field element can be handled. All the control operations, as well as the byte-based operation in the AES PU, require three clock cycles. The control-based operations require one clock cycle to decode the microinstruction, one clock cycle to execute and request the next microinstruction, and another clock cycle for the remaining execution/storage procedures. The byte-controlled instructions require one clock cycle to read from the data RAM, one clock cycle to execute, and one final clock cycle to write the results in the data RAM. The limb-based microinstructions of the EC PU operate in a way similar to the AES byte-based operations given the schedule in Tables 14.1 and 14.2. The LMUL microinstruction of the EC PU requires more than three clock cycles of latency since the execution takes longer and the same limbs need to be available from the data RAM in two consequent clock cycles (see Table 14.2). Note that the request of the next microinstruction is issued by the PU during the execution of each microinstruction so that the next microinstruction is readily available in the PU after the current microinstruction is complete.

The EC PU does not support a native microinstruction that allows implementing the field multiplication and inversion. Hence, the available microinstructions have to be used to program these more complex operations. The

TABLE 14.4. Microinstructions' Latency in Clock Cycles for the EC PU and the AES PU

(a) EC PU											
ADD	SQR	RED	READ	WRITE	KEY	LADD	LMUL	COST	COSTE	JMP	END
13	14	14	1	1	4	3	5	3	3	3	3

(b) AES PU									
AB	ALB	LAB	LB	READ	WRITE	JMPSET	JMPINC	JMPDEC	END
3	3	3	3	1	1	3	3	3	3

PU supports two different granularities in the operations supported and the COST and COSTE microinstructions are used to change between granularities. The field multiplication can be programmed with the limb-based operations with a sequence of the LADD and LMUL microinstructions delimited by the COST and COSTE operations. The Karatsuba–Ofman multiprecision multiplication algorithm (applying Eq. 14.9 recursively) can be used for this purpose. With this algorithm, the field multiplication (with no reduction) can be accomplished with 1 COST, 9 LMUL, 40 LADD, and 1 COSTE microinstructions, resulting in a total latency of 171 clock cycles. The field inversion can be accomplished with field multiplications and other field operations already supported by dedicated microinstructions (see Eq. 14.11). The algorithm used to implement field inversions requires 9 field multiplications, 9 RED, and 162 SQR, a total of 3933 clock cycles. Given the field multiplication and inversion, it is now possible to determine the total latency of the EC point addition described in Equation (14.13) and of the multiplication described in Algorithm 14.2. The point addition requires 3 field multiplications, 3 RED, 1 SQR, 8 ADD, 1 field inversion, and 1 END microinstruction, a total of 4609 clock cycles. Algorithm 14.2 requires 982 field multiplications, 982 RED, 651 SQR, 496 ADD, 163 JMP, 1 field inversion, and 1 END microinstructions, a total of 201,657 clock cycles.

On the other hand, the AES PU already contains all the microinstructions required to program the ciphering and key expansion procedures. Therefore, the ciphering requires 1 JMPSET, 9 JMPINC, 160 AB, 448 ALB, 144 LB, and 1 END microinstruction, a total of 2289 clock cycles. The key expansion requires 1 JMPSET, 10 JMPINC, 130 AB, 40 ALB, 20 LB, and 1 END microinstruction, a total of 606 clock cycles. The AES deciphering has exactly the same latency as the ciphering, given that both have the exact inverse operations of the other. The microcoded programs for all the applications supported by the PUs fit into a single RAM with 1024 entries of 36 bits that can be easily implemented with two block RAMs natively available in the FPGA device, each having 18 kbits available. Concerning the data RAM in each one of the PUs, a single block RAM is used to implement it, providing enough storage resources for the addressed applications.

TABLE 14.5. FPGA Implementation and Related State-of-the-Art Results for a Single AES PU and a Single EC PU

(a) AES PU					
Reference	Device	Key Size	Area (Slices)	Block RAMs	Throughput (Mbit/s)
Lim and Benaissa [45]	XCV800	128	2,330	–	3.8
Good and Benaissa [40]	XC2S15	128	124	2	2.2
Chaves et al. [41]	XC2VP20	128	515	12	2,300
This chapter	XC4VSX35	128	1,243	4	8.8

(b) EC PU					
Reference	Device	m	Area (Slices)	Block RAMs	Throughput (Op/s)
Lim and Benaissa [45]	XCV800	163	>2,330	–	188
Ernst et al. [16]	XC4085XLA	191	2,634	–	862
Leong and Leung [4]	XCV1000E	155	1,868	–	120
Saqib et al. [23]	XCV3200E	191	18,314	12	17,857
Chelton and Benaissa [5]	XC4VLX200	163	16,209	–	50,000
This chapter	XC4VSX35	163	2,120	3	756

Different levels of security are specified as the key size for the AES and m of the underlying $GF(2^m)$ for the EC results.
Op, operations.

Table 14.5 presents the results after place and route of the addressed PUs while computing an AES ciphering/deciphering or an EC point multiplication. Also, some results were selected after a thorough review of the related state of the art for the same applications. The selected related state-of-the-art implementations refer either to high throughput or highly compact approaches. The aim of Table 14.5 is to contextualize the designed PUs among other solutions, rather than perform a direct comparison between all of them. Note that the resources presented in Table 14.5 for the processor addressed in this chapter include not only the PUs but also all the resources required to operate with a single PU, namely the resources to communicate with the host, the RNG, the microinstructions arbiter, and the RAM. The throughput results for the presented PUs were measured at the maximum operating frequency, which is 153 and 158 MHz for the EC and AES PUs, respectively.

In Table 14.5a, the design in Lim and Benaissa [45] is the only one capable of supporting both the AES and EC cryptography. This implementation aims to be a compact design and is supported by a Xilinx Virtex *XCV800*. The maximum operating frequency is 41 MHz and relies on several logical units that support the basic field operations over $GF(2^8)$ that are organized in two possible configurations. One configuration supports the AES by applying a single instruction, multiple data (SIMD) approach, whereas the other configuration supports the EC arithmetic by a single instruction, single data (SISD) approach. However, the AES and EC arithmetic is not supported simultaneously because the logical units require reconfiguration to reuse resources when changing between the two functionalities. As described in Section 14.2, despite the similarity of the field $GF(2^m)$ arithmetic for the AES and EC, the values of m are significantly different, turning the efficient implementation of each operation, such as the reduction, significantly different as well. Hence, sharing the datapath between the AES and EC by splitting an operation in $GF(2^m)$ with a large value m into smaller ones with small values of m, as in Lim and Benaissa [45], limits the performance, given that these operations could be more efficiently computed in dedicated hardware or using LUTs. Note that the resource values for Lim and Benaissa [45] in Table 14.5a comprise a conversion of the provided number of gates to slices. Also, the values for the EC implementation refer to a $GF(2^{64})$ finite field being the real values for the slice resources greater than 2330 for a $GF(2^{163})$ finite field.

The implementations in Good and Benaissa [40] and Chaves et al. [41] refer to AES-only implementations. A compact design for the AES is proposed in Good and Benaissa [40]. This design runs at 67 MHz on a Xilinx *XC2S15* FPGA. A multiply–accumulate and a byte substitution unit to support the nonlinear function required in the AES are the two main arithmetic units underlying this design. Similarly to the design herein proposed, the design in Good and Benaissa [40] is controlled by microinstructions and a microprogram counter controlling the program flow and branches. The design in Chaves et al. [41] aims for high throughput and uses several possibilities in the hardware design such as the datapath pipelining and data flow parallelization efficiently exploited in a Xilinx *XC2VP20* FPGA. Furthermore, several block RAMs are employed either to store independently the round keys, allowing getting a complete round key in a single clock cycle, or to implement the *S-box* lookup in parallel and in a pipeline fashion. To the best of the authors' knowledge, the design in Chaves et al. [41] provides the highest throughput for an AES FPGA accelerator to date.

Concerning the EC point multiplication-only implementations, the works of Leong and Leung [4] and Ernst et al. [16] refer to compact implementations, and the works of Chelton and Benaissa [5] and Saqib et al. [23] to high throughput ones. In Ernst et al. [16] the underlying finite field elements are described as terms of a normal basis instead of a polynomial basis as employed in the herein addressed processor. Therefore, hardware structures optimized for this kind of basis, such as the so-called Massey–Omura multiplier and rotational shifts for squaring, are employed. The implementation in Ernst et al. [16] is also controlled with microinstructions that directly instantiate the multiplication, addition, or squaring of field elements. The inversion is constructed from the later microinstructions as in the herein addressed EC PU. Another property of the processor in Ernst et al. [16] is the fact that it has no storage for microinstructions, being the program feed from outside the device each time the program is to be run.

In Leong and Leung [4] an optimal normal basis is also used to implement an EC accelerator for a Xilinx *XCV1000E* FPGA. The area and throughput in Table 14.5b for this implementation are estimations provided by the authors [4], since there is no optimal normal basis for the $GF(2^m)$ finite field with $m = 163$ defined in the standards [8]. The estimation of the area considers a linear increase with the field size m while the latency values are obtained by fitting a curve to the results provided in Leong and Leung [4]. This implementation also uses a microcoded program stored in the RAM available in the FPGA device. The processor contains a register file and an arithmetic unit which contains a field multiplier that benefits from the optimal normal basis' properties that allow minimizing the fanout and the interconnections between the logic elements.

The design in Saqib et al. [23] uses a polynomial basis and proposes the utilization of two completely parallel hardwired Karatsuba–Ofman multipliers to achieve a high throughput figure. Furthermore, this design also uses hardwired reduction and squaring units. The data storage is assured by 24 block RAMs that allow storing and loading a complete field element in a single clock cycle. The EC point multiplication algorithm employed is similar to the one herein addressed. However, the processor presented in Saqib et al. [23] is not designed to support the inversion. Thus, the latency in Table 14.5b does not include the final projective to affine conversion of the resulting EC point.

To the best of the authors' knowledge, the design in Chelton and Benaissa [5] presents the highest throughput of the EC point multiplication over a binary finite field to date. The processor in Chelton and Benaissa [5] is a fully dedicated structure that implements the same point multiplication algorithm supported by a polynomial basis as employed in the processor herein presented. The design in Chelton and Benaissa [5] executes the EC point multiplication by pipelining the field operations involved. The core of this processor is the field multiplier that uses the property in Equation (14.6) and consecutive additions to obtain the reduced field multiplication in several steps, also pipelined. Squaring and addition logic is also inserted in the pipeline as well as data forwarding, avoiding pipeline exposure during the computation. Regardless of the microinstructions used to control the processor in Chelton and Benaissa [5], this implementation is similar to a dedicated one, since the processor is designed considering a predetermined program to be fed into the processor. For example, the data forwarding is designed knowing that specific data dependencies exist in the program.

The remaining component that requires individual evaluation is the RNG. The RNG follows the same structure presented in Figures 14.6 and 14.7. The randomness source is the jitter in the oscillators which requires the number of inverters and the number of oscillator rings to be experimentally dimensioned to assure that the randomness harvested from the oscillators is enough to obtain the required randomness properties for the output bitstream. In the design of the RNG the implementation details in Section 14.3.3 were taken into account. Also, the number of inverters per oscillator is the same for all the oscillators, given that the related results in Wold and Tan [43] suggest that it has a reduced impact in the randomness of the bitstream. The criteria used for the number of inverters is a compromise between the required resources and the wider variation of the jitter for longer series of inverters. Experimental evidence suggests that, for the FPGA device supporting the presented prototype, it is possible to obtain an RNG that successfully passes all the main statistical randomness tests [37, 38], with a combination of 20 oscillators with three inverters each sampled at a frequency of 100 MHz.

14.4.2 Complete Processor Evaluation

In this section, the behavior of the complete processor comprising several PUs is evaluated. For this purpose, two main components are fundamental: the arbiter and the FPGA configuration port. The FPGA configuration port is the component that allows exploiting the reconfiguration capabilities of the processor, that is, the runtime replacing of the EC by the AES PUs and vice versa. FPGAs support several methods for reconfiguration. Among these methods is the possibility to feed the reconfiguration bitstream in the device's pads reserved for this purpose, or the FPGA internal reconfiguration port often called the internal configuration access port (ICAP). The latter method relies on the instantiation of a special component which the remaining hardware of the processor communicates with in order to dynamically reconfigure itself. In the presented processor, this method can be easily accommodated in the design by providing the processor's host with the possibility to send the reconfiguration bitstream used to reconfigure the hardware at runtime. With this purpose, the static hardware resources in the FPGA comprise the possibility to redirect the reconfiguration bitstream to the ICAP. Concerning the arbiter, this component assures the correct behavior of all the PUs in the processor by regulating the requests to the microinstructions RAM. The arbiter must have a priority scheme to solve these conflicts. As discussed in Section 14.3.4, the priority should be related with the average number of clock cycles per microinstruction in the microcoded program running in the PUs.

In order to evaluate the processor and the effect of the arbiter and the dynamic reconfiguration, the layout depicted in Figure 14.10 is used. This layout consists of up to six PUs attached to a port of the microinstructions RAM intermediated by an arbiter. Three of the PUs connected to one arbiter are static and the remaining three are reconfigurable. The static PUs were defined as EC PU while the other three can support either the AES or EC. Note that the AES microinstructions require three clock cycles to complete;

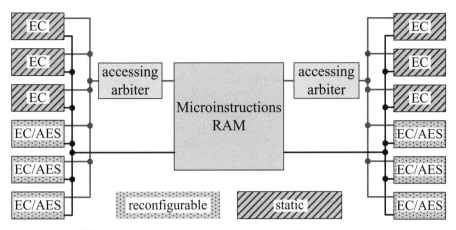

FIGURE 14.10. Macroview of an AES and EC enabled system.

thus, given the priority scheme, assigning more than three AES PUs to one arbiter would not be profitable since the fourth PU would have to wait for any of the first three to finish computing. For the reconfigurable PU, a reconfigurable region with 21 configurable logic blocks (CLBs) high and 13 CLBs wide was set, containing 1092 slices. The reconfigurable region coordinates were judiciously chosen so that the reconfiguration frames of the several regions do not overlap, hence each region can be reconfigured without interfering with the others. The resources used by the reconfigurable PU are of 943 ± 7 and 157 ± 2 slices (86% and 14%) for the EC and AES PU, respectively. Considering that all PUs correspond to EC configurations, the total amount of resources is 14,092 slices and 11 block RAM (two for the microinstructions storage, one for each PU, and three for the LUTs needed by the AES PU implemented in the static regions). The processor operating frequency for this evaluation prototype was set to 100 MHz. Notwithstanding, this frequency is limited by the maximum operating frequency of the PU, which are higher than 150 MHz.

The obtained configuration bitstreams for the reconfigurable regions vary from 30,662 to 31,067 words of 32 bits. Given that the maximum frequency of the ICAP is 100 MHz for the selected FPGA device, up to $32{,}067/100 \text{ MHz} = 310 \ \mu s$ are required to perform the reconfiguration of a single region.

In order to evaluate the efficiency of the processor, the PUs attached to a single arbiter were isolated and the results were obtained for the processor operating frequency (100 MHz) as presented in Table 14.6. The throughput metric should be multiplied by 2 so as to obtain the maximum throughput with the two processor arbiters. Table 14.6 also presents an efficiency metric. This efficiency metric is obtained by computing several consecutive EC point multiplications and AES ciphers in a time window T. Considering that n_{EC} and

TABLE 14.6. Performance Metrics for Different Combinations of Simultaneously Working PUs per Arbiter

Number of ECC PUs	Number of AES PUs	Latency (K *clk* cycles)	ms	ECC Throughput (Op/s)	AES Throughput (Mbit/s)	Efficiency (%)
0	0	–	–	–	–	–
1	0	201.7	2.02	496	–	100.00
2	0	201.7	2.02	992	–	100.00
3	0	201.7	2.02	1,488	–	100.00
4	0	342.3	3.42	1,169	–	82.50
5	0	344.9	3.45	1,450	–	71.60
6	0	390.5	3.91	1,536	–	61.67
0	1	201.5	2.02	–	5.59	99.98
1	1	206.9	2.07	483	5.44	99.08
2	1	223.5	2.24	895	5.04	96.61
3	1	348.8	3.49	860	3.23	81.80
4	1	354.5	3.55	1,128	3.18	71.24
5	1	391.4	3.91	1,278	2.88	61.59
0	2	201.5	2.02	–	11.18	99.98
1	2	208.1	2.08	481	10.83	98.57
2	2	348.7	3.49	574	6.46	79.27
3	2	350.3	3.50	856	6.43	70.72
4	2	385.9	3.86	1,037	5.84	61.20
0	3	201.5	2.02	–	16.77	99.98

Op, operations.

n_{AES} operations were computed in the time window T, the efficiency metric E is obtained as:

$$E = \frac{n_{\text{EC}}T_{\text{EC}} + n_{\text{AES}}T_{\text{AES}}}{n_{\text{PU}}T}, \tag{14.30}$$

where $n_{\text{PU}} = n_{\text{EC}} + n_{\text{AES}}$ and T_{AES} and T_{EC} are the computation times of the AES and EC programs, respectively, if no conflicts occur. The results in Table 14.6 show that for $n_{\text{PU}} \leq 3$, the efficiency is very close to 100% since 3 is the minimum latency for a microinstruction. For larger values of n_{PU}, the efficiency decreases, being above 60% for $n_{\text{PU}} = 6$, which means that more than 60% of the throughput is obtained regardless of the selected PUs.

14.5   CONCLUSIONS

In this chapter, a complete and fully functional processor to accelerate the computation of cryptographic procedures is presented and discussed. In particular, public- (EC cryptography) and private-key (AES) algorithms are able to be accelerated as well as the generation of true random numbers. Furthermore, the design of this processor is accomplished, with emphasis not only on the efficiency but also on the flexibility of the implementation. This flexibility allows complying with the constant evolution of the cryptographic algorithms and with the runtime requirements of the system which is hosting the processor. A microcoded approach and modular design are the key guidelines to meet this flexibility. The utilization of a microcode eases the modification of the processor behavior by loading the required microcoded program, whereas the utilization of a modular structure for the processor enables the utilization of the dynamic reconfiguration available on FPGA devices. With this, it is possible to replace the processor modules at runtime by reconfiguring the FPGA accordingly, resulting in the constant conformation of the computational resources with the runtime demands.

A prototype was developed and thoroughly evaluated for a Xilinx Virtex 4 FPGA. Experimental results suggest that regardless of the provided flexibility, the performance figures are competitive regarding the related state of the art. At a frequency of 100 MHz, throughputs of up to 3072 EC point multiplications and up to 33.5 Mbit/s for the AES block ciphering are obtained. The runtime reconfiguration of a single module (which supports either the EC point multiplication or the AES ciphering) of the processor can be accomplished in 310 μs.

REFERENCES

[1] D.R. Stinson, Cryptography: Theory and Practice, 3rd ed., Chapman & Hall/CRC, Boca Raton, FL, 2005.

[2] N. Koblitz, "Elliptic curve cryptosystems," Mathematics of Computation, 48(177), 1985, pp. 203–209.

[3] V.S. Miller, "Use of elliptic curves in cryptography," in H. Williams, ed., Lecture Notes in Computer Science: Advances in Cryptology—CRYPTO 1985 Proceedings, Springer Berlin/Heidelberg, Berlin and Heidelberg, Germany, 1985, pp. 417–426.

[4] P.H.W. Leong, I.K.H. Leung, "A microcoded elliptic curve processor using FPGA technology," IEEE Transactions on Very Large Scale Integration (VLSI) Systems, 10(5), 2002, pp. 550–559.

[5] W.N. Chelton, M. Benaissa, "Fast elliptic curve cryptography on FPGA," IEEE Transactions on Very Large Scale Integration (VLSI) Systems, 16(2), 2008, pp. 198–205.

[6] S. Antão, R. Chaves, L. Sousa, "Compact and flexible microcoded elliptic curve processor for reconfigurable devices," in IEEE Symposium on Field Programmable Custom Computing Machines—FCCM, IEEE, Napa, CA, 2009, pp. 193–200.

[7] N. Koblitz, A. Menezes, S. Vanstone, "The state of elliptic curve cryptography," designs," Codes and Cryptography, 19(2), 2000, pp. 173–193.

[8] National Institute of Standards and Technology (NIST), "Digital Signature Standard (DSS)—FIPS PUB 186-3," 2009.

[9] IEEE Computer Society, "IEEE Std 1363-2000: IEEE Standard Specifications for Public-Key Cryptography," 2000.

[10] IEEE Computer Society, "IEEE Std 1363a-2004 (Amendment to IEEE Std 1363-2000): IEEE Standard Specifications for Public-Key Cryptography- Amendment 1: Additional Techniques," 2004.

[11] Certicom Research, "Standards for Efficient Cryptography 1 (SEC 1) : Elliptic Curve Cryptography, Version 2," 2009.

[12] Certicom Research, "Certicom Research, Standards for Efficient Cryptography 2 (SEC 2) : Recommended Elliptic Curve Domain Parameters," 2010.

[13] Ç.K. Koç, ed., Cryptographic Engineering, Springer Science+Business Media, LLC, New York, 2009.

[14] B. Sunar, Ç.K. Koç, "An efficient optimal normal basis type II multiplier," IEEE Transactions on Computers, 50(1), 2001, pp. 83–87.

[15] J.K. Omura, J.L. Massey, "Computational Method and Apparatus for Finite Field Arithmetic," US Patent 4587627, 1986.

[16] M. Ernst, B. Henhapl, S. Klupsch, S. Huss, "FPGA based hardware acceleration for elliptic curve public key cryptosystems," Journal of Systems and Software, 70(3), 2004, pp. 299–313.

[17] S.S. Erdem, Ç.K. Koç, "A less recursive variant of Karatsuba-Ofman algorithm for multiplying operands of size a power of two," in IEEE Symposium on Computer Arithmetic—ARITH, IEEE, Santiago de Compostela, 2003, pp. 28–35.

[18] H. Brunner, A. Curiger, M. Hofstetter, "On computing multiplicative inverses in $GF(2^m)$," IEEE Transactions on Computers, 42(8), 1993, pp. 1010–1015.

[19] U. Daepp, P. Gorkin, "Fermat's little theorem," in S. Axler, F.W. Gehring, K.A. Ribet, eds., Reading, Writing, and Proving: A Closer Look at Mathematics (Undergraduate Texts in Mathematics), Springer New York, New York, 2011, pp. 315–323.

[20] T. Itoh, S. Tsujii, "A fast algorithm for computing multiplicative inverses in $GF(2^m)$ using normal bases," Information and Computation, 78(3), 1988, pp. 171–177.

[21] F. Rodríguez-Henríquez, A.D. Pérez, N.A. Saqib, Ç.K. Koç, Cryptographic Algorithms on Reconfigurable Hardware, Springer Science+Business Media, LLC, New York, 2006.

[22] W. Stallings, Cryptography and Network Security, 4th ed., Pearson Prentice Hall, Upper Saddle River, NJ, 2006.

[23] N.A. Saqib, F. Rodríguez-Henríquez, A. Díaz-Pérez, "A parallel architecture for fast computation of elliptic curve scalar multiplication over $GF(2^m)$," in International Parallel and Distributed Processing Symposium—IPDPS, Santa Fe, NM, 2004, p. 144.

[24] M. Bednara, M. Daldrup, J. von zur Gathen, J. Shokrollahi, J. Teich, "Reconfigurable Implementation of elliptic curve crypto algorithms," in International Parallel and Distributed Processing Symposium—IPDPS, IEEE, Ft. Lauderdale, FL, 2002, pp. 157–164.

[25] J. López, R. Dahab, "Improved algorithms for elliptic curve arithmetic in $GF(2^m)$," in S. Tavares, H. Meijer, eds., Lecture Notes in Computer Science: Selected Areas in Cryptography, Springer Berlin/Heidelberg, Berlin and Heidelberg, Germany, 1999, pp. 201–212.

[26] P.L. Montgomery, "Speeding the Pollard and elliptic curve methods of factorization," Mathematics of Computation, 48(177), 1987, pp. 243–243.

[27] J. López, R. Dahab, "Fast multiplication on elliptic curves over $GF(2^m)$ without precomputation," in Ç.K. Koç, C. Paar, eds., Lecture Notes in Computer Science: Advances in Cryptology—CRYPTO 1999 Proceedings, Springer Berlin/Heidelberg, Berlin and Heidelberg, Germany, 1999, pp. 316–327.

[28] National Institute of Standards and Technology (NIST), "Specification for the Advanced Encryption Standard (AES)—FIPS PUB 197," 2001.

[29] National Institute of Standards and Technology (NIST), "Data Encryption Standard (DES)—FIPS PUB 46-3," 1999.

[30] D. Coppersmith, "The Data Encryption Standard (DES) and its strength against attacks," IBM Journal of Research and Development, 38(3), 1994, pp. 243–250.

[31] J. Daemen, V. Rijmen, "AES Proposal: Rijndael," 1999. National Institute of Standards and Technology (NIST). Available: http://csrc.nist.gov/archive/aes/rijndael/Rijndael-ammended.pdf.

[32] J. Daemen, V. Rijmen, "The block cipher Rijndael," in J.-J. Quisquater, B. Schneier, eds., Lecture Notes in Computer Science: Smart Card Research and Applications, Springer Berlin/Heidelberg, Berlin and Heidelberg, Germany, 2000, pp. 277–284.

[33] R.C. Tausworthe, "Random numbers generated by linear recurrence modulo two," Mathematics of Computation, 19, 1965, pp. 201–209.

[34] M. Matsumoto, T. Nishimura, "Mersenne twister: a 623-dimensionally equidistributed uniform pseudo-random number generator," ACM Transactions on Modeling and Computer Simulation, 8(1), 1998, pp. 3–30.

[35] C. Gabriel, C. Wittmann, D. Sych, R. Dong, W. Mauerer, U.L. Andersen, C. Marquardt, G. Leuchs, "A generator for unique quantum random numbers based on vacuum states," Nature Photonics: Letters, 4(10), 2010, pp. 711–715.

[36] B. Sunar, W.J. Martin, D.R. Stinson, "A provably secure true random number generator with built-in tolerance to active attacks," IEEE Transactions on Computers, 26(1), 2006, pp. 109–119.

[37] G. Marsaglia, "Diehard Battery of Tests of Randomness," 1995. Available: http://www.stat.fsu.edu/pub/diehard/.

[38] National Institute of Standards and Technology (NIST), "A Statistical Test Suite for Random and Pseudorandom Number Generators for Cryptographic Applications—Special Pub. 800-22—Rev. 1," 2010.

[39] Xilinx, Inc., "Virtex-4 FPGA User Guide," 2008. Available: http://www.xilinx.com/support/documentation/user_guides/ug070.pdf.

[40] T. Good, M. Benaissa, "AES on FPGA from the fastest to the smallest," in J. Rao, B. Sunar, eds., Lecture Notes in Computer Science: Advances in Cryptology—Cryptographic Hardware and Embedded Systems CHES 2005, Springer Berlin/Heidelberg, Berlin and Heidelberg, Germany, 2005, pp. 427–440.

[41] R. Chaves, G. Kuzmanov, S. Vassiliadis, L. Sousa, "Reconfigurable memory based AES co-processor," in International Parallel and Distributed Processing Symposium—IPDPS, IEEE, Rhodes Island, Greece, 2006, p. 9.

[42] S. Antão, R. Chaves, L. Sousa, "AES and ECC cryptography processor with runtime configuration," in International Conference on Advanced Computing and Comunications—ADCOM, IEEE, Bangalore, India, 2009.

[43] K. Wold, C.H. Tan, "Analysis and enhancement of random number generator in FPGA based on oscillator rings," International Journal of Reconfigurable Computing, 2009, Article ID 501672, 8 pages.

[44] Annapolis Micro Systems, Inc., "WILDCARD 4 with Xilinx Virtex 4 FPGA," 2007. Available: http://www.annapmicro.com/wc4.html.

[45] W.M. Lim, M. Benaissa, "Subword parallel $GF(2^m)$ ALU: an implementation for a cryptographic processor," in IEEE Workshop on Signal Processing Systems—SIPS, IEEE, Seoul, South Korea, 2003, pp. 63–68.

INDEX

Embedded Systems: Hardware, Design, and Implementation, First Edition.
Edited by Krzysztof Iniewski.
© 2013 John Wiley & Sons, Inc. Published 2013 by John Wiley & Sons, Inc.